Industrial
Design A-Z

Front Cover: Rolls-Royce *Trent* fanjet engine, 1995
(Science & Society Picture Library, London)
Spine: Ettore Sottsass & Perry King, *Valentine* typewriter for Olivetti, 1969
(Archivo Storico Olivetti, Ivrea)

TS
17/
.F54
2000

© 2000 TASCHEN GmbH
Hohenzollernring 53, D–50672 Köln
www.taschen.com

© 2000 for the works by
Marianne Brandt: Bauhaus-Archiv GmbH, Berlin
Charles & Ray Eames: Eames Office, Venice, CA, www.eamesoffice.com

© 2000 for the works by Josef Albers, Peter Behrens, Lucian Bernhard, Max Bill,
Ludwig Hohlwein, Bruno Ninaber van Eyben, Jacobus Johannes Pieter Oud,
Sigurd Persson, Wilhelm Wagenfeld, Frank Lloyd Wright: VG-Bild Kunst, Bonn

Editing and coordination: Susanne Husemann, Cologne
Production: Martina Ciborowius, Cologne
Design: UNA (London) designers

Printed in Italy
ISBN 3–8228–6310–6

Industrial Design A-Z

Charlotte & Peter Fiell

TASCHEN

KÖLN LONDON MADRID NEW YORK PARIS TOKYO

CONTENTS

For over 200 years, the products of mechanized industrial production have shaped our material culture, influenced world economies and affected the quality of our environment and daily lives. From consumer goods and packaging to transportation systems and production equipment, industrial products encompass an extraordinary range of functions, techniques, attitudes, ideas and values and are a means through which we experience and perceive the world around us.

The nature of industrial products and how they come to be is determined by an ever more complex process of design that is itself subject to many different influences and factors. Not least of these are the constraints imposed by the social, economic, political, cultural, organizational, and commercial contexts within which new products are developed, and the character, thinking and creative abilities of the individual designers or teams of designers, aligned specialists and manufacturers involved in their realization.

Industrial design – the conception and planning of products for multiple reproduction – is a creative and inventive process concerned with the synthesis of such instrumental factors as engineering, technology, materials and aesthetics into machine-producible solutions that balance all user needs and desires within technical and social constraints. Engineering – the application of scientific principles to the design and construction of structures, machines, apparatus or manufacturing processes – is an essential and defining aspect of industrial design. While both disciplines are concerned with finding optimum solutions to specific problems, the primary distinguishing characteristic of industrial design is its concern for aesthetics.

The origins of industrial design can be traced back to the Industrial Revolution which began in Great Britain in the mid-18th century, and which heralded the era of mechanization. Prior to this, objects were craft-produced, whereby both the conception and the manufacture of an object were the work of a single individual. With the development of new industrial manufacturing processes and the division of labour, design (conception and planning) was progressively separated from the act of making. At this early stage, however, design had no intellectual, theoretical or philosophical foundation and was considered just one of the many interrelated aspects of mechanical production. Thus the industrial goods of the years up to the 19th century were created by specialists from the technical, materials and production spheres rather than by an industrial designer. Towards the end of the 19th century, however, manufacturers began to realise that they could gain a critical competitive advantage by improving the constructional integrity and aesthetic appearance of their products. As a consequence, they began to invite specialists from other spheres– most notably, architects – to contribute to the design process.

Industrial design subsequently became a fully-fledged discipline in the early 20th century, when design theory was integrated into industrial methods of production. Among the first professional industrial design practitioners was the German architect Peter Behrens, who was recruited by AEG in 1907 to improve the company's products and corporate identity. Since then, industrial design has become an increasingly important factor in the success of industrial products and the companies that manufacture them.

This survey of industrial design encompasses all aspects of the subject – from heavy industrial products to sports equipment, and medical devices to aerospace vehicles – and is intended as a companion work to our earlier *Design of the 20th Century*. The manufacturers, industrial design consultancies, individual designers and inventors we feature – in entries organized in alphabetical order – are those who have developed some of the most important and influential products and technologies of the last 200 years. Their innovative solutions strike the best possible balance between the intellectual, functional, emotional, aesthetic and ethical expectations of the user/consumer and the influences and factors bearing upon the design process. Other related topics that have been crucial to the development of industrial design – from Aerodynamics to Utilitarian Design – are outlined in the "Themes and Materials" section. Several case studies have also been included to demonstrate the frequently evolutionary nature of industrial design. Cross-references appear in the text in bold type so as to reveal the many illuminating interrelationships between designers, design consultancies and manufacturers, while a time-line at the back of the book places key industrial design developments within the context of world events.

As more and more countries are drawn into the global free-market economy, so industrial design has become an increasingly vital means of competing on a global scale. By reflecting on the history of industrial design and focusing in particular on its successes, this book hopes to demonstrate how industrial design has consistently sought to de-mystify technology and to deliver it in accessible forms to the greatest number of people.

A further aim of the book is to highlight the phenomenal extent to which manufacturing companies themselves have shaped the history of industrial design. Without their willingness to risk the necessary and sometimes massive investments demanded by the development of new products, there would be very little industrial design. Innovative and commercially-motivated yet socially-minded manufacturing companies are too often the unsung heroes of our material culture.

"Our capacity to go beyond the machine rests upon our power to assimilate the machine. Until we have absorbed the lessons of objectivity, impersonality, neutrality, the lesson of the mechanical realm, we cannot go further in our development toward the more richly organic, the more profoundly human."
Lewis Mumford, *Technics and Civilization*, 1934

TASCHEN

A-Z

DESIGNERS AND FIRMS

AEG

FOUNDED 1883
BERLIN, GERMANY

Peter Behrens,
Turbine hall for AEG,
1908–1909

→ Peter Behrens,
Light fitting, 1907

Peter Behrens, AEG
advertisement, 1907

In 1881 Emil Rathenau (1838–1915) visited the "Exposition Internationale d'Electricité" in Paris, where he saw **Thomas Alva Edison**'s light bulb. He was so impressed that he bought the patent licence and in 1883 founded the Deutsche Edison Gesellschaft (German Edison Company for Applied Electricity). The company subsequently changed its name to the Allgemeine Elektricitäts Gesellschaft, or AEG. Its catalogue for the 1900 Paris "Exposition Universelle" was designed by the Jugendstil artist Otto Eckmann (1865–1902), who also designed an Art Nouveau-style logo for the company. In 1907, however, AEG appointed as its artistic director the architect and designer **Peter Behrens**, who proceeded to create the first wholly integrated **corporate identity** for the company. Having re-invented AEG's logo, Behrens went on to design not only a unified range of electrical goods such as kettles, clocks and fans, but also the factory buildings required for their production. AEG's production engineer, Michael von Dolivo-Dobrowolsky, realised that the key to the successful mass production of high quality goods lay in the **standardization** of interchangeable components, which would allow them to be used for several products rather than just one. This type of product standardization and the modern methods of manufacture employed by AEG reflected the ideals of the **Deutscher Werkbund**, of which Behrens had been a founding member in 1907. Hermann Muthesius (1861–1927), another central figure in the formation of the Deutscher Werkbund, had urged German companies to

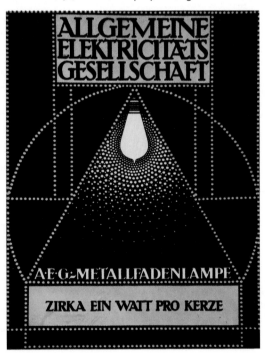

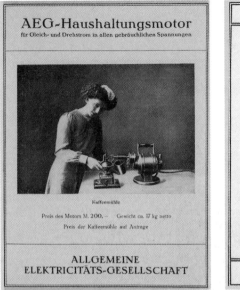

AEG-Haushaltungsmotor
für Gleich- und Drehstrom in allen gebräuchlichen Spannungen

Kaffeemühle

Preis des Motors M. 200, — Gewicht ca. 17 kg netto
Preis der Kaffeemühle auf Anfrage

ALLGEMEINE
ELEKTRICITÄTS-GESELLSCHAFT

BERLIN **AEG** 1913

Arbeiter-Kontroll-Uhren

Allgemeine Elektricitäts-Gesellschaft

Peter Behrens,
Electric coffee
grinder, c. 1911

↗Peter Behrens,
Clocking-in machine,
c. 1910

establish a national aesthetic of "types" and "standards" that would result in "a unification of general taste". AEG was one of the first companies to embrace and implement Muthesius' ideas and to assemble household products from standardized components, as was the case with Peter Behrens' well-known range of kettles and his electric table fan (1908). With a strong brand identity forged by Behrens' comprehensive design programme, AEG went on to become a leading manufacturer of electrical products. In 1927 the company took part in the Deutscher Werkbund's *Die Wohnung* (The Dwelling) exhibition held in Frankfurt, where it displayed electric fans, water heaters, lamps and kettles, among other products. In the 1960s, AEG became primarily known for its almost puristic high quality white goods. In the 1980s, however, AEG entered a financial crisis and was taken over, first by Daimler-Benz, and in 1994 by the Swedish Company **Electrolux**. Today their concept of high quality and restrained form is very successful and AEG remains a large and prosperous manufacturing company.

Peter Behrens,
Electric desk fan,
1908

Hans Ehrich and
Tom Ahlström with
Clean shower and
toilet chair, 1999

→*Fresh* bath boards
for Etac, 1998

Jordan 1230
Dishwashing brushes
for Jordan, 1974

One of the leading industrial design consultancies in
Sweden, A&E Design was founded in 1968 by Tom
Ahlström (b. 1943) and Hans Ehrich (b. 1942), both of
whom had previously studied design and metalwork at
the Konstfackskolan, Stockholm. From its outset, this
Stockholm-based practice researched the application
of **plastics** while specializing in **design for disability**.
A&E Design has since developed numerous aids and
implements for use by the elderly and handicapped –
from a lift-and-ride wheelchair for transporting patients
to the shower, to bath seats and locking armrests for
toilets. The consultancy has also developed other er-
gonomically-led designs including tools for Anza and
dishwashing brushes for Jordan. A&E Design is renowned for innovative product development including meticulous working models. Unusually, around 20% of
the work undertaken at A&E Design is dedicated to the
development of its own products, which "seek an optimal balance between aesthetics, function and market".
In 1982, Ahlström and Ehrich founded a sister design
consultancy, **InterDesign**, that similarly manufactures
models – from mock-ups to fully functional prototypes
– in its own workshop. Although A&E Design is one of
the leading Swedish consultancies, many of its product
designs have attributes more closely associated with
Italian rather than Scandinavian design. Its founders
readily acknowledge the influence of Italian designers,
most notably **Joe Colombo**, and have always maintained
close links with Milan and the design journal *Domus*.
A&E Design has received 14 Excellent Swedish Design
Awards and no less than three Roter Punkt awards
for Highest Design Quality.

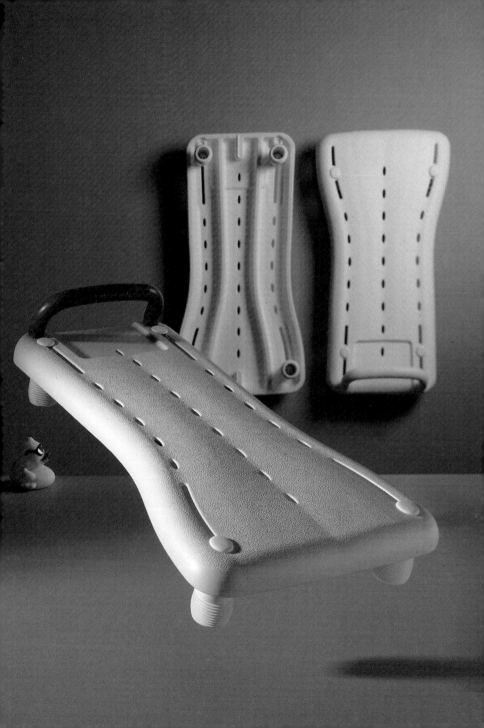

Wally Byam

AIRSTREAM

FOUNDED 1934
LOS ANGELES, USA

The history of the iconic Airstream caravan can be traced to plans for a trailer which were submitted to a DIY magazine published by Wally Byam (1896–1962) in the 1920s. Byam received several complaints from readers that the design did not actually work, so he set himself the task of designing his own trailer that would actually perform properly. By lowering the floor between the wheels and raising the roof height, Byam designed the first travelling trailer in which one could stand upright. Byam eventually abandoned publishing in 1929 in order to concentrate on perfecting the revolutionary design of his trailers, in which he applied techniques borrowed from the aircraft industry, including riveted **aluminium** monocoque structures and **streamlining**. In 1934, he conceived the name "Airstream" for his trailers, which travelled along "like a stream of air". Two years later, he introduced the *Clipper* built along aircraft principles. Throughout the Depression of the 1930s, Byam's trailers bucked the trend with demand outstripping supply despite their relatively high price. Following the outbreak of the Second World War, however, Byam's growing busi-

Bambi Airstream trailer (re-edition of the 1961/1962 original)

ness was abruptly halted by the US Government's decision in 1941 to
restrict the use of aluminium to weapons manufacturers. Byam spent the
war years working for a Los Angeles aircraft manufacturer, where he undoubtedly gleaned new ideas for his trailers. After the war, Airstream was
reborn and by 1952 increased demand for Byam's high quality trailers had
led to the construction of new manufacturing plants in Jackson Centre, Ohio
and Santa Fe Springs, California. The gleaming Airstream trailer came to
symbolize the American Dream; just like the earlier covered wagons of the
pioneers, these futuristic silver bullets promoted the spirit of freedom and
adventure.

Following Byam's death in 1962, a foundation was created to host annual
caravan meetings across America in order to enhance cross-cultural relationships. 1978 saw the introduction of the first Airstream motor homes, and
today the "Wally Byam Caravan Club, International" boasts some 19,000
member trailers/motor homes. The sleek good-looks of the Airstream trailer
have bred much fanaticism among its enthusiasts and Wally Byam's dream
of some 65 years ago has truly become an international phenomenon. In the
words of Byam, Airstream strives "endlessly to stir the venturesome spirit
that moves you to follow a rainbow to its end ... and thus make your travel
dreams come true."

ALESSI

FOUNDED 1921
OMEGNA, ITALY

Group of Alessi
workers outside the
factory, early 20th
century

→Luigi Massoni &
Carlo Mazzeri,
Model No. 870
cocktail shaker, 1957

Carlo Alessi, *La
Bombé* tea and
coffee service, 1945

Giovanni Alessi founded FAO (Fratelli Alessi Omegna) in Omegna, Italy, in 1921. According to a sign which it displayed at trade fairs during the 1920s, FAO was a "Workshop with foundry specializing in working brass and nickel silver sheet". From its earliest beginnings, FAO was highly innovative and ready to experiment. From 1929, the company concentrated on producing catering and domestic tableware in either nickel-plated or silver-plated brass. Around 1935, Giovanni's son Carlo Alessi (b. 1916) was appointed principal designer to the firm. The ensuing years saw FAO making the transition away from craft-based methods to industrial production. In 1945 – the year in which he also became general manager– Carlo conceived his *Bombé* coffee set. This enduring design, produced in four different sizes, was an unashamedly industrial product.

Also in 1945, another member of the family, Ettore Alessi, joined the firm as a technical expert in cold metal pressing – Alessi had started using pressed stainless steel as a consequence of the general shortage of nickel-silver and brass after the Second World War. From around 1955, the company began

Guido Venturini,
GV08 Okkio clothes
brushes, 2000

inviting outside designers such as Luigi Massoni (b. 1930) to contribute
to its programme. Many of the resulting designs, which tended to be rather
forward-looking, were consequently mass-produced – 1.5 million *Model
No. 870* cocktail shakers by Massoni, for example, have been sold to date. In
the late 1970s, with Carlo now president, the firm began producing limited-
edition, signed ranges by internationally celebrated designers and architects,
such as **Ettore Sottsass** and **Richard Sapper**. In 1983 Alessi registered a new
trademark, *Alessi Officina*, for products of experimental design that could be
developed outside the constraints of industrial production. That same year,
Carlo's son, Alberto, initiated the *Tea and Coffee Piazza* project in which
eleven architects designed limited edition services. This publicity-motivated
"architecture in miniature" project brought Alessi international recognition
and made the company one of the leading exponents of Post-Modernism
within the decorative arts during the 1980s. In the 1990s Alessi introduced
other product lines, such as *Twergi*, which included characterful wooden
household items designed by Ettore Sottsass and Andrea Branzi (b. 1938)
and a range of affordable yet high-style plastic wares by Stefano Giovannoni
(b. 1954) and Alessandro Mendini (b. 1931). Designs such as these are the
outcome of Alberto Alessi's belief that "design is a global creative discipline
with a strictly artistic and poetic matrix, and not simply one of the many
tools at the service of marketing and technology to ensure improved production methods and better sales".

Alejandro Ruiz,
Parmenide parmesan
cheese graters
(*Memory Container*
collection), 1994

ALFA ROMEO

FOUNDED 1910
PORTELLO/MILAN, ITALY

Super Sport, 1947

In 1910 the company known as Anonima Lombarda Fabbrica di Automobili (ALFA) was founded in the Portello district of Milan, having emerged from the defunct Società Italiana Automobili Darracq, which had been established in 1906 to produce low-cost cars. The new company decided to espouse a distinctly Milanese image and adopted a red cross and the Visconti dragon for its trademark. Its first venture was a production run of 300 *24 HP Torpedo* cars (1910) designed by Giuseppe Merosi. Three years later, the company produced its first sports car, the *40/60 HP Corsa* (1913), which had a top speed of 147 km per hour. Around this period, Nicola Romeo, a Neapolitan entrepreneur, took over the company and added his surname to its title. He also ensured that cars began rolling off the lines in Portello in significant numbers. During this era, the company produced several remarkably progressive cars, including the tear-drop shaped *40/60 HP Aerodynamica* (1914), which could reach an impressive top speed for its day. The company went into full-scale industrial production with Giuseppe Merosi's *RL* (1922–1923). Described by Alfa Romeo as its first masterpiece, the *RL* was also produced in sports and touring versions. The fast and reliable racing version of the *RL*, known as the *Targa Florio*, with its characteristic "prow-shaped" nose, placed second, third and fifth in the famous Sicilian 1,000 mile road race – the Mille Miglia

↘6C 2500 Super Sport, 1947

Giulietta Spider, 1955

– and brought the world's attention to the fledgling Milanese car industry. At this stage, the Carrozzeria Castagna also designed several car body variations including a cabriolet version of the four-cylinder *RM* and an elegant

open touring version of the *RM*. In 1923, the designer Vittorio Jano joined the firm and his *P2* won Alfa Romeo's first world title. His *8C-2300*, which won Le Mans in 1931 and 1934, ensured that the marque became synonymous with racing during the early 1930s. Although this was the golden age of Alfa Romeo, the company was nationalized in 1933. Sadly, the Italian Government did not provide the company with sufficient funding and so it moved away from customized prize-winning racing cars in order to concentrate on general use vehicles, although it did continue producing its sleek, high-quality sports cars. After the Second World War, Alfa Romeo mass-produced elegant road cars such as the *6C 2500 Super Sport* (1947) with super-lightweight coachwork. Incorporating the latest aerodynamic research, the *Super Sport* and the similar *Freccia d'Oro* possessed a streamlined beauty and a sense of poise. While the market in the 1950s was demanding more functional and less luxurious models, Alfa Romeo was disinclined

↑ *Spider Duetto*, 1966

Alfa *GTV*, 1995

Alfa 166, 1998

Interior of Alfa 166, 1998

to completely sacrifice its sporting heritage and decided instead to produce affordable performance cars. By the mid-1950s and early 1960s, the company had increasingly industrialized its production methods and was manufacturing classic models such as the *Giulietta Spider* (1955) styled by **Pininfarina** and the *Giulietta Sprint* (1954) styled by **Bertone**. Under the directorship of Orazio Satta, the company had come up with the *Giulietta* project so as to meet the growing demand for smaller and cheaper private vehicles. The *Giulietta* managed to compress Alfa Romeo's high performance technology into an engine of only 1300 cubic centimetres. In 1962, the slightly larger *Giulia* was launched and the company began targeting the European market with classics such as the *Giulia Sprint* styled by Bertone. Later, Pininfarina styled the streamlined *Duetto* (1966), which with a few minor changes remains in production today. In 1971, **Giorgetto Giugiaro** styled the groundbreaking *Alfasud* that helped to considerably increase the company's market share. From the post-war period until 1986, Alfa Romeo was under the control of the Italian Government, which oversaw its operations through the Istituto per la Ricostruzione (IRI). Even today, Alfa Romeo designs such as the Alfa 156 (1997) and Alfa 166 (1998) acknowledge their legendary sporting ancestry and project a strong and purposeful aesthetic.

EMILIO AMBASZ

BORN 1943
RESISTENCIA, ARGENTINA

Emilio Ambasz studied architecture at Princeton University, taught for a
year at the Hochschule für Gestaltung in Ulm and held a professorship at
Princeton until 1969. In 1967 he also co-founded the avant-garde Institute
of Architecture and Urban Studies in New York. From 1970 to 1976 he was
Curator of Design at New York's Museum of Modern Art, where in 1972 he
organized the groundbreaking exhibition "Italy: The New Domestic Landscape – Achievements and Problems of Italian Design". He subsequently
founded his own design studio, Emilio Ambasz & Associates, in 1977 and
the Emilio Ambasz Design Group in 1981 – both in New York. Between 1981
and 1985, Ambasz was president of the Architectural League and taught
extensively at Princeton and other American universities. While Ambasz is
widely celebrated for his design teaching and writing, he is also acclaimed

Cummins *Signature*
600 engine,
1996–1997

Polyphemus
flashlight, 1985

for several notable seating and lighting designs, such as the highly success-ful seating systems *Vertebra* (1977) with Giancarlo Piretti and *Dorsal* (1981), and the lighting range for Logotec (1981). His architectural projects include the Center for Applied Computer Research and Programming, Las Prome-sas, Mexico (1975), the Grand Rapids Art Museum, Michigan (1975), the Museum of American Folk Art, New York (1980) and the San Antonio Bo-tanical Garden Conservatory, Texas (1982). Ambasz also won first prize and a gold medal in a competition to design the master plan for Expo 1992 in Seville. Since 1980 Ambasz has been chief design consultant to Cummins Engine Co. and has received numerous awards for his lighting and seating designs – his *Polyphemus* flashlight (1985) was nominated for both a Com-passo d'Oro and an IDSA (Industrial Designers Society of America) award in 1987, his *Qualis* seating system (1991) was awarded a Compasso d'Oro in 1991, and his *Soffio* modular lighting system won him the Industrial Design Excellence award by the IDSA. He has stated that the methodological princi-ple guiding his work "is the search for basic principles and prototypical or pilot solutions which can first be formulated into a general method and then applied to solve specific problems". Ambasz believes that design should not just fulfil functional requirements but should also take poetic form in order to satisfy our metaphysical needs. He maintains that designers must learn to reconcile the past and the future in their work and give "poetic form to the pragmatic".

APPLE COMPUTER INC.

FOUNDED 1976
PALO ALTO, USA

Steve Jobs

→Apple IIc, 1984

↘Lisa, 1983

High school friends Steven Wozniak (b. 1950) and Steven Jobs (b. 1955) shared an interest in electronics but were both perceived as "outsiders" by their peers. After graduation, they dropped out of further education and found employment in Silicon Valley – Wozniak at Hewlett-Packard and Jobs at Atari. In 1976 Wozniak, who had dabbled in the development of computers for some time, designed what would eventually become the *Apple I* computer. Jobs, who was more business-orientated than his friend, insisted that this product should be developed and marketed. As a result, Apple Computers was founded in Palo Alto on 1st April 1976 in the hope that the new personal computer, the *Apple I*, would revolutionize computers just as **Ford**'s *Model T* had revolutionized cars. The venture did not take off until a year later, when the *Apple II* – the first ever commercial personal computer – was launched at a local trade fair. This was the first personal computer to have a plastic casing and to include colour graphics. 1978 saw the introduction of the *Apple Disk II*, which at the time was the cheapest and easiest-to-use floppy drive available. Increasing sales drove the company's rapid growth and by 1980, the year in which the *Apple III* was launched, the business was employing a workforce of several thousand. In 1981 Apple hit the first of several rocky periods in its history; market saturation meant fewer sales, **IBM** released its first PC and Wozniak, the creative force behind the business, was injured in a plane crash and only returned to the company for brief periods of time. Undeterred, Jobs began working on the *Apple Macintosh*, which had initially been conceived as a $500 computer but in the end far exceeded that amount. Its subsequent launch in 1984 marked a real breakthrough in personal computing. With its user-friendly interface, high-definition screen and mouse, the *Mac* heralded the true advent of the home computer. Unlike the *Apple II*, the *Apple Mac* featured an integrated monitor and disk drive. Designed

Apple I, 1976

by **Frogdesign**, its **styling** was radically different from its more angular competitors – it had softer, flowing lines that gave it a sleeker aesthetic. Although the smaller, cream-coloured *Mac* initially sold well, users became increasingly frustrated by its small amount of RAM and its lack of hard drive connectivity. Problems also arose through personal differences between Jobs and the then CEO of Apple, John Scully, which eventually resulted in Jobs being ousted in 1985 from the company he had originally created. Bill Gates' introduction of *Windows 1.0*, which had many similarities to the *Mac*'s GUI interface, brought further difficulties. In particular, while Gates had signed a legal statement to the effect that Microsoft would not use *Mac* technology for *Windows 1.0*, it did not extend to future versions of Windows. This resulted in Apple's losing the exclusive rights to its revolutionary, user-friendly icon-based interface. Although the late 1980s were a prosperous time for Apple, by 1990 the market was swamped with cheaper PC clones that could run the newly launched *Windows 3.0* operating system. While watching its market share erode, Apple nevertheless remained the design in-dustry's computer system of choice because of its superior graphics and publishing applications. This was especially true after the launch of the *Powerbook* in 1991 and the *PowerMac* in 1994. By the mid-1990s, Apple was yet again on the brink of collapse with $1 billion of back orders that it could not fulfil because it did not have the parts to make up the products. In 1996,

Apple logo

Jobs returned to Apple and began making significant structural changes to the company, after which Apple sold its products directly over the Internet and a cross-licensing agreement was drawn up with its one-time rival Microsoft. In 1998 Jobs oversaw the launch of the company's new landmark product, the *iMac* – an-other design from Apple that re-defined the personal computer. Importantly, the affordable, fun-coloured and translucent *iMac*, the design of which had been over-seen by the brilliant young indus-trial designer **Jonathan Ive**, tapped into the lucrative educational and home markets to the extent that,

by the autumn of 1998, it had become the best-selling computer in America. As Ive notes: "One of the primary objectives for the design of the *iMac* was to create something accessible, understandable, almost familiar." With the success of the *iMac* and its family of products, including the *iBook* and *G3* (1993), the death knell of the anonymous and essentially alienating grey box was finally tolled. Significantly, the *iMac* must also be the first computer to have stylistically influenced the design of other products – from tasking lamps to desk accessories. With its commitment to design excellence and new paradigm innovation, there is little doubt that Apple Computer Inc. will continue to challenge and redefine the personal computer market.

Apple Macintosh I
28 K, 1984

APRILIA

FOUNDED C. 1950
NOVALE, ITALY

Detail of Aprilia *SR
Stealth*, 1999

→Aprilia *SR Stealth*,
1999

The origins of Aprilia can be traced to a family-run company that manufactured motorised bicycles from around 1950. In 1968, Ivano Beggio joined the firm and from the late 1970s, the company committed itself to design and technological innovation. The company also established a networked production system, with smaller companies producing components for assembly at Aprilia's main factory in Scorzé, nearby its headquarters in Novale, just outside Venice. Beggio's love of racing inspired him to invest heavily in the company's motorcycle racing programme. For Aprilia, the racing circuit is an important testing ground for new developments, which are later incorporated into standard road models. In 1991 Aprilia became the first Italian motorcycle manufacturer to produce a scooter with plastic bodywork and a liquid cooling system – the *Amico 50*. A year later the company won its first World Motorcycle Championship and also produced 55,000 motorcycles and scooters. Within six years this number had soared to 300,000, making Aprilia Europe's second largest manufacturer of motorcycles and scooters, with five factories based in Italy. By commissioning designs from **Philippe Starck**, the company has also kept pace with the latest trends in stylistic taste. Its design centre at Novale maintains close links with the Institute of Applied Art and Design in Turin and the Faculty of Engineering at Padua University, where it has jointly set up a postgraduate course in Motorcycling Theory and Technology. Since 1997, Aprilia has also collaborated with the Art Center College of Design in Pasadena, California – one of the foremost styling and design research centres in America. Through these contacts

Aprilia *SR125*, 1999

with young designers, Aprilia has managed to forge a distinctive house style that, although centred on the idea of "creating emotions", never loses sight of the fundamental values of function and durability. Today, with its high-performance and exquisitely styled product range, Aprilia has captured some 60% of the European sportbike market and in 1999 its awesome *1000 cc RSV Mille* motorcycle was voted Motorbike of the Year by the German magazine *Motorrad*.

RON ARAD

BORN 1951
TEL AVIV, ISRAEL

Ron Arad studied at the Jerusalem Academy of Art from 1971 to 1973, and continued his training under Peter Cook at the Architectural Association in London. In 1981, Arad and Caroline Thorman established a design studio, workshop and showroom known as One Off London, which functioned as a forum for Arad's own furniture designs as well as those of other avant-garde British designers. Combining materials associated with High Tech, scaffolding poles and *objets trouvés*, Arad's early furniture resembled "readymade" art works. His *Concrete Stereo* (1984) was consciously distanced from the precision of standardized hi-fi equipment and accepted notions of "good form". Arad's later limited-edition mild steel furniture and lighting designs of the late 1980s, which required extremely costly labour-intensive techniques to produce, brought him widespread publicity and the attention of established manufacturers. Having moved into industrial production during the 1990s, Arad designed several commercially successful products, including his *Book Worm* shelving (1997), of which 1000 kilometres has been produced by **Kartell**. He also created innovative designs that exploited new materials and production techniques, such as his *Tom Vac* vacuum-formed **aluminium** chair (1998), which was later produced in plastic by Vitra, and his *Fantastic Plastic Elastic* chair (1998) for Kartell. In 1997 Arad was appointed Professor of Furniture Design at the Royal College of Art, London, and two years later Professor of Industrial Design. For his sculptural furniture, Arad now uses **computer-aided design** (CAD) systems, which allow him to translate sketches into three-dimensional models, thus considerably easing the transition from initial concept to finished products.

Book Worm for Kartell, 1997

BOOK WORM

Kartell

Packaging design for
White Rose, c. 1951

Having worked as a sports editor in 1916 for the *Tribune-Citizen* newspaper in Albuquerque, New Mexico, in 1917 Egmont Arens opened his own bookstore in New York. A year later he began printing newspapers under the imprint "Flying Stag Press" and edited the magazines *Creative Art* and *Playboy* (a journal specializing in modern art, not the Hugh Hefner publication of today). Arens later became editor of *Vanity Fair* magazine and also worked at the advertising agency of Earnest Elmo Calkins, where he began his career as an industrial designer by establishing an industrial **styling** department. Like other well-known industrial designers who gained prominence in America during the 1930s, Arens styled products for manufacturers so as to make them more alluring to consumers, for example the **packaging** for the grocery chain A&P (his *8 o'clock coffee* package is still in use). He termed this practice "consumer engineering" and wrote extensively on the relationship between design and marketing, most notably in his book *25 Years in a Package*. Arens' many products in the streamline style, including a range of spun **aluminium** saucepans, were an attempt to "design" America out of the Great Depression.

Egmont Arens &
Theodore Brookhart,
Meat slicer, *Model
No. 410 Streamliner*
for Hobart Manu-
facturing Company,
1941

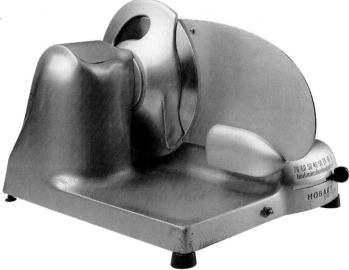

RICHARD ARKWRIGHT

BORN 1732 PRESTON, ENGLAND
DIED 1792 CROMFORD, ENGLAND

Patent drawing for
the *Water Frame*,
1769

Experimental
model of Spinning
Machine, c. 1769

The famous British inventor Richard Arkwright initially worked as a wig-maker – a vocation that allowed him to travel throughout Britain. His remarkable appetite for learning led him, however, to an interest in spinning technology. In 1764 he began designing and building his first spinning machine, which was patented five years later and became known as Arkwright's *Water Frame*. This water-powered device produced a yarn stronger than that spun by **James Hargreaves'** *Spinning Jenny*, making it suitable for warp weaving. Arkwright subsequently established factories in Nottingham and Cromford that made use of his revolutionary design. A few years later, he expanded his operations by opening several other factories that were able to carry out the complete textile manufacturing process – from carding to spinning. In 1773 Arkwright began manufacturing an all-cotton calico cloth, which revolutionized the British textile industry. As a result of his success, by 1782 Arkwright had a workforce of 5,000 and capital of £200,000. Four years later he was knighted for his spinning innovation that had led Britain to become the leading textile manufacturing nation in the world. Arkwright's *Water Frame* is thus recognized as one of the quintessential designs of the **Industrial Revolution**.

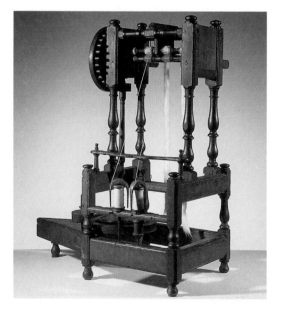

Emma Schweinberger-
Gismondi, *Dedalo*
umbrella stands, 1966

Vico Magistretti, *Eclisse*
lamp, 1965

ARTEMIDE

FOUNDED 1959
MILAN, ITALY

Artemide was founded by Ernesto Gismondi (b. 1931) in 1959. Dedicated to the manufacture of furniture and lighting products, it was quickly successful with designs by Vico Magistretti (b. 1920), such as the *Demetrio 45* table (1966), *Stadio* table (1967), and *Selene* and *Gaudí* chairs (1969 & 1970), which were initially manufactured in reinforced plastic and later in injection-moulded ABS. High quality designs such as these assisted in the general acceptance of **plastics** as noble materials and helped to bring increasing international attention to Italian design. Artemide's many landmark lighting products include Gianfranco Frattini (b. 1926) and **Livio Castiglioni**'s snake-like *Boalum* lamp (1969), Vico Magistretti's *Chimera* floor lamp (1966) and **Richard Sapper**'s *Tizio* task lamp (1972). Always at the forefront of progressive design, Gismondi provided exhibition space in the Artemide showroom for the design group Memphis during the early 1980s. Artemide has also produced lighting designs by **Michele De Lucchi**, **Enzo Mari**, **Ettore Sottsass** and Santiago Calatrava (b. 1951), and its products are represented in over 100 museums worldwide. During recent years, Artemide has manufactured lighting by **Norman Foster** and Ernesto Gismondi, whose *e.light* (2000) was designed to compliment **Apple Computer**'s *iMac* range.

ARZBERG

FOUNDED 1887
ARZBERG, GERMANY

Michael Sieger, Jug
and glass from the
Tric series, 1997

←Hermann Gretsch,
1382 service, 1931

The industrial manufacture of porcelain in the Bavarian town of Arzberg dates back to 1887, although it was only in 1928 that the Porzellanfabrik Arzberg – by now well established as a leading producer of high quality porcelain – assumed its official name. During the 1930s the factory manufactured functionalist wares, most notably Hermann Gretsch's *Form 1382* service. This design, which reflected the Modern Movement ethos of "Less is More", was described by one commentator in 1931 as "practical, unadorned, and timeless ... something good that is above the whims of fashion". Occasionally, however, this white dinner service was decorated by Arzberg with a pale blue, sprig-like floral pattern or with a simple dark red line, both of which somewhat detracted from the design's pure utilitarian aesthetic. Subsequent designs produced by Porzellanfabrik Arzberg (now a subsidiary of Winterling AG) also show a commitment to essential forms and practical use, as for example **Heinrich Löffelhardt**'s *Form 2000*, *Form 2025* and *Form 2050* tableware from the 1950s, **Hans Theo Baumann**'s *Form 3000* service from the 1970s and Michael Sieger's *Tric* range from the 1990s.

Ulrike Bögel,
Teaworld service,
1984

CARL AUBÖCK

BORN 1924 VIENNA, AUSTRIA
DIED 1993 VIENNA, AUSTRIA

Neopan microscope
for C. Reichert
Optische Werke,
1961

Through the activities of his father's workshop Carl Auböck was introduced to metal-casting, leather-working and wood-working at an early age. He later went on to study architecture at the Technische Hochschule in Vienna and subsequently trained at the Massachusetts Institute of Technology in Cambridge, Massachusetts. He also worked as an assistant lecturer under Professor Merinsky at the Institut für Baukunst und Ingenieurwesen in Vienna from 1950 to 1955. After leaving this post, he began working as an independent industrial designer developing a diverse range of products for industrial manufacture: steel cutlery (1957) and steel cookware (1962) for Neuzeughammer Ambosswerk; glassware services for Burg (1961 & 1966); an electric food-warmer for Ultra (1961); ski bindings for Tyrolia (1964 & 1968); school and institutional furniture for Wiesner-Hager (1965); **packaging** for Radio Köck (1975); cutlery for **Rosenthal** (1974); and the *Neopan* research microscope for C. Reichert Optische Werke (1963). Alongside this work, he designed numerous shops, factories and apartment buildings, mainly in Vienna, and created collections of women's fashions and ski-wear. He also designed glassware and cutlery in non-industrial materials such as brass, horn and wood for his family's workshop. Auböck regarded the role of the designer as that of problem-solver and believed that design practice should be broadly generalized, covering a multiplicity of disciplines, rather than becoming narrowly focused on particular areas of specialization.

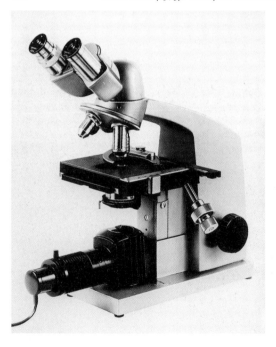

Poster announcing
the change from
Horch to Audi, 1910

AUDI

FOUNDED 1909 ZWICKAU, GERMANY
SINCE 1965 INGOLSTADT, GERMANY

In 1896 August Horch (1868–1951) was hired to supervise the mass production of cars at the factory owned by Karl Benz (1844–1929), who had patented his designs for a vehicle powered by an internal combustion engine ten years earlier. After three years with Benz, in 1899 Horch established his own automobile factory in Cologne, Horch & Cie. His first car, designed with a front engine and a new type of gearbox, proved successful and production capabilities rapidly expanded – by 1908 over 100 cars were being manufactured per annum. In 1909, when Horch left the company, Fritz Seidel was appointed its chief designer and Heinrich Paulmann its technical director. Now no longer associated with the company that bore his name, Horch set up another factory in Zwickau in 1909 and employed August Hermann Lange (1867–1922) as his technical director. The venture began producing cars in 1910 bearing the new Audi marque (the latin word for Horch). Audi 4-cylinder cars driven by Horch triumphed at the Austrian Alpine Runs during the

Wanderer W3
automobile, 1913

1910s, winning successive Alpine Trophies from 1912 to 1914 and bringing valuable publicity to the new company. Prior to the First World War, Audi had produced five passenger models and two truck models and had a manufacturing capability of 200 vehicles a year. After the war, German society was on the brink of economic collapse. Its currency had gone into free-fall and there was little if any demand for obsolete cars produced with out of date machinery. Between 1925 and 1929, however, German manufacturers began adopting factory-line assembly systems, which allowed the industry to double its productivity. During the 1920s, smaller cars for ordinary people were manufactured. Audi's first post-war model, the *Type K*, was developed by Hermann Lange. In Germany, this was the first car to be offered with left-hand drive and a central gear shift as standard. In 1932, the Audi and Horch companies, together with the vehicle manufacturer DKW and the automotive division of the Wanderer company, joined forces and formed Auto Union AG of Chemnitz – its logo of four rings symbolizing the four entities. The centralization of these companies brought about greater productivity and increased efficiency. Later in 1936, a Central Experimental Department was formed allowing Auto Union to become the first German manufacturer to undertake crash tests. At the end of the Second World War, Auto Union had lost whole factories – its Audi plant was now in the Soviet Military Zone. By 1949, with the help of bank loans, Auto Union re-emerged at a new production site and began producing vehicles, including those with the Audi marque, from its new assembly plant in the Bavarian town of Ingolstadt. By 1965, however, the "Four Rings" had become a wholly owned subsidiary of **Volkswagen**. A year later, a new range of Audis was produced, including the

Audi 920, 1939

well-equipped Audi *Super 90*. More important models in re-establishing the marque, however, were the Audi *100* (1970), which was developed clandestinely by Ludwig Kraus, and the Audi *80* (1972) by Nuccio **Bertone**. From 1979, the company decided to move Audi up-market and this resulted in the appearance of the high performance Audi *Quattro* in the spring of 1980. This powerful and solidly built car not only epitomized German engineering excellence, but also revolutionized the international rallying scene. These three Audi models served as design blueprints for subsequent models until the introduction of the Audi *A4*, *A6* and *A8* and sporting *S* versions in 1994. Recently, Audi has developed the light weight, safe and economical *A2* – the world's first volume-production **aluminium** car. Audi sees itself as defining the future with designs such as these, which are born out of its commitment to "cutting edge technology and visionary design". In 1996 Audi announced its sponsorship of the International Audi Design Award, with the objective of creating a forum for the inter-disciplinary exchange of ideas. According to Audi, "progress is not a matter of abolishing yesterday, but of retaining its essence ... to create a better today". Its famous motto, "Vorsprung durch Technik" (Advancement through Technology) succinctly sums up the company's guiding principles.

Audi *TT Coupé*
Quattro, 1999

AUTHENTICS

FOUNDED 1980
HOLZGERLINGEN NEAR STUTTGART, GERMANY

Hansjerg Maier-
Aichen, *Rivista*
magazine rack, 1997

Authentics was established in 1980 to manufacture "products of everyday life". It was founded by Hansjerg Maier-Aichen (b. 1940), who studied interior design in Wuppertal and painting at the Staatliche Hochschule der Bildenden Künste, Munich, as a subsidiary of Artipresent GmbH, of which Maier-Aichen was managing director. Using state-of-the-art **plastics** moulding technology and an essentialist approach to design, Authentics has attempted to raise the status of synthetic materials and to "create sensible and appropriate solutions". Its products – CD-holders, magazine racks, laundry baskets, waste bins, buckets, lamps, soap dispensers and so forth – have been designed by an international array of talented young designers, including Matthew Hilton (b. 1957), Konstantin Grcic (b. 1965) and Sebastian Bergne (b. 1966). Through the quality of their products and the intelligent exploitation of the inherent characteristics of synthetic materials, such as translucency, texture and colour, Authentics have "restored the dignity" of thermoplastics, and in particular polypropylene. While its designs are often copied, Authentics remains at the forefront of the rapidly growing home accessories market, having succeeded in creating a powerful brand identity through its promotion of "minimal, economical and ecological design concepts".

Konstanin Grcic,
2 Hands baskets,
1995 & 1997

CHARLES BABBAGE

BORN 1791 LONDON, ENGLAND
DIED 1871 LONDON, ENGLAND

→ Detail of Charles
Babbage's *Difference
Engine No.1*, begun in
1824 and assembled
by Joseph Clement in
1832

The calculating engines which the mathematician Charles Babbage developed have led him to be called "The Father of Computing", in recognition of his remarkable contribution to the basic design of the computer. Around 1812/13 Babbage came up with the concept of mechanically calculating mathematical tables. After co-founding the Analytical Society with Herschel and Peacock in 1820, in 1823 Babbage began work on his *Difference Engine No.1* with funding from the British Government. His initial scheme called for 6 decimal places and a 2nd-order difference. While only approximately 1/7th of the the *Difference Engine No.1* was actually assembled in 1832, it was the first successful automatic calculator. In 1834 Babbage began developing a more ambitious version, the *Analytical Engine*. This improved device, the design of which called for 20 decimal places and a 6th-order difference, was intended to be capable of any mathematical operation. Despite being appointed Lucasian Professor of Mathematics at Cambridge University –

Engraving showing
section of Charles
Babbage's *Difference
Engine* built in 1832

a chair held earlier by Isaac Newton – Babbage could not persuade the British Government to finance the construction of the *Analytical Engine*. While his calculating engines were never fully completed, they were remarkably early precursors of the digital calculator and, ultimately, the modern-day computer – arithmetic problems were solved through instructions introduced to the machines via punched cards. They also had a memory unit that could store numbers and a sequential control. Babbage's extraordinary development of the world's first universal digital computer remained largely overlooked, however, until his unpublished notebooks were rediscovered in 1937.

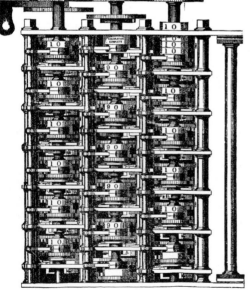

B. H. Babbage

Land Rover *Long 109-inch* 10-seater station-wagon, 1958

DAVID BACHE

BORN 1926 BIRMINGHAM, ENGLAND
DIED 1995 SOLIHULL, ENGLAND

During the late 1940s, David Bache began his apprenticeship as a designer for the motor manufacturer Austin, which merged with Nuffield in 1949. Bache initially worked under the Italian designer Riccardo Burzi, who had designed for Austin since the 1920s and who had previously been employed by Lancia. As an assistant, Bache designed the instrument panel for the Austin *A30*. In 1954 he moved to Rover, where his first vehicles to go into production were the re-styled landmark *Series II* Land Rover (1958) and the Rover *P5* (1959). Bache redesigned the Land Rover from the original farm vehicle by altering its proportions and cut-lines and by adding flash-gaps. The *Series II* has been described as the definitive Land Rover and is a classic British design. In 1963 Bache designed the sleek and sophisticated Rover *2000*, but due to the turmoil brought on by nationalization of the British car industry, some of his best designs, many of which were produced with the assistance of his engineering partner Spen King, were never realized. These included the *P9*, which in its day was compared favourably to the **Porsche** *911*. Bache did, however, see his highly influential *Range* Rover (1970) and *P10* (1975) go into production and achieve much success.

Land Rover *Series II,* 1958

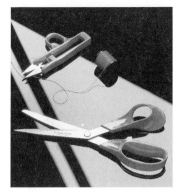

OLOF BÄCKSTRÖM

BORN 1922 FINLAND
DIED FINLAND

O-Series scissors for
Fiskars, 1967

Olof Bäckström trained as an electronic engineer, a
career he pursued for eleven years before focusing his
attention on woodcarving and design in 1954. He initially produced wooden domestic ware, some of his
designs winning a silver medal at the XI Milan Triennale
in 1957. In 1960 Bäckström designed a range of cutlery
for camping that was also awarded a silver medal at the XII Milan Triennale
that same year.

In 1958 he became an industrial designer for the Helsinki-based manufacturer **Fiskars**, where his first project was a range of tableware made of
melamine. Between 1961 and 1967, Bäckström developed his groundbreaking *O-Series* scissors. Launched in 1967 (a left-handed variation came out
in 1971), the scissors had distinctive bright orange ABS handles that ergonomically conformed to the shape of the hand. As a highly proficient wood-
carver, Bäckström initially prototyped the design in wood before it was
translated into plastic. Fiskars continue to manufacture a modified version
of Bäckström's scissors, which demonstrate so clearly that **ergonomics** can
dramatically enhance comfort and performance.

Fiskars scissors
closely based on
Olof Bäckström's
earlier design, 1990s

First image of a
human face received
by a Baird *Televisor*
in 1926

→Baird TV, 1936

JOHN LOGIE BAIRD

BORN 1888 HELENSBURGH, SCOTLAND
DIED 1946 BEXHILL-ON-SEA, ENGLAND

John Logie Baird studied electrical engineering at the
University of Glasgow. In 1922 he began researching
and developing relatively primitive television equipment
By 1924 he was able to transmit images of objects in
outline and a year later managed to transmit images of
human faces which, although quite blurred, were just
about recognizable. On 26 January 1926, at the Royal Institution in London,
Baird televised the first images of objects in motion. Three years later, the
German post office provided him with the facilities to develop a television
service. Apart from black-and-white images, Baird also demonstrated the
transmission of colour images and achieved transatlantic television transmission in 1928. In 1930 he showed his first big-screen television and, two
years later, demonstrated ultra-short wave transmission. When the British
Broadcasting Corporation (BBC) began broadcasting its first regular high-
definition television service in 1936, however, John Logie Baird's system
was being seriously rivalled by **Marconi**'s EMI system, which was eventually
adopted by the BBC in 1937. As the inventor of modern television, John
Logie Baird continued to pioneer the design of television equipment during
the 1940s and shortly before his death in 1946, completed research into
stereoscopic television.

John Logie Baird
adjusting the
receiving apparatus
on his early wireless
vision apparatus,
1926

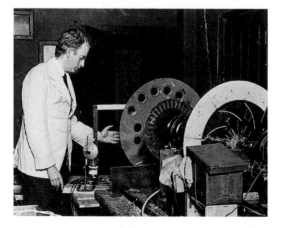

BeoSound 9000 CD
player, 1996

BANG & OLUFSEN

FOUNDED 1925
QUISTRUP NEAR STRUER, DENMARK

As a child, Peter Bang (1900–1957) experimented with
radio equipment and built a mains receiver that required
neither batteries nor an accumulator. He later trained
as an engineer at the Århus Electrotechnical School and
graduated in 1924. While there, Bang studied alongside
Svend Olufsen (1897–1949), who subsequently conducted radio experiments at his family's estate in Quistrup. While attempting to build a mains
receiver, he turned for help to Bang, who had just returned from America
where he had been researching the latest developments in radio technology.
In November 1925, Bang & Olufsen was officially founded and the two friends
began working on an aggregate to connect battery receivers to the mains
power supply – the *B&O Eliminator*. In 1929 they introduced their own
mains-powered radio known as the *Five Lamper*, which was criticized by the
architect Ole Wanscher for its old-fashioned aesthetic. During the Second
World War, B&O established an illegal radio service for communications
between the British and the Danish Freedom Fighters. Because of the company's links with the resistance, B&O's factory was bombed in 1945. The
factory was quickly rebuilt and production of radios resumed, but again
the company was criticized, this time by the celebrated designer Poul Henningsen (1894–1967), who said its products appeared as "a mass of in-
competence". To justify the company's by-line – "The Danish Hallmark of
Quality" – in the mid-1950s B&O began working with some of Denmark's
leading architects and designers,
such as **Sigvard Bernadotte** and
Acton Bjørn. These collaborations
produced spectacular results –
the company's new-generation,
high-quality products were completely integrated and, through
their refined Modern forms, communicated a sophisticated design
rhetoric to a new and highly receptive design-conscious audience.
In the 1960s B&O pioneered the

Jakob Jensen,
Beogram 1200 record
player, 1969

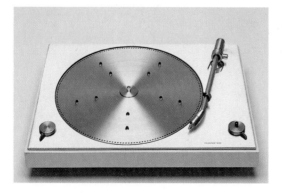

David Lewis,
Beosystem 2500
music system,
1991

design of integrated stereo record players, "for those who discuss taste and quality before price". Having become a pioneer of audio technology, the company later launched several landmark designs by **Jakob Jensen**, which married superlative quality with sleek good looks. Jensen's "easy touch" layouts and "flush-designed" systems quite literally redefined the design vocabulary for hi-fi equipment. During the 1980s, however, the company suffered badly from Asian competition and eventually formed an alliance with **Philips** in order to survive. In the early 1990s B&O changed its approach to manufacturing, moving strategically from mass production to order-production. There was also an important shift in terms of design, implemented by David Lewis, who had been Jensen's assistant. Lewis' revolutionary *Beosystem 2500* (1991) with its "all in one system" set itself apart from the anonymous stacked black boxes produced by B&O's competitors. Believing that aesthetics were as important to consumers as function and ease of use, Bang & Olufsen skillfully balanced external design with internal technological considerations in order to create distinctive products that powerfully communicated the company's values and vision about technology and concepts. Today, Bang & Olufsen continues its commitment to design and remains a distinguished leader in its field.

OSCAR BARNACK

BORN 1879 LYNOW, GERMANY
DIED 1936 BAD NAUHEIM, GERMANY

Oscar Barnack's design for the **Leica** *I* of 1925 was nothing short of a photographic revolution. Prior to Barnack's designs for the manufacturer Ernst Leitz Wetzlar (later renamed Leica), 35 mm cameras had been relatively bulky and hence difficult to use. Barnack's first 35 mm Leica camera, the *UR-Leica*, was designed and prototyped in 1913 and was eventually put into production around 1918 as the *Leica A*. Rather than using rolls of perforated cine-film, this groundbreaking camera utilized a ridge film cassette, which allowed it to be portable and light. Both the *Ur-Leica* and the later *Leica A* of 1925 (which had a film winder and controls) were highly convenient cameras that had optically superior *Elmar* lenses developed by Max Berck. The *Leica I*, with its distinctive "hockey stick" element on the front of the camera, was launched to great acclaim at the 1925 Leipzig Fair. It set the standard for subsequent Leica designs and ultimately changed the art of still photography. The *Leica I* was designed to be extremely versatile and could be used to photograph anything from microscopic specimens, portraits and landscapes to even the outer limits of the visible solar system. Barnack's designs, which were completely functionally conceived, possessed a very pure quintessentially Machine Age aesthetic.

Leica I camera, 1930

Josef Albers,
Teaglass, 1926

BAUHAUS

1919–1933 WEIMAR,
DESSAU & BERLIN, GERMANY

The Weimar State Bauhaus was born in April 1919 out of the merger of the city's School of Arts & Crafts and Academy of Art into one new interdisciplinary school of craft and design. For **Walter Gropius**, the school's first director, construction was an important social, symbolic and intellectual endeavour and this sentiment pervaded Bauhaus teaching. While based in Weimar, the Bauhaus promoted an inter-disciplinary approach to the arts and was fundamentally craft-based. The curriculum included a one-year preliminary course where students were taught the basic principles of design and colour theory. After completing this foundation year, students entered the various workshops and trained in at least one craft. These workshops were intended to support themselves financially through private commissions. During the earliest period of the Bauhaus, it was Johannes Itten (1888–1967) who played the most important role within the teaching staff. He taught theories of form, colour and contrast as well as the appreciation of art history. Itten believed that spatial composition was governed by natural laws and students were taught the importance of simple geometric forms such as the circle, square and cone. These elemental shapes came to exemplify Bauhaus design, which was characterized by unadorned sur-

Marianne Brandt,
Tea and coffee
service, 1924
(re-edited by Alessi,
1985)

faces and functional forms. Eventually, conflict arose between Gropius and Itten and the latter left in March 1923, marking the end of the Bauhaus' Expressionist period. Josef Albers (1888–1976) and László Moholy-Nagy (1895–1946) were appointed as Itten's successors and pursued a more industrial approach, with students being taken on factory visits. In 1923 the Bauhaus staged a landmark exhibition highlighting the full scope of its activities. One development seen at this exhibition was the new image that the Bauhaus had forged for itself – the graphics from this period were self-consciously Modern, incorporating sans serif "New Typography". The exhibition, which also featured De Stijl designs, won the school international critical acclaim. This success was short-lived, however, when the right-wing parties won the majority vote in the State elections in Thuringia in 1924, the Bauhaus' budget was slashed by half. In 1925 the Masters voted to dissolve the school in Weimar and Gropius relocated the Bauhaus to Dessau, where the ruling Social Democrats and the liberal mayor were far more receptive to

its ideals. This industrial city offered the Bauhaus the financial support it so desperately required on the understanding that the school would part-fund itself through the manufacture and sale of the designs it produced. The money it put up was sufficient to allow a new purpose-built school to be constructed, and in 1926 the Bauhaus moved into its newly-completed Dessau headquarters designed by Walter Gropius.

The Dessau Bauhaus, with its highly rational pre-fabricated building, marked an important turning-point in the school's evolution away from craft and towards industrialized production. By now, Gropius had become disillusioned with socialism; he believed that Henry **Ford**'s type of industrial capitalism could benefit workers and that, in order to survive, the Bauhaus needed to adopt an industrial approach

→Adolf Meyer &
Walter March
(construction);
Georg Muche
(design), Model
kitchen for the Haus
am Horn, Weimar,
1923

to design. In the conviction that a better society could be created through the application of functionalism, Bauhaus designs were now conceived for industrial production and a machine aesthetic was consciously adopted. **Marcel Breuer**'s revolutionary series of furniture, commencing with his *B3* chair (1925–1927), which was designed for Wassily Kandinsky's staff quarters at the Dessau Bauhaus, employed state-of-the-art **tubular metal** and utterly transformed the language of chair design and production. This group of furniture was later mass-produced by Standard-Möbel and Gebrüder **Thonet** from around 1928. In November 1925, with the financial support of Adolf Sommerfeld, Gropius realized his long-held ambition of establishing a limited company to promote and retail the school's designs. Bauhaus GmbH duly produced a catalogue, which illustrated Bauhaus products, but sales of these items were far from overwhelming. For the most part, this was no doubt due to the severity of their aesthetic. Furthermore, although they looked machine-manufactured, the majority of Bauhaus GmbH products were in fact unsuitable for industrial production. A few licensing agreements were drawn up between the Bauhaus and outside manufacturers, but these did not bring in the revenues that Gropius had hoped for. Eventually, the Swiss architect Hannes Meyer (1889–1954) took over from Gropius as director of the school, which by now bore the subtitle "Hochschule für Gestaltung" (Institute of Design), reflecting its shift away from the arts and crafts towards functionalism. Meyer, who was a Communist, believed that form had to be governed by cost as well as function, so that products would be not just practical but also affordable for working-class consumers. During his tenure, the Bauhaus' approach to design became more scientific. At this time, the Bauhaus also became more politicized with the school site being used as the focus for the political activities of a group of Marxist students. In 1930 Ludwig Mies van der Rohe (1886–1969) took over as director of the Bauhaus. Under pressure to de-politicize the school for its own survival, he established a new curriculum with the preliminary course becoming non-compulsory. The study of architecture was given greater importance and although the applied art workshops continued, their remit was to supply only products that could be industrially manufactured. In October 1931, however, the National Socialists swept to power in Dessau, too, winning 19 out of 36 seats, and in August 1932 their motion proposing the closure of the school was passed. Briefly re-established by Mies as a private school in Berlin, in July 1933 the Masters themselves voted to dissolve the Bauhaus – formally marking the end of this remarkable institution. As one of the first art schools to pioneer the teaching of industrial design, the Bauhaus had a fundamental impact on subsequent industrial design education and practice.

HANS THEO BAUMANN

BORN 1924
BASEL, SWITZERLAND

Hans Theo Baumann studied painting, sculpture and design in Dresden from 1943 to 1945 and interior design, graphics and stained glass at the Kunstgewerbeschule in Basel from 1946 to 1950. In 1953 Baumann worked under Egon Eiermann (1904–1970) on stained glass panels for the Pforzheimer Kirche and the Gedächtniskirche in Berlin. Two years later he established his own product design office in Schopfheim. Since then he has specialized in the design of **ceramics** and glassware. Using simple, unadorned geometric forms, which he softens by rounding off edges sculpturally, Baumann's designs combine the visual qualities of refinement and sturdiness. His clients have included, most notably, **Arzberg**, **Rosenthal**, KPM, Thomas, Daum, Süssmuth and Schönwald. The Baumann Product Design studio was founded in Basel in 1971 and another office established later in Gaillan en Médoc, France. His ceramic and glassware designs, such as the *Brasilia* cof-

Form 5500, Brasilia service for Arzberg, 1975

fee and tea service (1975) for Arzberg, have brought him significant international recognition. Baumann has also designed textiles, furniture and lighting that express a similar pared-down aesthetic. In 1970, Baumann helped establish a chair for ceramics teaching at the National Institute of Design in Ahmedabad, India, and from 1983 he was a professor at the Hochschule der Künste in Berlin. Baumann's elegant yet robust forms are born out of a deep understanding of materials and their sculptural potential. He believes that "the importance of a design does not spring from the formulas of mathematical calculation – it is solely the result of the ability of the human being to add or take away. It is possible to bring weight, material, purpose, and function into such a relationship with one another that the object which results is beautiful."

Delta service for
Arzberg, 1979

TREVOR BAYLIS

BORN 1937
LONDON, ENGLAND

Trevor Baylis grew up in Southall and left school at the age of 15. During his national service in the 1950s he built himself a go-kart in the barracks. After a succession of careers, including that of an underwater stunt artist, and with many years of invention behind him, Baylis developed a clockwork generator that could power an **Apple** laptop computer for 16 minutes before crashing. While watching a BBC television programme on AIDS in Africa in 1991, Baylis had a "eureka moment" and realised that a clockwork-powered radio could bring practical help to third world countries that did not have access to stocks of inexpensive batteries. Having successfully developed and patented the clockwork radio, his requests for manufacturing assistance were turned down by various established companies, including **Philips,**

Freeplay torch, 1999

Marconi and BP as well as the **Design Council** of Great Britain. Baylis' luck changed, however, when his revolutionary design was featured on the BBC television programme, "Tomorrow's World". With the necessary investment subsequently secured and with the assistance of the industrial designer, Andy Davey (b. 1962), Baylis was able to develop the first production version of the *Freeplay* wind-up radio, which was awarded first prize in the Products Category in the 1996 BBC Design Awards. Since then, Baylis has developed a wind-up torch and is currently developing, among other things, a wind-up computer. The manufacturer of the self-powered radios and torches, Freeplay, now sells over 100,000 units per month, which are manufactured in two plants in Cape Town that are run by a disabled workforce. Since the launch of Baylis' ingenious radio in 1994, nearly two million wind-up products have been sold. Baylis hopes to found an Academy of Invention with the proceeds gained from the sale of his products, which will encourage what he describes as "thinking with the hands".

Freeplay radio, 1999

PETER BEHRENS

BORN 1868 HAMBURG, GERMANY
DIED 1940 BERLIN, GERMANY

Peter Behrens studied at the Kunstgewerbeschule in Hamburg, the Kunstschule in Karlsruhe and at the Düsseldorfer Akademie from 1886 to 1889.
In 1890 he began working as a painter and graphic designer in Munich,
where he came under the influence of the Jugendstil movement. During this
period he produced colourful woodcuts, illustrations and book-bindings in
the Jugendstil style and in 1892 became a founding member of the Munich
Secession, a progressive group of exhibiting artisans. In 1897, in a joint venture with Hermann Obrist (1863–1927), August Endell (1871–1925), Bruno
Paul (1874–1968), Richard Riemerschmid (1868–1957) and Bernhard Pankok
(1872–1943), Behrens co-founded the Vereinigte Werkstätten für Kunst im
Handwerk (United Workshops) in Munich, dedicated to the mass production of handcrafted everyday objects. From 1899 to 1903 Behrens was an
active member of the Darmstädter Künstlerkolonie (Darmstadt artists'
colony), which had been initiated by Grand-Duke Ernst-Ludwig of Hesse-
Darmstadt, the grandson of Prince Albert. In Darmstadt, Behrens designed
his first building, the Behrens House. This project was conceived as a *Gesamtkunstwerk* (complete-art-work), with furniture and even glassware being
specially designed for it. The house marked an important point of departure
for Behrens – away from Jugendstil and towards a more rational approach
to design. Between 1902 and 1903, Behrens gave master classes at the Bay-
erisches Gewerbemuseum in Nuremberg and exhibited at the "Esposizione
Internazionale d'Arte Decorativa Moderna" in Turin in 1902. From 1903 to

Kettles illustrated in
an AEG catalogue
from 1909/1910

1907 he was director of the Kunstgewerbeschule in Düsseldorf. The
commercial logic of the concept of
"industrial art" led the founder of
AEG, Emil Rathenau (1838–1915) –
at the instigation of Paul Jordan
(director of the AEG factories) – to
appoint Behrens artistic director to
the company in 1907. This was the
first time a company had employed
a designer to advise on all aspects
of design. In this capacity, Behrens

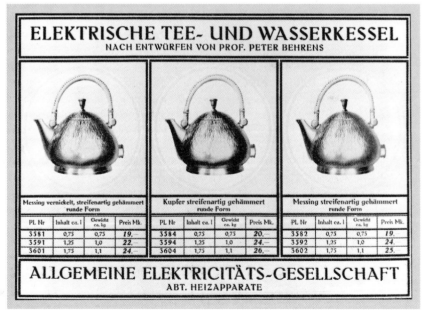

ELEKTRISCHE TEE- UND WASSERKESSEL
NACH ENTWÜRFEN VON PROF. PETER BEHRENS

AUSFÜHRUNG:

Leicht auswechselbare Patronen-heizkörper, komplett, mit Anschlußstöpsel, 2 m Litze und Stecker Pl. Nr 725

Die 0,75 l Kocher haben einen Stromverbrauch von 275 Watt, die 1,25 und 1,75 l Kocher einen Stromverbrauch von 440 Watt

Messing vernickelt, glatt ovale Form

Pl. Nr	Inhalt ca. l	Gewicht ca. kg	Preis Mk.
3585	0,75	0,75	19,—
3595	1,25	1,0	21,—
3605	1,75	1,1	23,—

Messing glatt, matt ovale Form

Pl. Nr	Inhalt ca. l	Gewicht ca. kg	Preis Mk.
3586	0,75	0,75	19,—
3596	1,25	1,0	22,—
3606	1,75	1,1	24,—

ALLGEMEINE ELEKTRICITÄTS-GESELLSCHAFT
ABT. HEIZAPPARATE

ELEKTRISCHE TEE- UND WASSERKESSEL
NACH ENTWÜRFEN VON PROF. PETER BEHRENS

Messing vernickelt, streifenartig gehämmert runde Form

Pl. Nr	Inhalt ca. l	Gewicht ca. kg	Preis Mk.
3581	0,75	0,75	19,—
3591	1,25	1,0	22,—
3601	1,75	1,1	24,—

Kupfer streifenartig gehämmert runde Form

Pl. Nr	Inhalt ca. l	Gewicht ca. kg	Preis Mk.
3584	0,75	0,75	20,—
3594	1,25	1,0	24,—
3604	1,75	1,1	26,—

Messing streifenartig gehämmert runde Form

Pl. Nr	Inhalt ca. l	Gewicht ca. kg	Preis Mk.
3582	0,75	0,75	19,
3592	1,25	1,0	24,
3602	1,75	1,1	25.

ALLGEMEINE ELEKTRICITÄTS-GESELLSCHAFT
ABT. HEIZAPPARATE

produced designs for workers' housing and factories, including the concrete, steel and glass AEG Turbine Factory (1908–1909), which was one of
the first true expressions of Modern industrial architecture. Beyond buildings, Behrens also undertook the design of electrical products such as
kettles, fans and clocks, which incorporated standardized components
that were interchangeable between products so as to rationalize production
methods. Behrens was also responsible for the graphics used by the company and created a strong and highly unified **corporate identity** for it. In
October 1907, shortly after his appointment at AEG, Behrens co-founded
the **Deutscher Werkbund** together with Peter Bruckmann (1865–1937),
Josef Maria Olbrich (1867–1908), Fritz Schumacher (1869–1947), Richard
Riemerschmid and Hermann Muthesius (1861–1927). The Deutscher
Werkbund was inspired by the Arts & Crafts Movement in Britain and
attempted to revive the status of craftsmanship and apply it to industrial
production. Its members actively promoted the unification of art and industry and Behrens contributed regularly to its journal, *Die Form*. As a pioneer
of the Modern Movement and a high-profile member of the Deutscher
Werkbund, Behrens realised that **standardization** and, thereby, a rational
approach to design had to be adopted if industrially produced goods were
to achieve the high level of quality found in handcrafted products. In 1907
Behrens also established a major architectural and design office in Berlin,
working alongside **Walter Gropius**
from 1907 to 1910, Ludwig Mies
van der Rohe (1886–1969) from
1908 to 1911 and Le Corbusier
(1887–1965) from 1910 to 1911. This
highly prolific practice undertook
numerous architectural commissions, including the German Em-
bassy in St Petersburg (1911–1912),
and much industrial design work
over the following years. After the
First World War, Behrens' architectural style moved away from "scrap-
ed classicism" and became influenced by Expressionism, as can
be seen in his design for the Frankfurt office building of I. G. Farben
Hoechst (1920–1925). In 1926
Behrens designed New Ways, a

Electric clock for
AEG, 1929

→Electric light fitting
for AEG, 1907

house in Northampton for the British industrialist Wynne Bassett-Lowke, which was the first complete example of Modern Movement architecture in Great Britain. By the 1930s, Behrens was working in the International Style, as evidenced by his State Tobacco Administration warehouse in Linz (1930). Behrens also designed porcelain for Manufaktur Mehlem Gebrüder Bauscher in Weiden, glassware for Rheinische Glashütten in Cologne and geometrically patterned linoleum for Delmenhorster Linoleum Fabrik. From 1922 to 1936, he was director of architecture at the Akademie der Bilden-den Künste in Vienna, where he gave master classes. In 1936 he became director of the architecture department at the Preußische Akademie der Künste in Berlin, a position he held until his death. As one of the very first industrial designers, Behrens was the most influential German practitioner of design in the 20th century. His simple, practical and rational design solutions were hugely important in the formation and dissemination of Modernism. Behrens believed that: "Technology has thus far contrived to reach a peak only in terms of material existence, for the unity of the material and the spiritual – that is, the values of culture – has never achieved formal expression. Indeed, nobody has given a moment's thought to it ... Only through industry have we any hope of fulfilling our aims. It alone can save us from economic misery." Behrens helped to promote the practice of industrial design consultancy and through his commitment to standardization and technological progress, managed to foster greater professionalism within this fledging field of activity. His pioneering work in the realm of corporate identity, through the implementation of a coherent house design strategy at AEG, was also of much influence upon later companies such as **Braun**.

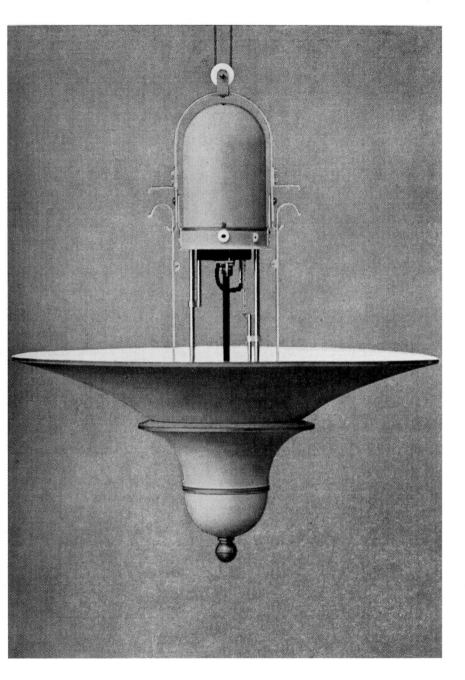

The American audiologist Alexander Graham Bell is best remembered as the inventor of the telephone. He was the third generation of his family, who originated from Edinburgh, to be recognized as a leading authority in elocution and speech therapy. His father's publication, *Standard Elocutionist*, was so popular that it went into nearly 200 editions. Trained by his father, Alexander initially worked as an elocution and music teacher in Elgin, County Moray. In 1864 he became a residential master at the Weston House Academy in Elgin, and it was there that he began his first researches into the nature of sound. Four years later he moved to London, where he became his father's assistant. Following the untimely death of his eldest brother from tuberculosis and the ill health of his younger brother, who had also contracted the disease, in 1870 the surviving Bell family, including Alexander, emigrated to Ontario, Canada. A year later, in Boston, Alexander lectured on and demonstrated his father's system of teaching people with impaired hearing how to speak. In 1872 he founded his own school for training teachers of the deaf, and a year later was appointed professor of vocal physiology at Boston University. While there, Bell began experiments with a young mechanic named Thomas Watson, which resulted in the development of

Alexander Graham Bell speaking for the first time on the newly established telephone connection between Chicago and New York, 18th October 1892

an apparatus that could transmit sound electrically. Bell was granted a patent for a telegraph machine that could send multiple messages, and began outlining the specifications for his new invention, the telephone, which was patented in 1876. That same year, Bell set up the **Bell Telephone Company** to commercialize his invention. Almost immediately, this attracted the attention of market rivals such as the Western Union Telephone Company, which attempted unsuccessfully to infringe Bell's patent. In France in 1880, Bell was awarded

Early Bell telephone and terminal panel, 1877
This example was used by Queen Victoria at Osbourne Cottage in Southampton to communicate with Alexander Graham Bell in London on 14th January 1878.

the Volta prize of 50,000 francs. This enabled him to establish the Volta Laboratory, where with Charles Sumner Tainter and his cousin Chichester Bell, he developed the "Graphophone" in 1886 – a precursor of **Emil Berliner**'s Gramophone. After 1900, Bell experimented with man-carrying kites and undertook research into a number of diverse fields, from sonar detection to hydrofoil watercraft. Bell's revolutionary designs quite literally changed the world of communications and entertainment and in so doing heralded the birth of the Modern Age.

BELL HELICOPTER TEXTRON

FOUNDED 1941
GARDENVILLE, NEW YORK, USA

UH-1D helicopters
coming into a
forward landing
base in Vietnam,
1967

Immediate-response
combat unit of the
US Army unloading
from a Bell "Huey"
helicopter in Vietnam,
late 1960s

In the early 1930s, a brilliant young inventor named
Arthur Young produced and demonstrated a model
of a helicopter that could fly. Impressed by the design,
Larry Bell, the founder of the Bell Aircraft Company, set
up a workshop for Young in Gardenville in 1941. Bell
had already earned considerable recognition for its conventional aircraft,
including the *P-39 Airacobra* and the *P-59*, America's first jet-powered aeroplane. Bell later went on to develop the first supersonic aircraft, the *X-1*. In
1945, with Young's helicopter about to enter production, Larry Bell incorporated Young's unit into the Bell Aircraft Corporation and moved his main
plant to a larger site in Niagara Falls. Bell's first helicopter prototype, the
Model 30, could reach speeds of 100 mph in level flight. This was succeeded
in 1946 by the *Model 47*, which was granted the world's first helicopter licence and heralded the beginning of a brand new industry. That same year,
Bell began supplying helicopters to the US military. In 1957 Bell's helicopter
division was restructured as the Bell Helicopter Corporation and this was
eventually purchased by Textron in 1960. The *Model 47* was widely used during the Korean War, and over the succeeding years other, more advanced
military and commercial rotorcraft were developed, including the success-

ful *206 JetRanger* (1960–1965) and
the *UH-1D Iroquois*, nicknamed the
"Huey" (1960–1963). This legendary
helicopter was used extensively in
Vietnam for medical evacuations,
gun-carrying and anti-tank missile
firing and became an iconic symbol
of the war. Since 1946 Bell Helicopter has produced more than 34,000
military and commercial helicopters. Today, the company is recognized as the worldwide leader in the
design and manufacture of rotary-
wing aircraft, and continues to develop highly innovative rotorcraft.

BELL TELEPHONE COMPANY

FOUNDED 1876
NEW YORK, USA

Henry Dreyfuss,
Model 300 telephone
for the Bell Telephone
Laboratories, 1937

Bell Telephone
Company
advertisement, *Look*
magazine, 1948

Alexander Graham Bell established the Bell Telephone Company in New York in 1876 as a means of commercializing his telephone patent of the same year. In 1881 Bell purchased Gray & Barton, the manufacturing company of his competitor, Western Electric. With this acquisition, Bell secured a virtual monopoly of both equipment and transmission technology within the fledgling telephone industry. In 1900 Bell formed the American Telephone & Telegraph Company (AT&T), which in 1925 set up a research division known as Bell Telephone Laboratories. Since then, Bell Laboratories have developed literally thousands of revolutionary inventions and associated product designs, such as the world's first synchronized sound-and-motion picture system (1926), a pioneering electrical relay digital computer (1937) and the transistor (1947). Researchers working for this prestigious research and development facility were awarded Nobel Prizes for physics in 1937, 1947 and 1978 for their innovative contributions to the communications field. The Bell Telephone Company also used the skills of independent design consultants, such as **Henry Dreyfuss**, designer of the historic *Model 300* telephone. AT&T's Bell Laboratories continue to innovate new technology applications and commission design consultancies such as **Frog-design** to assist in the development of state-of-the-art communication tools such as the *Personal Communicator* (1992).

GOOD NEWS about Long Distance

Today, Long Distance calls go through in about two minutes on the average. Now and then there are delays, but we're handling nine out of ten calls while you hold the line.

We've added many new circuits and switchboards and more are coming along. Our operators know their job.

Our aim is to put your out-of-town calls through faster than ever before. And we're working hard to do it.

BELL TELEPHONE SYSTEM

MARIO BELLINI

BORN 1935
MILAN, ITALY

Mario Bellini studied architecture at the Politecnico di Milano, graduating in 1959. From 1959 to 1962 he was design director at La Rinascente, the influential chain of Italian department stores that did much during the post-war years to promote industrial design. In 1962 he founded an architectural office with Marco Romano and in 1973 established Studio Bellini in Milan. As chief design consultant to **Olivetti**, a post he has held since 1963, he designed the *Divisumma 18/28* calculators (1973) and the *Praxis 35* and *Praxis 45* typewriters (both 1981). From 1969 to 1971 he was president of ADI (Associazione per il Disegno Industriale) and in 1972 he showed a mobile micro-living environment entitled *Kar-a-Sutra* at the "Italy: The New Domestic Landscape – Achievements and Problems of Italian Design" exhibition held at the Museum of Modern Art in New York. This led to his appointment as a research and design consultant to the car manufacturer **Renault** in 1978. During the 1970s, Bellini organized workshops to explore the complex relationships between humans and their man-made environment – a theme that has informed virtually all his work. From 1986 to 1991 Bellini was editor

Logos 50/60 adding machine for Olivetti, 1972

→ *Divisumma 18* adding machine for Olivetti, 1973

UHF 33 adding machine for Olivetti, 1981–1982

→ *ET 55P* adding machine for Olivetti, 1987

of *Domus* and since 1979 he has been a member of the Scientific Council for the Milan Triennale's design section. He has held professorships at the Istituto Superiore del Disegno Industriale, Venice (1962 to 1965), the Hochschule für angewandte Kunst, Vienna (1982 to 1983) and at the Domus Academy, Milan (1986 to 1991). Bellini has also lectured at many other design colleges including the Royal College of Art, London. His most notable furniture designs include the *Le Bambole* seating system for B&B Italia (1972), the *Cab* seating for Cassina (1977) and the *Figura* office seating programme codesigned with Dieter Thiel for Vitra (1985). He has also designed lighting for **Flos, Artemide** and **Erco** and audio equipment for **Yahama** and **Brionvega**. Bellini's designs have won numerous prizes, including seven Compasso d'Oro awards. Bellini believes that "an effective improvement in the quality of our environment may be obtained through the reunification of design standards ...[and].... that such a reunification can be accomplished only by superseding the compartmentalization of sectors and moving towards a more anthropocentric view of man's environment, at least as far as disciplinary and didactic, or cultural, aspects are concerned."

EMIL BERLINER

BORN 1851 HANOVER, GERMANY
DIED 1929 WASHINGTON, USA

Francis Barraud,
His Master's Voice,
painting featuring
Berliner's gramo-
phone, which
became the HMV
trademark

Emil Berliner emigrated to the United States from Germany in 1870. Then in 1877, a year after **Alexander
Graham Bell** had invented and patented the telephone, Berliner developed a
much-improved telephone receiver out of earlier experiments with transmit-
ters that utilized loose metal contacts. After a series of later inventions relating to the development of the telephone, Berliner patented his invention of
the gramophone in 1887. Unlike **Thomas Alva Edison**'s contemporaneous
Perfected Phonograph or Tainter and Bell's slightly earlier graphophone,
Berliner's recording machine utilized rotating flat discs (records) instead of
cylinders. The stylus of the gramophone cut a horizontal spiral groove into
the wax coating of the disc according to the modulation of the sound being
recorded. The grooves made on the discs had a deeper profile than those
used on wax-coated cylinders, which enabled the stylus to track better without the need of a feed screw. As Berliner's design did away with this element

Berliner
gramophone, c. 1891

it could be manufactured more cheaply than either the phonograph or the
graphophone. An added advantage
of Berliner's design was the fact
that surface pressings, or "recordings" were also easier to make.
Early recordings had a duration
of only about one minute because
of the small capacity of Berliner's
13 cm discs, which were spun by
hand at 70 rpm. Early recordings
included nursery rhymes, bugle
calls and imitations of farmyard
animals and were listened to via
the means of rubber tubes. Although in its day Berliner's gramophone was considered little more
than a novelty, it was an important
precursor of later more sophisticated recording machines.

SIGVARD BERNADOTTE

BORN 1907
STOCKHOLM, SWEDEN

Facit PI typewriter for Åtvidabergs Industrie, 1958

The son of King Gustavus VI of Sweden, Count Sigvard Bernadotte (later Prince Sigvard Bernadotte) studied at the University of Uppsala (1926–1929), at the Kungliga Konsthögskolan, Stockholm (1929–1931), and at the Staatsschule für angewandte Kunst, Munich, in 1931, prior to joining the Georg Jensen workshop. He was the first designer there to adopt geometric forms rather than naturalistic forms, as his *Bernadotte* flatware of 1939 demonstrates. During the late 1930s he also designed a volumetric metal and **Bakelite** cocktail shaker for an English manufacturer. After serving as a director of the Georg Jensen silversmithy, in 1949 he established, a Copenhagen-based design consultancy with the Danish designer **Acton Bjørn**. This multi-disciplinary studio later expanded, opening branches in Stockholm and New York, and Bernadotte and Bjørn co-designed several notable industrial products, including the melamine *Margrethe* stacking bowls for Rosti (1950) and the compact *Facit PI* typewriter (1958). In 1964 Bernadotte independently founded the Bernadotte Design studio, whose output was distinguished by its use of strong geometric forms and interesting colour combinations. Sigvard Bernadotte is an important pioneer of industrial design in Europe and is also the first European designer to have become a member of the American Designers' Institute.

Margrethe mixing bowl with salad servers for Rosti, 1950 (co-designed with Acton Bjørn)

BERTONE

FOUNDED 1912
GRUGLIASCO, ITALY

Guilietta Spider for
Alfa Romeo, 1954

The Carrozzeria Bertone was founded in 1912 by the cartwright Giovanni Bertone (1884–1972). It was not until 1921, however, that the bodywork company received its first order from SPA to build automobile coachwork for its car, the *Torpedo*. With the industrialized approach to automobile manufacture beginning to emerge, coachwork building methods had to follow suit quickly – wooden bodies were now bonded to the metal chassis using assembly-line techniques. Bodywork needed ever-greater strength owing to the increasing speeds that vehicles could now achieve. In response to this, Bertone constructed a number of hard-top-like canopies to transform models such as Vincenzo Lancia's *Lambda* into safer and more robust sedan versions. The following decade, the Bertone company was transformed from a craftsman's workshop into a coachwork factory. Giovanni Bertone's son, Guiseppe "Nuccio" Bertone (1914–1997), joined the family business in 1934 and three years later the company's streamlined bodywork for the **Fiat** *1500* won the Turin **styling** competition. As its manager, Nuccio Bertone transformed the business into one of the world's premier automotive design consultancies. In 1954 the **Alfa Romeo** *Giulietta Sprint* was launched at the Turin Motor

Carabo for Alfa
Romeo, 1968

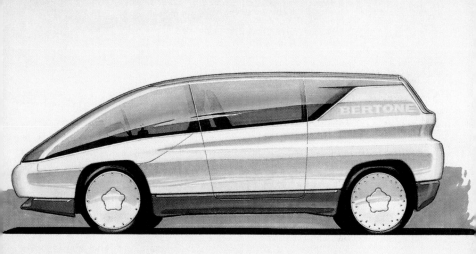

Sketch of *Genesis* concept car, 1988

Show and Bertone's body styling for this vehicle became a hallmark of Italian post-war design. During the 1960s Bertone styled cars for Alfa Romeo, Fiat, **BMW**, Aston Martin and **Ferrari**, and received widespread acclaim for its **Lamborghini** *Miura*, launched in 1966. Two years later, Bertone's futuristic styling of the Alfa Romeo *Carabo* signalled the beginning of the so-called "wedge-line" trend in automobile styling that remained fashionable for many years. In 1971 the first prototype of the Bertone-designed Lamborghini *Countach* was unveiled to much praise, but owing to the oil crisis of 1972, the studio was forced to concentrate on more utilitarian vehicles, such as the **Volkswagen** *Polo* (1975). In recognition of his immense contribution to car styling, Nuccio Bertone was recently nominated for the prestigious title of "Car Designer of the Century". Many well-known car designers, including Giovanni Michelotti (1921–1980), **Giorgetto Giugiaro** and Sergio **Pininfarina**, served apprenticeships at the Bertone Style Centre. The forward-looking and often angular formal vocabulary of the Bertone studio, which has distinguished such disparate vehicles as the Ferrari *Dino 308* GT4 (1973), the Lancia *Stratos* (1973) and the **Volvo** *780* Coupé (1985) is born out of the firm's philosophy that design should not only anticipate future trends but should also give style to functional form.

Max Bill trained as a silversmith at the Kunstgewerbeschule in Zurich from 1924 to 1927 and while there was influenced by trends in contemporary fine art such as Cubism and Dada. From 1927 to 1929 he studied art at the Dessau **Bauhaus** and fully embraced the school's functionalist approach to design. Upon completion of his studies, Bill returned to Zurich and worked as a painter, architect and graphic designer. In the 1930s he became the leading exponent of Constructivism within the Swiss School of graphics, and during this period designed graphics for the Wohnbedarf store in Zurich. He established his own architectural practice in 1930 and, as a member of the Swiss Werkbund, designed the Neubühl estate near Zurich in the Modern style (1930–1932). In 1931 Bill adopted Theo van Doesburg's theory of "concrete art", which argued that universality could only be achieved through clarity. From 1932 Bill also worked as a sculptor and became a member of various art organizations, including the Abstraction-Création group in Paris, the Allianz (Association of Modern Swiss Artists), the CIAM (Congrès International d'Architecture Moderne) and the UAM (Union des Artistes Modernes). In 1944 Bill turned his attention to industrial design. His **aluminium** wall clock of 1957 was manufactured by **Junghans**, for whom he also designed a range of wristwatches characterized by a similarly rational layout and a strong industrial aesthetic. Bill also designed a minimalist stool, the *Ulmer Hocker* (1954), one of his best-known products. Bill was responsible for the founding of the Bundespreis award and the exhibition "Die gute Industrieform", which sought to promote excellence in industrial design. Importantly, he also co-founded the in-

Kitchen clock for Junghans, 1956

Wristwatches for
Junghans, 1962
(re-edited by
Junghans, 1993)

fluential Hochschule für Gestaltung in Ulm in 1951 and was the school's
rector and head of its architecture and product design departments from
1951 to 1956. At Ulm, Bill championed Bauhaus-type geometric formalism,
believing that products based on mathematical laws had an aesthetic purity
and thereby a greater universality of appeal. This approach to industrial design was continued by **Hans Gugelot** when he took over the product design
department. Upon leaving Ulm, Bill set up his own Zurich-based studio in
1957 and concentrated on sculpture and painting. He was the chief architect
of the "Educating and Creating" pavilion at the Swiss National Exhibition of
1964 and was made an honorary member of the American Institute of Architects that same year. As much a theorist as a designer, Max Bill believed that
"good design depends on the harmony established between the form of an
object and its use".

Biró logo

British patent
for László Biró's
ballpoint pen, 1938

LÁSZLÓ BIRÓ

BORN 1899 BUDAPEST, HUNGARY
DIED 1985 BUENOS AIRES, ARGENTINA

László Jozsef Biró, a Hungarian-born journalist, created the first ballpoint pen, which he patented in 1938. The principle of the ballpoint pen, however, dates back to 1888, when a patent was filed by the American John J. Loud for a device to mark leather. This design had a number of failings and the patent was not commercially exploited. During the 1930s, while working as an editor of a magazine in Hungary, Biró became aware of an ink used for printing that was quick-drying and smudge-free and so decided to develop a pen that could use the same ink. As this thicker ink would not flow from a regular pen nib, Biró devised a point by fitting his pen with a tiny ball bearing at its tip. As the pen moved along the paper, the ball rotated, picking up ink from the ink cartridge and leaving it on the paper. In 1940 Biró, who was a fervent Communist, left Hungary when it entered into an alliance with Nazi Germany and subsequently emigrated to Argentina. Biró became an Argentinian citizen and adopted a new Spanish-sounding name, Ladislao José Biró. By 1942 his ballpoint pen had been perfected as a result of a special ink formulated by his chemist brother, Georg. After filing a fresh patent in 1943, Biró raised US$ 80,000 worth of backing to allow his pen to go into full-scale mass production. In 1944 Henry George Martin, an English accountant who had invested heavily in Biró's company, Eterpen Co., brought the fully patented ballpoints back to England

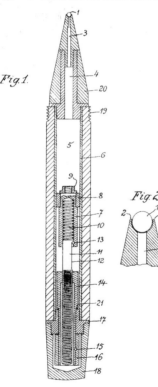

Fig. 1

Fig. 2

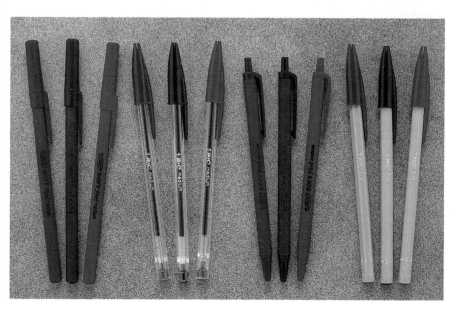

and offered them to the British and Allied armed forces. Unlike regular fountain pens, Biró's ballpoints were especially useful in unpressurized aircraft
because they did not leak. Over 30,000 ballpoints were manufactured for
the RAF by the Miles Martin Pen Company in Reading, and an even greater
number were produced for the American forces. At the end of the war, the
innovative and robust ballpoint pen went on sale to the general public
around the world. It was initially seen as a status symbol, no doubt due
to its high cost: in America the pen retailed at $12.50. Nevertheless, an in-
credible 10,000 were sold on its first day of sales. In Britain the pen was
priced at 55 shillings (£2.74) – the equivalent of a secretary's weekly wage
packet. In 1949 Marcel Bich (1914–1994), an expert in **plastics** machining,
perfected his own much less expensive ballpoint pen, which sold under
the name "BIC Point", a shortened version of his name. Backed by an innovative advertising campaign, the newly founded Société BIC was selling 42
million units annually after only three years. In 1957, Société BIC purchased
Biró Swan, the descendent of the Miles Martin Pen Company. Today, BIC
markets a range of some 20 writing instruments that are direct descendants
of the original Biró pen. Astonishingly, over 15 million BIC pens are sold
every day.

Magic Wand hand-mixer for ESGE, 1955

ACTON BJØRN

BORN 1910 COPENHAGEN, DENMARK
DIED 1992 DENMARK

Acton Bjørn studied architecture and town planning and began working as an industrial designer during the Second World War. In 1949 Bjørn and **Sigvard Bernadotte** established Bernadotte & Bjørn, one of the first Scandinavian industrial design consultancies. This Copenhagen-based practice not only designed metalwork and **ceramics** but also industrial machines. Among their most notable work was office equipment for Facit and plastic kitchenware for Rosti during the 1950s. In 1966 Bjørn won a Danish ID Award for his *Beolit 500* transistor radio for **Bang & Olufsen**. This design was one of the first radios to achieve a functional simplicity comparable to that of a telephone. The same year, he opened an office in Copenhagen specializing in household appliances, office furniture and **packaging**. He also received recognition for his **corporate identity** work, executed mainly for banking and insurance companies. Bjørn's designs succeed in balancing practical function with simple sculptural forms.

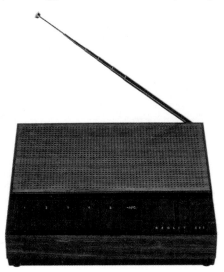

Beolit 500 transistor radio, 1965

MISHA BLACK

BORN 1910 BAKU, AZERBAIJAN
DIED 1977 LONDON, ENGLAND

Misha Black studied at Central School of Arts & Crafts in London for a short while, but was essentially self-taught as a designer and architect. In 1930 he founded Studio Z and three years later began working with Milner Gray (1899–1994) at the Bassett-Gray Group of Artists and Writers. In 1935 this became the Industrial Design Partnership, the first inter-disciplinary design consultancy in Britain, whose designs most notably included television sets and radios for Ekco. From 1940 to 1945 he worked as an exhibition designer for the Ministry of Information. In 1943 Black and Milner Gray founded the **Design Research Unit** in London, followed two years later by the Design Research Group. The DRU was the most important design consultancy in post-war Britain and produced numerous industrial designs – from **Kodak** cameras and Watney beer **packaging** to trains and electric heaters – all of which were distinguished by a clarity of lay-out and a no-nonsense, engineering approach. Black also contributed to major exhibitions, including the 1946 "Britain Can Make It" exhibition. Throughout his career he tirelessly promoted the Modern design cause.

Misha Black &
J. Beresford Evans,
E3001 locomotive
for British Railways,
c. 1956

BLACK & DECKER

FOUNDED 1910
BALTIMORE, USA

Workmate, the first version was designed by Ron Hickman in 1972

Black & Decker was founded in Baltimore in 1910 by Duncan Black (1883–1951) and Alonzo Decker (1884–1956) for the manufacture of specialist machinery. Early products included milk-bottle cap machines, letter-graphs, pocket-sized adding machines, candy-dipping machines and machinery for the U.S. Mint. In 1914 the company filed a key patent for a revolutionary electric drill that had a pistol grip and a trigger-style switch. This first ever portable power tool and its later successors brought the company huge success – within a decade of its foundation, Black & Decker's annual sales exceeded US$ 1 million. In 1935 the company commenced manufacturing in the UK. It was not until 1950 that Black & Decker began making tools for home use. Its first domestic product was the famous "Little Red Drill", which had a 1/4" capacity. In 1964 the company developed the *Cordless Zero Torque* tool for **NASA**'s Gemini Space Project. In 1971 Black & Decker designed the *Apollo Moondrill* for the Apollo 15 Mission – a cordless zero-torque tool that could remove core samples from the lunar surface. One of the company's greatest innovations was the *Workmate*, designed by Ron Hickman (b. 1932) in 1972. This multi-use bench was the blueprint for many subsequent variations. In 1979 Black & Decker launched its highly successful hand-held cordless *Dustbuster* vacuum cleaner, and a decade later introduced its *Multi Tools*, including the multiple-use *Multisander*. The development of multi-purpose tools eventually resulted in the launch, in 1998, of the four-in-one *Quattro* tool, which incorporates a jigsaw, drill, screwdriver and sander. Driven as much by state-of-the-art technology as by design innovation, Black & Decker has managed to straddle both the industrial and domestic markets with its extensive range of tools for building, gardening (it launched its first range of lawn-mowers in 1969) and cleaning.

PHH-I drill, c. 1999

BMW

FOUNDED 1916
MUNICH, GERMANY

BMW 3/15 PS,
1929–1932

Bayerische Motoren Werke was founded as an aircraft engine manufacturer in 1916. Within just three years Franz Zeno Diemer had set the world altitude record in an aircraft powered by BMW engines. In 1923 the company began producing motorcycles. Five years later, it acquired a car factory in Eisenach and with it, a licence to manufacture a small car, the *Dixi*, which became known as the BMW 3/15 PS. The first car to be designed in-house was the BMW 3/20, which was launched in 1932. The company designed several streamlined cars over the course of the 1930s, most notably the BMW 331 and the classic BMW 328 sports car. BMW also dominated car racing during those years, with its aerodynamic 328 roadsters winning their class at the 1938 Mille Miglia. In 1940 the company began mass-producing the 801 aircraft engine – some 30,000 were manufactured over the succeeding five-year period – and also won the Mille Miglia again with a modified "aerodynamic coupé" version of the 328. In 1941, car production was suspended and resources were directed to the design and construction of rockets and aircraft engines, including one of the world's first jet engines, the 003. By 1945, however, BMW's factories in Eisenach and Berlin had been lost and the remaining plant in Munich was dismantled. After a three-year production ban, BMW resumed the manufacture of motorcycles in 1948 with the single-cylinder BMW R24. In 1951 the company produced its first post-war car, the 501, but

BMW 331, 1930s

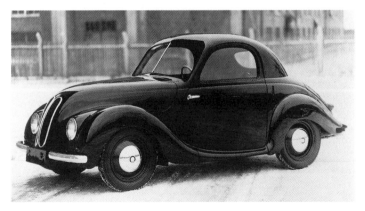

BMW *Isetta 600,*
1955

BMW *1602–2002*
models and the
succeeding *3 Series*

it was a commercial failure. Four years later BMW launched the diminutive *Isetta* (an italian licence), which bridged the motorbike and automobile markets. This hybrid design was extremely popular in Germany with around 200,000 examples being produced. Having decided to concentrate on the small car market, BMW launched the *Model 700* in 1959 and the *Model 1500* in 1962, which established the design vocabulary of its later compact touring cars. Despite this strategy, the company faced bankruptcy in the late 1960s. Critically, BMW was only able to emerge from the financial doldrums by introducing a range of high-quality, conventionally styled road cars that drove like sports cars. In 1973 the more aggressively styled BMW *2002* became the first mass-produced turbo-powered car in the world. Two years later the *3 Series* was introduced, followed by the *6 Series* coupé and *7 Series* in 1976 and 1977 respectively. During the 1980s and 1990s the company continued to expand; in 1997 it launched its retro-designed *M Roadster*, and a year later introduced the fifth generation of the highly successful *3 Series*. The achievements of BMW have largely rested on the continuity of its brand values, which are based primarily on a commitment to technological innovation and design.

No. 1208 Santos
coffee maker, 1958

BODUM

FOUNDED 1944
COPENHAGEN, DENMARK

The Bodum domestic products company was established in Copenhagen in 1944 by Peter Bodum (1910–1967) to import glassware from Eastern Europe to sell in Denmark. In 1958 it introduced its first branded product, the *Santos* coffee maker designed by Kaas Klaeson. One of the company's most famous products, the *Bistro* coffee maker (1974), was the outcome of a collaboration between Peter Bodum's son, Jørgen (b. 1948), and the industrial designer Carsten Jørgensen (b. 1948). Both believed in simple, effective design without unnecessary adornment and in the use of appropriate Modern materials to create useful everyday products at a reasonable price. In 1983 Pi Design, located in Switzerland, was founded as the Product & Design development company of the Bodum Group. Jørgensen was appointed a director of Pi and made responsible for the design and **corporate identity** of Bodum worldwide. Since then, Pi has designed a broad range of distinctive and highly rational products for Bodum that eloquently express the company's philosophy, namely that "some everyday things are at their best when they merely constitute a plain function and are simple in shape and proportions". In 1986 Bodum established its first retail outlet in London and later went on to open other shops in Paris, Copenhagen, Lisbon, Oporto, Zurich and Lucerne.

⟍*No. 1508 Bistro*
coffee maker, 1974

No. 1846 Assam tea
maker, 1992

← *No. 1928*
Chambord coffee
maker, 1958

→ *No. 5670 Antigua*
coffee grinder, 1996

← *No. 5500 Ibis*
kettle, 1992

→ *No. 5600 Le Curl*
kettle, 1998

Boeing *B17 Flying Fortress*, 1934–1935

Boeing *B29 Superfortress*, 1939–1940

BOEING

FOUNDED 1916
SEATTLE, USA

William E. Boeing

William Boeing (1881–1956) studied engineering at Yale University. In the late 1900s he became fascinated with aircraft and, after a period of some research, concluded that he could build a better bi-plane than any then available. In 1915 he asked the designer, Westervelt, to develop a new aeroplane that he could manufacture. The resulting aircraft was the birch-and-canvas B&W twin-float seaplane. During the First World War, Boeing received its first production order from the US Navy for 50 *Model C* seaplanes. Around this time, William Boeing declared: "We are embarked as pioneers upon a new science and industry in which our problems are so new and unusual that it behoves no one to dismiss any novel idea with the statement, It can't be done." Under his careful guidance the company grew and produced a wide range of aircraft – from mail planes and flying boats to military observation planes and torpedo planes. While its competitor, Douglas, was winning international acclaim for its first around-the-world flight in 1924, Boeing was developing state-of-the-art pursuit fighters and was building more fighter aircraft than any other manufacturer. The company's series of fighter aircraft from the 1920s and 1930s included the Boeing *P-26 Peashooter* monoplane, which flew 27 mph faster than any of its bi-plane competitors

Boeing *B52 Stratofortress*, 1948–1952

Boeing 707,
1955–1957

and employed innovative production techniques in its construction, including the arc-welding of its fuselage frame. With the introduction of anti-trust legislation in 1934, William Boeing grew disheartened and left the firm. The company's new president, Claire Egtvedt, determined that the future lay not just in the development of new military aircraft but also in the design and manufacture of large passenger aircraft, such as the long-range four-engine *Clipper* seaplane. During the Second World War, Boeing massively increased bomber production and by 1944 was building an astonishing 362 *B-17 Flying Fortresses* per month. The enormous *B-17* became legendary for its ability to remain flying even after sustaining severe damage. In total, Boeing plants built 6,981 *B-17*s in various models with another 5,745 being built under a nation-wide collaborative effort by **McDonnell Douglas** and **Lockheed**. In 1942 Boeing began production of the long-range heavy bomber, the *B-29 Superfortress*, which is mainly remembered as being the aircraft from which the world's first atomic bomb was dropped. During the post-war years Boeing reduced production but did continue developing military aircraft, most notably the *B52 Stratofortress*. Produced between 1952 and 1962, this devastating bomber first saw action in Korea. Remarkably, the last production variation of the aircraft, the *B-52H*, remains in service today – testifying to the superior design and strength of its airframe. During the 1950s the president of Boeing, William Allen, moved the company into the commercial

Boeing 747,
1963–1968

jet market with designs such as the 377 *Stratocruiser* – a luxurious civilian version of the *B-29*. In 1952 the company's net worth was risked so as to begin development of the *Dash-80* commercial jet. The company eventually had to bow to pressure from the airlines and re-design the *Dash-80* so as to match Douglas' *DC-8*. The resulting design was the landmark *707*, which featured more powerful engines, a wider fuselage, larger wings and a greater seating capacity than the *DC-8*. The company's family of commercial jets expanded over the years to include the *727*, the *737* and the epic *747* jumbo jet. One of the most significant technological achievements in the history of aviation, the *747* is the world's largest commercial aeroplane. Its first flight was in 1969 and it continues to be built at the Boeing factory – the world's largest building by volume in Everett, Washington – clearly demonstrating the aircraft's versatility, popularity, longevity and value. Boeing has also contributed significantly to the American Space Programme, including the famous Apollo missions.

Advertisement for
Boeing 707, 1950s

Bombardier *B7*
snowmobile, 1937

BOMBARDIER

FOUNDED 1942
VALCOURT, QUEBEC, CANADA

Joseph-Armand Bombardier (1907–1964) was born near Valcourt, Quebec. At the age of 19 he began working as a mechanic in his own garage, and over the next decade researched and experimented with vehicles that could be driven over snow-covered terrain. Between 1926 and 1935 he developed several prototypes and, in 1936, designed and patented a seven-passenger tracked vehicle known as the *B7*. This early snowmobile allowed previously snow-bound areas to be accessed, and by 1940 Bombardier was producing 200 units per annum. In 1941 he developed a larger and more powerful snowmobile, the *B12*, and a year later founded the L'Auto-Neige Bombardier Limitée Company to manufacture these vehicles. A military version of the *B12* was commissioned by the Canadian Government and in 1943 Bombardier developed an armoured snowmobile known as the *Kaki*. After the war, the company diversified its product range to include a device developed by Joseph-Armand Bombardier and his brother, Gérard, for use on tractors. Known as the *TTA* (Tractor Tracking

Advertisement for
the Bombardier *B12*
snowmobile, 1941

Attachment), this traction mechanism improved the performance of tractors on muddy terrain and sold by the thousands around the world. The development in 1953 of an unbreakable warp-proof all-rubber sprocket and the design of a new vulcanizing machine that allowed the production of seamless shock-resistant tracks enabled Bombardier to produce the *Muskeg* multiple-transportation all-terrain vehicle. Between 1957 and 1959 Germain Bombardier, the eldest son of Joseph-Armand, designed and perfected the first ever *Ski-Doo*, a fun sports vehicle that instantly found approval among outdoor enthusiasts. Some 60,000 *Ski-Doos* were being sold annually by the mid-1960s, and by 1972 this figure had risen to 495,000. The oil crisis in the early 1970s, however, dramatically slowed demand for *Ski-Doos* and a year later Bombardier moved strategically into rail transit systems and manufactured the rolling stock for Montreal's newly expanded subway. The company continued to grow through acquisitions – including the aircraft manufacturer Canadair in 1986, the French railway manufacturer ANF-Industrie in 1989, the business jet manufacturer **Learjet** in 1990, and the famous aircraft manufacturer **de Havilland** in 1992. Guided by its motto, "strength in diversity", and its commitment to innovative design solutions, which have resulted in several new vehicle typologies, Bombardier has become one of Canada's largest and most prestigious manufacturing companies.

BOSCH

FOUNDED 1866
STUTTGART, GERMANY

Robert Bosch

→Lucian Bernhard,
advertisement for
the first Bosch spark
plug, 1902

Robert Bosch (1861–1942) served an apprenticeship as a precision instrument maker with Wilhelm Maier in Ulm from 1876 to 1879, and spent a year in America, working for Sigmund Bergmann and **Thomas Alva Edison**. On his return to Europe in 1885, he stayed briefly in London, where he worked for **Siemens** Brothers. In November 1886 Bosch opened his own workshop, the Werkstätte für Feinmechanik und Elektrotechnik in Stuttgart, which produced the *Low-Voltage Magneto* for stationary gas engines (1887) and a table-top telephone. By 1898 the company had opened a subsidiary in London, the Compagnie des Magnétos Simms-Bosch, and a year later sales offices were established in Paris and Budapest. In 1901 Robert Bosch's manu-

Early advertisement
for Bosch *Magneto*
spark plugs

facturing operation moved into its first factory, built on the site of the company's existing premises. 1902 saw the launch of the crucial high voltage *Magneto* spark plug, for which the graphic designer Lucian Bernhard (1883–1973) was commissioned to create bold **packaging** and advertising posters, which helped the company establish its powerful brand identity. In 1909 another factory was built in Feuerbach to increase production capacity. Over the succeeding years, the company diversified its product line by introducing various automotive innovations, including headlamps (1913), the first electric starter motor (1914), car horns (1921), the first electrically powered windscreen wipers (1926) and the first standard diesel injection pump for trucks

At Last You Can Get Them,

THE LONG MISSED AND LONG WANTED

Bosch-Magnetos

For all kinds of Combustion Engines.

AGENCY FOR THE UNITED STATES

of the world-famous Bosch-Magnetos, made by Robert Bosch, Stuttgart, Germany, the inventor of the Bosch Ignition Apparatus, and sole manufacturer of all machines sold as Bosch or Simms-Bosch Magnetos,

HAS NOW BEEN OPENED

Bosch ·Ignition is used by all the most important motor car and cycle manufacturers. Some of them are :

UNITED STATES

E. R. THOMAS MOTOR CO.
ROYAL MOTOR CAR CO.
THE GARFORD CO.
OLDS GASOLINE ENGINE WORKS
AMERICAN LOCOMOTIVE AUTOMOBILE CO.
DAIMLER MFG. CO.
ETC.

FRANCE	GERMANY	GREAT BRITAIN
Berliet	Adler	Daimler
Charron, Girardot & V.	Horch	Napier
Clément-Bayard	Neckasulm	Singer
Darracq	N. A. G.	Westinghouse
De Dietrich	Mercedes	Wolseley
Delahaye	Wanderer	Etc., etc.
Léon Bollée	Etc., etc.	**ITALY**
Peugeot		Fiat
Renault Frères	**BELGIUM**	Itala
Richard-Brasier	Fabrique Nationale	Isotta Fraschini
Rochet & Schneider,	Pipe	Rapid
Etc., etc.		Etc., etc.

WRITE, OR BETTER STILL, *CALL* AND OBTAIN PARTICULARS

ROBERT BOSCH, 1947 BROADWAY, New York City
Telephone, 4255 Columbus

(1927). In 1932 Bosch took over the Idealwerke AG and produced televisions
and radios under the label Blaupunkt. Bosch continued its commitment to
research and development throughout the 1930s, introducing the first serial
car radio in 1932 and the first standard diesel injection pump for cars in 1936.
During that decade the company also began manufacturing white goods
and presented its first refrigerator in 1933. Bosch expanded rapidly over the
following decades, opening foreign subsidiaries and numerous overseas
plants. It also continued to launch innovative products, amongst them the
first fuel injection system for cars (1951), the first automatic dishwasher
(1964) the first electronic fuel injection system for cars (1967) and the original ABS anti-blocking system (1978). Renowned for its white goods and electric power tools designed by the German industrial designer **Hans Erich
Slany** (b. 1926), Bosch maintains its tradition of producing well-designed
and well-built products that epitomize the high quality of German industrial

design. Today the Bosch Group is one of Germany's largest companies. Its scientists, engineers, designers and technicians are dedicated to the improvement of the function and reliability of existing products and to the development of new products and systems across all areas of its operations, including electrical and electronic automotive equipment, private and public communications technology, power tools, household appliances and thermotechnology as well as automation technology and **packaging** machinery.

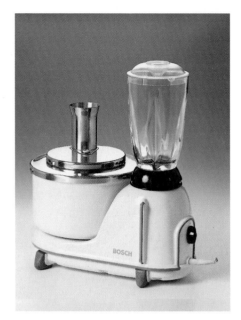

MARIANNE BRANDT

BORN 1893 CHEMNITZ, GERMANY
DIED 1983 KIRCHBERG, GERMANY

Marianne Brandt studied at the School of Fine Art (Großherzoglich-Sächsische Hochschule für Bildende Kunst) in Weimar from 1911 to 1917. She subsequently established her own studio in 1917 and worked as a freelance artist until she enrolled at the Staatliches **Bauhaus** in Weimar in 1923. After the preliminary course, she took an apprenticeship in the metal workshop, which was then directed by László Moholy-Nagy (1895–1946). Around this time she designed a coffee and tea service, which was based on simple geometric forms – the body of the diminutive teapot being hemispherical. Even at this early date, Brandt appeared more interested in functional form than in traditional craftsmanship and it is not surprising that her later designs became increasingly utilitarian in nature. After taking her journeyman's exam, Brandt became deputy director of the metal workshop in 1928 and organized projects in collaboration with the lighting manufacturers Körting & Mathiesen AG (Kandem) in Leipzig and Schwintzer & Gräff in Berlin. At the Bauhaus she worked alongside fellow metalworkers **Christian Dell** and Hans Przyrembel (1900–1945) and in 1928 co-designed the *Kandem* lamp with Hin Briedendieck (b. 1904), as part of a class project. Brandt's lamps were some of the most important designs to emanate from the Bauhaus because of their suitability for mass production. Although the Bauhaus' focus moved during the 1920s towards the realization of prototypes that were potentially suited to mass-manufacture, very few designs created at the school were successfully put into production. Brandt's lamps were a notable exception. She worked in the architectural office of **Walter Gropius** in 1929, and from 1930 to 1933 developed new design concepts for the Ruppelwerk factory in Gotha. In 1933 she returned to Chemnitz, where she took up painting and attempted to license some of her products to the Wohnbedarf department store. Brandt taught at the Staatliche Hochschule für Angewandte Kunst in Dresden from 1949 to 1950 and at the Institut für Angewandte Kunst in Berlin-Weißensee from 1951 to 1954. Although Brandt was an accomplished painter and received recognition for her photomontages, it is her industrial design work that is of greatest significance. Brandt also holds a special position as one of the very first women to design for industrial production – a field that has historically been dominated almost entirely by men.

702 Kandem bedside table lamp for Körting & Mathiesen, 1928

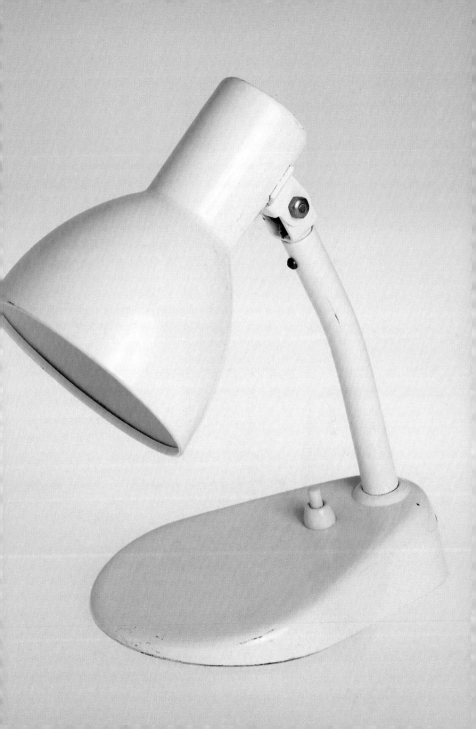

BRAUN

FOUNDED 1921
FRANKFURT/MAIN, GERMANY

Max Braun, *Model
S 50* electric shaver,
1950

→Gerd Alfred Müller
& Hans Gugelot,
SM3 electric shaver,
1960

Max Braun, *Model
S 50* electric shaver,
1950

In 1921 the engineer Max Braun (1895–1946) established a manufacturing company in Frankfurt to produce
connectors for drive belts and scientific apparatus. In
1923 he began producing components for the newly emerging radio industry. Following the advent of plastic pellets in 1925, he was quick to seize
upon this new material, using homemade presses to manufacture components such as dials and knobs. In 1928 the company moved into a functional modern factory building on Idsteiner Strasse in Frankfurt, and a year
later began producing its own radio sets, which were some of the first to
incorporate the receiver and speaker in a single unit. In 1932 the company
expanded its product range and became one of the first manufacturers to
introduce radio/phonograph combination sets. Braun developed a battery-
powered radio in 1936 and a year later won an award at the Paris "Exposition
Internationale des Arts et Techniques dans la Vie Moderne" for its "exceptional achievements in phonographs".

By 1947 the company was mass-producing radio sets,
albeit still styled as furniture rather than as Modern
electronic equipment. During this period Braun also
began production of its *Manulux* flashlights and in 1950
developed its first electric razor, the *S 50*. This shaver
incorporated an oscillating cutter-block screened by a
thin steel shaver foil – a system that is still used today.
In 1950 Braun also branched into domestic appliances
with the *Multimix*. After the death of Max Braun in 1951,
the firm was headed by his two sons, Artur (b. 1925) and
Erwin (b. 1921), who decided to implement a radical design programme that was both rational and systematic.
In 1953 Erwin identified a marketing opportunity for distinctive radios that were "honest, unobtrusive and practical devices" and embodied a Modern aesthetic. To this
end, **Wilhelm Wagenfeld** and designers associated with
the Hochschule für Gestaltung in Ulm, such as Fritz
Eichler (1911–1991), were commissioned in 1954 to redesign the company's radios and phonographs. This

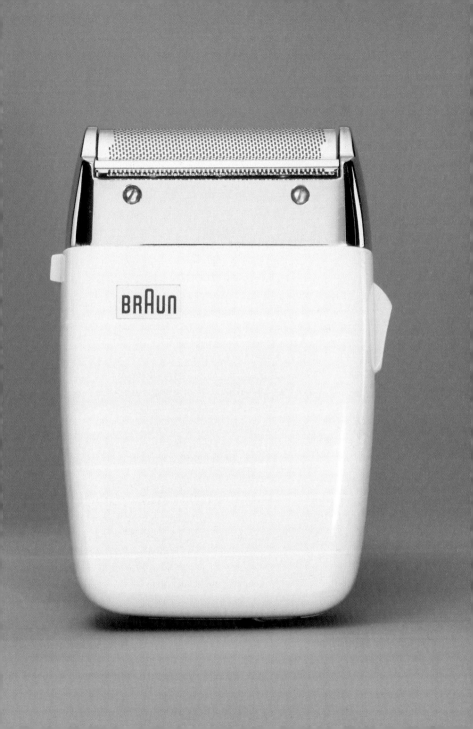

new Braun line was introduced at the Düsseldorf Radio Fair in 1955 and attracted international acclaim. 1956 saw the establishment of an in-house design department headed by Eichler, who proceeded to formulate a coherent corporate style based on geometric simplicity, utility and a functionalist approach to the design process. The Braun design vocabulary was not only used for products but was also applied to all areas of **corporate identity,** including **packaging**, logos and advertising. Eichler also commissioned other designers associated with the Hochschule für Gestaltung, such as Otl Aicher (1922–1991) and **Hans Gugelot,** to design sleek, unornamented products. Notable designs from this period include Eichler and Artur Braun's line of radios and phonographs (1955) and **Dieter Rams** and Hans Gugelot's *Phonosuper SK 4* radio-phonograph (1956), which was nicknamed "Snow White's Coffin". Rams also designed the *Transistor 1* portable radio (1956), the *T 3 T 4* pocket radio (1958) and the first component-based hi-fi system, the *Studio 2* (1959) – all of which helped establish Braun's international reputation. In 1955 **Gerd Alfred Müller** (b. 1932) joined the Braun design team and was responsible for some of the company's best-known designs from the late 1950s, including the *KM 3* multi-purpose kitchen mixer (1957), which embodied the austere rationalist aesthetic that became synonymous with German post-war design. In 1961 Dieter Rams was appointed head of the company's design department, and in 1968 overall director of design. Rams was to head the Braun design team for some 40 years, and his pared-down functionalist aesthetic permeated all the products it manufactured, from kitchen equipment to alarm clocks to electric shavers. During his tenure, Braun introduced a series of landmark designs including the *Permanent*

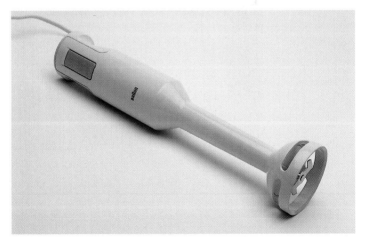

lighter (1966), which incorporated an electromagnetic device rather than a traditional friction cylinder, the *ET 22* electronic pocket calculator (1976) and the first radio-controlled clock (1977). In 1967 the Boston-based **Gillette** Company acquired a controlling stake in Braun AG. A year later, the International Braun Awards for design in engineering were established. In 1983 the company was itself awarded the first Corporate Design Award at the Hanover trade fair for its "exemplary conception of product design, information and presentation".

In 1990 Braun discontinued its hi-fi production so as to concentrate on the manufacture of personal grooming products, such as the *Silk-épil EE 1* depilator (1989), the highly successful *Flex Control* line of electric razors (1990) and the *Plak Control D 5* electric toothbrush (1991), as well as a range of hair-dryers. During the 1990s Braun also introduced innovative coffee machines, food processors, hand mixers, irons and alarm clocks. In 1996 Braun launched the *Thermoscan* infrared thermometer, which marked its entry into the personal diagnostic appliance market. Certainly, Braun's success stems from the fact that its products are jointly developed by designers, engineers and marketing experts in accordance with basic design principles. The company uses design innovation to achieve technical and functional innovation and has established a tradition of progressiveness within its

Ludwig Littmann,
K 750 Combi Max
food processor, 1997

design team. The strong aesthetic clarity of its products is the outcome of a logical ordering of elements and the quest for a harmonious and unobtrusive totality. Braun acknowledges that "integrated working methods are ultimately reflected in the obviousness of the product expression", and asserts that "Braun Design is the orientation towards lasting worthwhile values: innovation, distinctive, desirable, functional, clear, honest, aesthetic."

Roland Ullmann,
Flex Integral Colour Selection razors,
1997–1998

Carl Breer driving
a Chrysler *Airflow*

←Carl Breer with a
small-scale wooden
model of the
Chrysler *Airflow*,
1934

CARL BREER

BORN 1883 LOS ANGELES, CALIFORNIA, USA
DIED 1970 DETROIT, MICHIGAN, USA

Determined to increase the market share of his company, which at that time ranked only tenth in the automotive industry, in the early 1930s Walter Percy **Chrysler**
(1875–1940) decided to develop a totally new car. The result was the Chrysler
Airflow, designed by Carl Breer, chief engineer at Chrysler, in collaboration
with his staff engineers Owen Skelton and Fred Zeder. In order to improve
both the stability and the comfort of the final design, Breer had studied the
concept of **streamlining** by experimenting with a variety of automobile body
shapes in a wind tunnel. His idea was to reverse the lift effect normally associated with an aircraft wing by arriving at an aerodynamic form that would
press the car more firmly against the road at high speed. With its swept-
back "waterfall" nose and its headlamps and fenders partly absorbed into
the body, the *Airflow* was the first automobile production to employ leading-
edge aerodynamic **styling**. With this car, Breer also pioneered unitized body
construction and "cab-forward" positioning – an innovative approach to
weight distribution. Despite incorporating other significant advances, such
as Chrysler's first automatic overdrive transmission, the forward-looking
Airflow proved too radical in its appearance for the wider public and remained in production for only three years. While Breer's remarkable *Airflow* was
a commercial failure, its design was immensely influential for later car
designers, especially those at **Ford**, **General Motors** and **Porsche**.

Airflow, 1934

MARCEL BREUER

BORN 1902 PÉCS, HUNGARY
DIED 1981 NEW YORK, USA

In 1920 Marcel Lajos Breuer won a scholarship enabling him to study at the
Akademie der Bildenden Künste in Vienna. Dissatisfied with the institution,
he remained there for only a brief period before finding work in a Viennese
architectural practice. From 1920 to 1923 he studied at the Staatliches **Bauhaus** in Weimar, completing the preliminary course, the carpentry apprenticeship and his journeyman's exam. While still a Bauhaus student, Breuer
designed his *African* chair (1921) and his *Slatted* chair (1922–1924). In 1925,
when the Bauhaus moved to Dessau, Breuer joined the teaching staff as a
"Young Master" and was appointed head of the carpentry workshop. There
he designed his first **tubular metal** chair, the *B3* (1925) – this innovative
choice of material having been inspired, so it is said, by the Adler bicycle
he had recently purchased. Breuer subsequently designed a whole range
of tubular metal furniture, including chairs, tables, stools and cupboards,
which were manufactured and distributed by Standard-Möbel, Berlin from
1927, and by **Thonet** from 1928/1929. Tubular metal offered many benefits –
affordability, hygiene and an inherent resiliency that provided comfort without the need for springing. Breuer regarded his designs as essential equipment for modern living. At the Bauhaus, Breuer also designed the interiors
and furnishings for the school's new premises and for the Masters' houses.
His *B3* or *Wassily* chair was originally designed for Wassily Kandinsky's ac-
commodation. Breuer not only created standardized furnishings – in 1926
he also designed a small standardized metal house and, a year later, his
Bambos house. Breuer continued to teach at the Bauhaus until April 1928,
after which he directed his own architectural practice in Berlin from 1928 to
1931. Although his building projects remained unrealized, during this period
Breuer continued to design furniture, interiors and department stores. He
was commissioned by the **Deutscher Werkbund** to design interiors for the
German section at the "Société des Artistes Décorateurs Français" exhibition of 1930. In 1932 he completed his first architectural commission, the
Harnischmacher House in Wiesbaden, and designed the Wohnbedarf furniture store in Zurich. Two years later, in conjunction with Emil (1893–1980)
and Alfred Roth (b. 1903), he co-designed the Doldertal Houses, a pair of
experimental apartment blocks in Zurich for Sigfried Giedion (1888–1968),
the founder of the Wohnbedarf company. From 1932 to 1943, Breuer devel-

oped a range of pliant furniture using a patented method of construction
that incorporated flat bands of steel and **aluminium**. This range of metal furniture was manufactured by Embru and retailed by Wohnbedarf. In 1935
Breuer emigrated to London and initially worked in partnership with the
architect F. R. S. Yorke (1906–1962). Together they completed several architectural commissions, including houses in Sussex, Hampshire, Berkshire
and Bristol and the Gane Pavilion in Bristol (1936), which combined wood
and local stone (a far cry from the Bauhaus aesthetic of steel and glass).
Breuer and Yorke also designed a "Civic Centre for the Future", which remained unrealized. Between 1935 and 1937, as controller of design at Jack
Pritchard's company, **Isokon**, Breuer produced five **plywood** furniture designs, which were essentially translations of his earlier metal designs. These
Isokon designs reflected the popularity of the earlier plywood furniture by
Alvar Aalto (1898–1976) that had been exhibited in Britain in 1933. In 1937
Breuer moved to the USA, where **Walter Gropius** had offered him a pro-
fessorship at Harvard University's School of Design in Cambridge, Mas-

sachusetts. He also established an architectural practice in Massachusetts with Gropius, and together they designed the Pennsylvania Pavilion at the 1939 **New York World's Fair** and several private houses, including Gropius' own residence. In 1941 Gropius and Breuer dissolved their partnership and Breuer founded his own architectural practice, which he moved to New York in 1946. During the late 1940s and 1950s, Breuer designed some 70 private houses, mainly in New England, and in 1947 he built a home for himself in New Canaan, Connecticut. The Museum of Modern Art, New York, initiated a touring exhibition of his work in 1947 and the following year invited him to build a low-cost house in the museum's grounds, which would suit the needs of an average American family. He furnished this project with affordable plywood cutout furniture. In 1953 Breuer worked as part of a team on the new UNESCO building in Paris and he also designed the Bijenkorff department store in Rotterdam. In 1956 he founded Marcel Breuer and Associates in New York and around this time made concrete his material of choice. He used this medium in a highly sculptural and innovative way for his design of the monumental Whitney Museum of American Art, New York (1966). Breuer was one of the foremost exponents of the Modern Movement, and the enduring appeal of his highly democratic furniture designs testifies not only to his mastery of aesthetics but to his understanding of industrial methods of production.

Model B18 table (1st version), 1928, and *Model B55* armchair, 1928–1929

BRIONVEGA

FOUNDED 1945
MILAN, ITALY

Achille Castiglione,
RR 126 radio-gramo-
phone, 1965

The consumer electronics company, Brionvega, was founded by the Brion family initially for the production of radio sets. In 1952, after Italy's first television network had been established, the company launched the first all-Italian television sets. A decade later, it manufactured the first Italian fully transistorized television, the portable *Doney 14* (1962), which was designed by **Marco Zanuso** and **Richard Sapper**. The same designers also created the *TS 502* radio (1964) with a hinged ABS plastic housing, the *Algol 11* television with an angled screen and the cuboid *Black ST 201* television (1969) – all of which reflected the company's preference for neo-modern forms and primary colours.

Brionvega's product line was less austere than those of German and Danish companies and this helped the company to appeal to a younger market. **Achille** and **Pier Giacomo Castiglioni** designed several innovative products for Brionvega, including a wire broadcasting receiver and the famous *RR 126* radio-gramophone (1965). **Mario Bellini** also designed a number of products for the company, including the *Totem* radio and record player with integrated speakers (1971), as did Franca Helg (b. 1920) and Franco Albini (1905–1977).

Mario Bellini,
Totem hi-fi, 1971

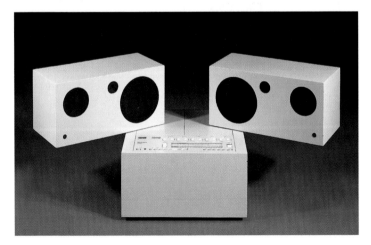

Marco Zanuso &
Richard Sapper,
Black ST 201
television, 1969

By this time however, the company was undoubtedly finding it difficult to compete with Japanese imports and at some point in the mid-to-late 1970s it ceased operating. Like its competitors **Braun** and **Bang & Olufsen**, Brionvega employed a Modern aesthetic for its product's casings, but the **styling** it promoted was more akin to trends within the contemporary fine arts.

Isambard Kingdom Brunel was the son of the French émigré Sir Marc Isambard Brunel (1769–1849), who invented a wide range of machines for printing, boot-making, stocking knitting and timber processing and who also worked as a civil engineer on tunnels and suspension bridges. Brunel senior was in charge of the building of the Thames Tunnel running between Rotherhithe and Wapping, and in 1825 appointed his 19-year-old son resident engineer for the project. Isambard Kingdom Brunel held this post for three years until he was severely injured in a flooding incident – an event that put the London Tunnel project on hold. While recovering his health, Brunel designed a suspension bridge to span the Avon Gorge, and this design was eventually chosen over that put forward by the renowned Scottish engineer, Thomas Telford (1757–1834). Later, as an engineer, Brunel implemented major improvements to the docks in Bristol and designed other docks at Monkwearmouth, Milford Haven, Plymouth, Brentford and Briton Ferry. He became chief engineer to the Great Western Railway (GWR) in 1833 and subsequently introduced the broad-gauge railway track (spanning seven feet), which made trains more comfortable, roomier and safer because of the engine's lower centre of gravity. Brunel's broad gauge also allowed trains

Great Western
engine, 1838

Clifton Suspension
Bridge, 1830–1863

to travel faster owing to the increased stability. The battle between the adoption of Brunel's gauge over standard gauge raged for many years and it was not until 1892 that the GWR, nicknamed "God's Wonderful Railway", eventually conceded to building only standard gauge. Brunel did, however, introduce more than 1000 miles of railroad in Britain and Ireland. He also built two Italian railway lines, was consulted on the construction of railway lines in Australia and India, and attempted to develop a system of pneumatic propulsion for the South Devon Railway in 1844 – a project that was not fully successful. Brunel is also remembered for his many bridge designs, including one in Maidenhead that remains the flattest arch made of brick in the world. He also worked as a marine engineer and designed three landmark vessels, which were each the largest in the world at their launching – the *Great Western* (1837), the *Great Britain* (1843) and the *Great Eastern* (1858). Brunel was knighted in 1841 for his construction of the Thames Tunnel, which although dogged by financial difficulties and technical setbacks, was a sheer marvel for its day. Brunel's extraordinary originality of thought and ability to put theory into practice against all the odds have ensured his enduring fame as a great pioneer of industrial design and the father of modern civil engineering.

Isambard Kingdom
Brunel, c. 1855

The *Great Eastern*
ready for launching

The *Great Eastern*
steamship, 1858

EDWIN BUDDING

BORN 1796 ENGLAND
DIED 1846 ENGLAND

Detail of Budding's
lawnmower

In 1830 Edwin Budding, an engineer from Stroud, Gloucestershire, invented the lawnmower. His inspiration for the basic mechanical principle came from a machine which he saw in a local cloth mill, which used a cutting cylinder mounted on a bench to cut the pile on textiles to make them smooth after weaving. Budding realized that he could adapt the concept for the cutting of grass if the mechanism were mounted in a wheeled frame in such a way that the blades rotated close to the lawn's surface. The subsequent lawnmower he designed in 1830 was constructed of cast iron and had spur gears connecting the main roller to the knife blades. The gears allowed the blades to rotate at a speed twelve times that of the large rear roller. In 1831 Budding went into partnership with John Ferrabee, who agreed to pay for the patents. Budding's lawnmower went into production and for many decades afterwards served as the essential prototype for subsequent models produced by other manufacturers.

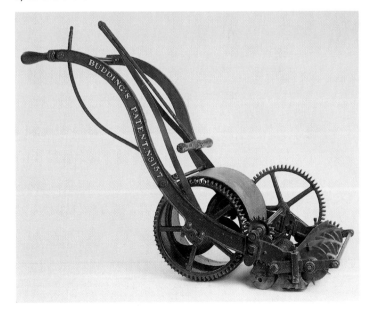

The first-ever
lawnmower, 1830

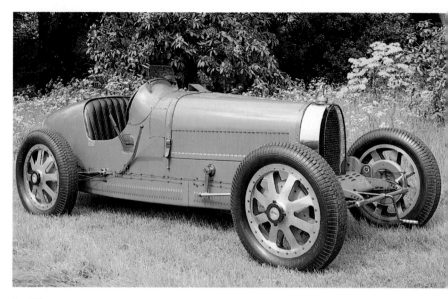

Bugatti *Type 35*, 1924

Ettore Bugatti with
the Bugatti *Type 35*

BUGATTI

ETTORE BUGATTI

BORN 1881 MILAN, ITALY
DIED 1947 PARIS, FRANCE

Ettore Bugatti established an automobile factory in 1909 in Molsheim, Alsace. He specialized in building racing cars such as the *Type 22* and the famous *Type 35* (1924), the latter proceeding to dominate motor racing through the mid-1920s by winning over 2,000 races. During the 1920s the Bugatti workshop also produced a handful (6–8) of the ultra-luxurious *Type 41* road cars. Sometimes referred to as the "Golden Bugatti" or "La Royale", the *Type 41* was possibly the most meticulously constructed car of all time. 1934 saw the introduction of the exceptionally beautiful *Type 57*. The streamlined design of this vehicle – its flush doors in particular – exerted significant influence on the car industry. After Ettore's death in 1947, the company's fortunes declined, but the marque was eventually revived by an Italian entrepreneur in 1987 and three new models were designed, including the *EB 110 GT* super car (1991) – the fastest car in the world at the time – and the luxurious *EB 112* sedan (1993) styled by **Giugiaro** (ItalDesign). In 1995 the company was bought by **Volkswagen**, and 1999 subsequently saw the launch of four powerful and elegant prototypes which reflected Bugatti's strong and remarkable heritage.

Bugatti advertisement by Gerold, 1932

Prototype Bugatti *EB 112*, styled by ItalDesign, 1993

Adjustable floor
lamp, c. 1929

EDOUARD-WILFRID BUQUET

ACTIVE 1920s–1930s FRANCE

In 1925 the French Art Deco designer Edouard-Wilfrid Buquet developed
an innovative adjustable lamp which predated **George Carwardine**'s more
famous articulated *Anglepoise* lamp by some seven years. Buquet's coun-
ter-balanced design was patented in Paris in 1927 and became one of the
most popular lighting designs of the period. Especially favoured by architects and interior designers, the chromed metal task lamp with its **aluminium** shade and lacquered wood base was often used by members of the
avant-garde, such as **Marcel Breuer** and Louis Sognot (1892–1970), for their
interior schemes. The lamp was exhibited at the 1929 Salon d'Automne in
Paris and widely illustrated in French magazines during the late 1930s, but
Buquet remained uncredited for its design until the 1970s. The lamp was
also produced in floor-standing and hanging versions, as well as in a variation with two articulated arms. While the lamp projects a Modern industrial
aesthetic, it was for the most part produced by hand. Despite this incon-
gruity, Buquet's lamp was highly influential for the design of subsequent
industrially manufactured tasking lights, such as Carwardine's *Anglepoise*
and the *Luxo L-1* by Jacob Jacobsen (b. 1901).

Adjustable
architect's lamp,
1927

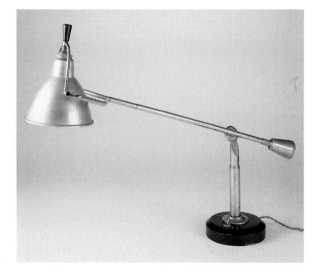

Stand selling products from the metal workshop of the Kunstgewerbeschule, Burg Giebichenstein, Halle, c. 1927

In 1915 the architect Paul Thiersch (1879–1928) became director of the Handwerkerschule in Halle and transformed the institution by adding courses and establishing two workshops – one for metalworking and the other for enamelling. When the school subsequently moved to new premises in Burg Giebichenstein in 1921/1922, it was renamed the Werkstätten der Stadt Halle, Staatlich-städtische Kunstgewerbeschule Burg Giebichenstein. The institution was one of the first to teach design students the importance of serial and industrial production. In 1923 the Berlin sculptor Karl Müller (1888–1972) was appointed director of the school's metal workshop, and during his tenure the Kunstgewerbeschule became one of the leading German design centres for metalware. Like the **Bauhaus**, the school in Halle made the philosophical transition from hand production to industrial manufacture, but did not depart completely from the idea of craftsmanship. Because the Halle school did not adhere to functionalism quite as strictly as the Bauhaus, the metalware products created there, ranging from tableware to lighting, were less severe and at times more expressive. Similarly, ceramic designs, whether porcelain dinner services or stoneware pots, were more often distinguished by softened elemental forms. Through its promotion of a subtler and more humanistic form of Modernism, the Halle school perhaps struck a better balance between art and industry than did its rival, the Bauhaus.

Marguerite Friedländer-Wildenhain, *Burg Giebichenstein* dinner service for KPM, Berlin, 1930

MIKE BURROWS

BORN 1943
ST. ALBANS, ENGLAND

Mike Burrows developed an interest in cycle design after his car "blew up" in 1976. He founded his own company in 1980 to produce his first bike design and this was soon followed by several recumbent models. Burrows originally conceived the recumbent cycle as a record-breaking high performance vehicle that was intended to go in a straight line only. In 1991 his most advanced monocoque superbike was seen by **Lotus** and adapted for the British cyclist, Chris Boardman, who went on to win a gold medal with it in the 400 metre pursuit event at the 1992 Barcelona Olympic Games. Having accepted an offer of a senior design post with the Giant Cycle Company, Taiwan in 1994, Burrows decided to license the manufacture of his most refined recumbent, the *Windcheetah*, to Advanced Vehicle Design in Altrincham, a manufacturer of radical, high-quality quadricycles. Also known as the "Speedy" because of its 50-mph capability and its 70-mph downhill potential, the *Windcheetah*, like the Boardman bike, was completely revolutionary and allowed "man to power a machine to new limits of speed and distance". The *Windcheetah* utilizes a highly sophisticated cruciform frame and a unique joystick steering system, which provides extraordinary cornering. Lightweight titanium, **aluminium** and kevlar components are used extensively in its construction and the combination of these high performance

Windcheetah T. I.
human-powered
vehicle, 1995

materials, together with wind cheating **aerodynamics**, delivers an outstanding rider experience. As the benchmark for recumbent performance, the *Windcheetah* is a tribute to Mike Burrows' genius.

JAKE BURTON

BORN 1954
NEW YORK, USA

As a child, Jake Burton experimented endlessly with a department store toy known as a *Snurfer* that was manufactured by Sherman Poppen in the mid-1960s. This child's plaything – a short fat ski without bindings or edges – allowed the user to surf on snow. No further development of this proto-snowboard took place over the next ten years, however. In 1977 Burton left the business world of Manhattan and moved to Vermont to become a snowboard "shaper". The same year, he founded the world's first snowboard factory, which in the beginning was a modest garage operation. The first snowboards he produced included the Burton *Backhill*, Burton *Backyard* and *Performer*, which were sold by Jake going from shop to shop. In these early years, snowboarding was very much an underground sport with sledging hills and even golf courses mainly being used. Burton knew that if snowboarders had to walk their way up a hill in order to board down it, the sport would not progress very far. He therefore lobbied local ski hills to allow snowboarders access to the lifts and slopes. Eventually, in the early 1980s, Stratton resort in Vermont acquiesced. As other ski resorts followed suit, Burton developed the *Performer Elite* specifically for hard-packed snow, as

Burton *Backhill* & *Backyard* snowboards, 1977

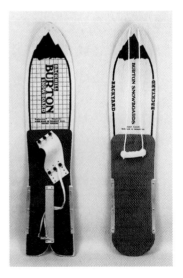

his earlier edgeless wooden boards were really only suited to powder snow. Burton's tireless efforts to promote snowboarding eventually ensured its widespread acceptance, but also turned snowboard manufacturing into a highly competitive business. Burton buckled down and focused on product development. In 1992 he moved his factory from Manchester, Vermont, to Burlington. While Jake Burton did not invent the first snowboard – there is an example from the 1920s in the Burton museum – he did invent the sport. Burton's enthusiasm and vision resulted in the creation of a remarkable industry that continues to seriously rival skiing. Not surprisingly, snowboarding has had a significant influence on ski design – the now common parabolic-shaped ski was born directly out of snowboard technology.

Burton *FL*
snowboard, 1998
(first manufactured
2000)

→Burton *Custom 56*
snowboard, 1998
(first manufactured
2000)

NICK BUTLER

BORN 1942
NORMANTON, ENGLAND

In 1967 Nick Butler founded BIB Design Consultants Ltd. and in 1975 was elected a fellow of the Chartered Society of Designers (CSD). He became a member of the judging panel for the **Design Council** Awards in 1975 and of the Council's Industrial Advisory Committee in 1978. Butler also chaired the British Design Export Group from 1980 to 1981 and was elected a fellow of the Royal Society of Arts in 1983. He served as an external examiner for the Royal College of Art from 1983 to 1986 and has lectured widely on industrial design both in Britain and abroad. In recognition of his contribution to this field, Butler received an OBE (Order of the British Empire) in 1988 and became Master of the RSA's Faculty of Royal Designers to Industry in 1995. During the mid-1990s, Butler consulted to the Bank of England and the Royal Mint on the design of the Euro currency and was a judge for the British Millennium Products programme. His office continues to offer a comprehensive product design service, which includes design audits, feasibility studies, product innovation and development, design engineering, human factors analysis, model-making and prototyping as well as productionization. Butler's product designs, such as the *Durabeam* flashlight (1982) which

Durabeam
flashlights for
Duracell, 1982

won a Design Council Award, are simple and intuitive to use and are guided by his belief that "design can make a real difference to how well a product is understood and therefore the efficiency of its usage". BIB has designed personal computers for **Apple**, fax machines for British Telecommunications and Panasonic, watches for Citizen, televisions for Thomson, a range of pens for Dunhill and kitchen equipment for Prestige.

CANON

FOUNDED 1947
TOKYO, JAPAN

The Precision Optical Instruments Laboratory was established in Tokyo in 1933 to conduct research into camera technology. A year later, the company developed a prototype of the *Kwanon*, Japan's first 35mm focal-plane shutter camera, and in 1940 designed Japan's first indirect X-ray camera. Introduced in 1946, the Canon *SII* camera was acclaimed by officers from the Occupation Forces and also by foreign purchasers. In 1947 the firm was renamed the Canon Camera Company and over the following years continued to produce innovations in camera technology, such as the world's first speed-light synchronized flash and shutter camera, the Canon *IVSb* (1952). The Canon *L1* still-camera and the Canon *8T* cine-camera became the first products to receive a "Good Design" designation from the Japanese Ministry of International Trade and Industry in 1957. During the 1960s and 1970s, Canon diversified its product line to include microfilm systems and calculators. In 1965 the company entered the copying machine industry with its introduction of the *Canofax 1000*. Three years later it also launched the first-ever four-track, four-channel recording head. Throughout the 1970s Canon continued developing

Kwanon camera – Japan's first 35mm focal-plane shutter camera, 1934

state-of-the-art cameras, including its top-end SLR, the Canon *F-1* (1971) and the landmark *AE-1* with a built-in microcomputer (1976), which triggered an *AE* SLR boom. In 1975 the company succeeded in developing a laser beam printer (LBP) and in 1976 introduced the world's first non-mydriatic retinal camera, the *CR-45NM*. This was followed in 1981 by the first bubble-jet printer and the new *F-1* SLR. In 1982 Canon launched a colour ink-jet printer and in 1986 the first multi-functional telephone with a built-in fax. Around this time, Canon also began producing digital camcorders, the *Optura* (1997) and the lighter and more compact *Elura* (1999). Over the company's 53-year history, the cumulative production of its cameras has exceeded 100 million units. From handshake-compensating binoculars to digital cameras to ultra-high-quality digital imaging equipment and high-definition television broadcasting lenses, all Canon products have a strong brand identity. Among the company's product design guidelines are: the use of advanced forms, materials and technology; the optimization of comfort with regard to human factors; the dedication to ease of operation; the consideration of the operational environment; and the creation of transcultural solutions.

G-Mark image stabilizer binoculars, 1995

C.BIO camera
designed by Luigi
Colani, 1982–1983

IXUS camera, 1996

GEORGE CARWARDINE

BORN 1887 ENGLAND
DIED 1948 ENGLAND

George Carwardine, a director of Carwardine Associates, Bath, was an automotive engineer who specialized in the design of suspension systems. In 1932 he patented the design of an articulated table lamp, the *Anglepoise*. This innovative design, which allowed flexible re-positioning, was based on the constant-tension principle of human limbs with the lamp's spring acting in much the same way as muscles. Carwardine balanced a weight against a spring through a linking element so as to allow the weight – and thereby, the arm – to remain stable in a variety of positions. Originally produced by the English spring manufacturer, Herbert Terry of Redditch, from 1933 the *Anglepoise* lamp was an instant success. It found applications in offices, factories and hospitals as well as in domestic environments and was manufactured in large numbers for over 50 years. In 1937 the patent for the lamp was acquired by the Norwegian lighting designer, Jacob Jacobsen (b. 1901), who had been influenced by it when he came to design his own

Anglepoise lamp, 1932 (produced from 1933)

Page from a brochure promoting the use of the *Anglepoise* lamp in the Birmingham Hospital Centre, mid-1930s

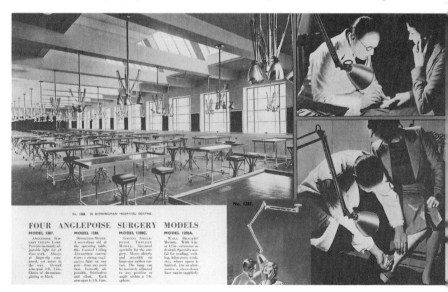

FOUR ANGLEPOISE SURGERY MODELS

MODEL 1287.
ANGLEPOISE SURGERY CEILING LAMP. Provides instantly adjustable light for all close work. Always at finger-tip command, yet never in the way. Overall arm-span 5 ft. 2 ins. Choice of chromium-plating or black.

MODEL 1288.
DISSECTING MODEL. A marvellous aid at the operating table, this Forward Ceiling Anglepoise concentrates a strong shadowless light on any part—from any position. Positively adjustable. Shadowless and silent. Each arm-span is 3 ft. 9 ins.

MODEL 1208C.
SPECIAL ANGLE-POISE TROLLEY MODEL. Intended specially for the surgery. Moves silently and smoothly on large-size rubber castors. The lamp can be instantly adjusted to any position or angle within a 7 ft. sphere.

MODEL 1208A.
WALL BRACKET MODEL. With 8-in. or 12-in. extension as desired. Specially useful for reading, writing, laboratory work, etc., where space is limited. (As an alternative a screwdown base can be supplied.)

Early example of an *Anglepoise* lamp for Herbert Terry & Sons, 1932 (produced from 1933)

highly successful *Luxo L-1* lamp the same year. Carwardine's *Anglepoise*, which was subsequently manufactured in Norway and retailed under a different name, was highly influential for later generations of task lighting. The fact that this classic design is still manufactured today testifies to its functionality and timeless Modern aesthetic.

Cover of *The Case Eagle*, September 1932 issue

CASE

FOUNDED 1842
ROCHESTER, WISCONSIN, USA

Jerome Increase Case (1819–1891) read an article in the *Genessee Farmer* about a machine that could be used to "thrash" wheat – a back-breaking chore that had always been done entirely by hand. The time-consuming nature of traditional threshing had hindered the expansion of farms in North America and Case saw the huge potential of a mechanical threshing machine. In 1842 he improved upon a primitive "ground-hog" threshing machine he had designed earlier and founded the J. I. Case Company in Rochester, Wisconsin. Two years later, he built a factory in Racine and became the first manufacturer to build mechanical threshers. His *Sweepstakes Thresher* of 1862 was capable of threshing 200–300 bushels a day, whereas one man could only manage six or seven bushels a day. Case soon became known as the "Threshing Machine King" and in 1865 he adopted the famous eagle trademark. Four years later he produced the first Case steam engine, known as the *Old No. 1*, which boasted 8 hp and which was loved by farmers as they could have it running all day, unlike animal-powered machinery. It was not until the mid-1870s, however, that sales of agricultural steam-powered machines really took off. 1880 saw the launch of the *Agitator* thresher, offering greater capacity and improved efficiency. By 1886 Case was the largest manufacturer of steam engines in the world and in 1892 introduced its first gasoline-powered tractor. By 1913 the company was producing smaller and faster tractors that were the antecedents of today's vehicles, and in 1919 Case added ploughs and tillage tools to its product line, thus enabling it to become a full-line farm equipment manufacturer. Case also dabbled in automobile manufacturing and even fielded a racing team for the first Indianapolis 500 motor race in 1911. In the 1930s, Case exported threshers as far afield as New Zealand and continued to introduce

Tresher/Stacken, 1908

Ground/tog thresher,
c. 1842

new products, from the *Case Hammer Mill*, which ground feed for livestock, to the *One-Man* combine. During the Second World War, Case produced more than 1.3 million shells for the American armed forces and manufactured parts for military trucks, Sherman tanks and fighter aircraft. After the war, Case returned to peacetime production of agricultural and construction equipment. Today the company remains a leader in its field, particularly in terms of innovative design and manufacture.

*Senty Next G11
Pocket Bell* pager,
1996

CASIO

FOUNDED 1957
TOKYO, JAPAN

*MG-800 calculator,
c. 1978 – one of
the first calculators
to incorporate a
primitive electronic
game*

In 1946 Tadao Kashio, an engineer, set up a company
called Kashio Seisakujo to manufacture aircraft parts
for the military. Acting on a suggestion from his brother
Toshio, Tadao and his three younger brothers also set about developing a
calculator. Unlike other calculators, which relied on electrically powered internal gears, the Kashio brothers' design ingeniously adapted the relay elements developed for telephone switching equipment. This new electronic
calculator was significantly more compact, since it did not need gears. 1957
saw the founding of Casio Computer Co. and the launch of the all-electric
14-A model. Ten years later, the company introduced the world's first programmable electronic desktop calculator, the *AL-1000*. At this stage, calculators were still extremely expensive and so Casio decided to develop a design
that could be afforded by individuals rather than just corporate consumers.
The resulting palm-sized *Casio Mini* (1972) was the world's first personal
calculator and helped to make the calculator part of everyday life. Using its
pioneering technology, Casio set new standards in miniaturization with the
first-ever electronic wristwatch, the *Casiotron* (1974) and the original card-
sized calculator, the *Casio Mini Card* (1978). The *Casiotron*, with year, month,
day, hour, minute and second display capabilities,
was based on the concept that "time is a continuous
process of addition". In 1981 the Casio design team
decided to develop a wristwatch that was as "strong
as a tank but also flexible", and which would last
10 years – the result was the *G-Shock* (1983), which
could endure the shock of hitting the ground from
a height of 10 metres. Many subsequent *G-Shock*
models followed, including a ladies' version, the
Baby-G (1994), which became an instant "must-
have" fashion accessory. Throughout the 1980s and
1990s, Casio continued to specialize in the design
and development of state-of-the-art calculators,
wristwatches and electronic musical instruments,
as well as hybrid products such as the *BM-100WJ*
(1989) digital wristwatch with a sensor for forecast-

← BG 100 OCL-7T
Baby G wristwatch,
1995

→ BG 100–3T Baby G
wristwatch, 1995

← AW 500–1E,
G-Mark, G-Shock
wristwatch, 1988

→ DW 003B-9
G-Shock wristwatch,
c. 1991

ing weather, and the *Pathfinder PAT1GP-1* (2000), the world's first wristwatch
with built in GPS for longitude and latitude display. Casio strives to make
original products with innovative capabilities that are easy to operate, versatile and affordable. Research and development of totally new products is a
central theme of its business philosophy.

LIVIO, PIER GIACOMO & ACHILLE CASTIGLIONI

| BORN 1911 MILAN, ITALY | BORN 1913 MILAN, ITALY |
| DIED 1979 MILAN, ITALY | DIED 1968 MILAN, ITALY |

BORN 1918 MILAN, ITALY

Livio Castiglioni studied architecture at the Politecnico di Milano, graduating in 1936. In 1938 he set up a studio in collaboration with his brother Pier Giacomo and Luigi Caccia Dominioni (b. 1913), designing silver and **aluminium** cutlery. Their most notable design, however, was the 547 radio (1938–1939) for Phonola. As the first Italian radio to be made in **Bakelite**, the 547 changed the face of radios, which up till then had mostly been cased in wooden boxes. The 547 was awarded a gold medal at the VII Milan Triennale of 1940, where the Castiglioni office also curated an exhibition of radios. Livio worked as a design consultant for Phonola from 1939 to 1960 and for **Brionvega** from 1960 to 1964. He was president of the ADI (Associazione per il Disegno Industriale) from 1959 to 1960 and designed many audio-visual presentations. He collaborated with his younger brothers on several lighting projects, although his best-known light design, the snake-like *Boalum* (1970), was executed in conjunction with Gianfranco Frattini (b. 1926). Like their elder brother, Pier Giacomo and Achille Castiglioni also graduated from the Politecnico di Milano, in 1937 and 1944 respectively. Achille joined

Livio Castiglioni,
Pier Giacomo
Castiglioni & Luigi
Caccia Dominioni,
547 radio for
Phonola,
1938–1939

his elder brothers' design studio in Piazza Castello and during the post-war years they undertook town-planning and architectural commissions as well as exhibition and product design. The brothers were extremely active and helped establish the Milan Triennale exhibitions, the Compasso d'Oro awards and the ADI. When Livio left the partnership in 1952, the two younger brothers continued to work together. They designed the "Colori e forme nella casa d'oggi" exhibition at the Villa Olmo in Como, where they first presented their radical "ready-made" designs of 1957 – the *Mezzadro* (Sharecropper's Stool), incorporating a tractor seat, and the *Sgabello per Telefono* (Telephone Stool), which had a bicycle seat. The brothers also produced less contesting designs, such as the Neo-Liberty style *Sanluca* armchair (1959) for the furniture manufacturer Dino Gavina, whose company offices in Milan they designed in 1963. Other notable designs by Pier Giacomo and Achille Castiglioni include the *Tubino* desk lamp (1950), the *Luminator* floor lamp (1955), the *Arco* floor lamp (1962) and the *Taccia* table lamp (1962) – all of which were manufactured in significant numbers. In 1966 they designed the *Allunaggio* seat, which was inspired by the first moon landing. Their prestigious client list included **Kartell**, Zanotta, Bernini, **Siemens**, Knoll, Poggi, Lancia, Ideal Standard and Bonacina. Apart from furniture and lighting, Pier Giacomo and Achille Castiglioni also designed audio equipment for Brionvega, including the *RR 126* radio-gramophone (1965). After Pier Giacomo's death in 1968, Achille continued to work in industrial design, creating such well-known pieces as the *Lampadina* table lamp (1972) for **Flos**, the sleek cruet set (1980–1984) for **Alessi** and the *Gibigiana* directional table lamp (1980) for Flos.

The Castiglioni brothers exerted much influence on succeeding generations of Italian designers not only through their activities and success as industrial designers, but also through their long-standing involvement in design education. Pier Giacomo taught at the Politecnico di Milano from 1946 to 1968, while Achille was professor of artistic industrial design from 1970 to 1977 and professor of interior architecture and design from 1977 to 1980 at the Politecnico di Torino. In 1981 Achille returned to the Politecnico di Milano as professor of interior design and since 1986 has been professor of industrial design there.

During his long career, which has spanned over half a century, Achille Castiglioni has been honoured with eight Compasso d'Oro awards as well as numerous other design prizes. While the language of design he and his brothers pioneered was grounded in rationalism, it was tempered with ironic humour and sculptural form – an unusual approach to industrial design that has been described as "rational expressionism". This, and the remarkable consistency of quality of their designs, which are both structurally inventive and aesthetically pleasing, make the Castiglioni brothers three of the most important figures in Italian design in the 20th century. Achille Castiglioni believes that "the need for a new and original approach in contemporary design is emerging more and more urgently". "Such an approach," he claims, "can be effected through design activity on two different levels. On the one hand, there is design for mass production; on the other, design for the production of a limited quantity." In both these areas of activity, Achille has not only experimented but also excelled, with his work communicating emotional values while at the same time "straining to achieve adequate results with minimum means".

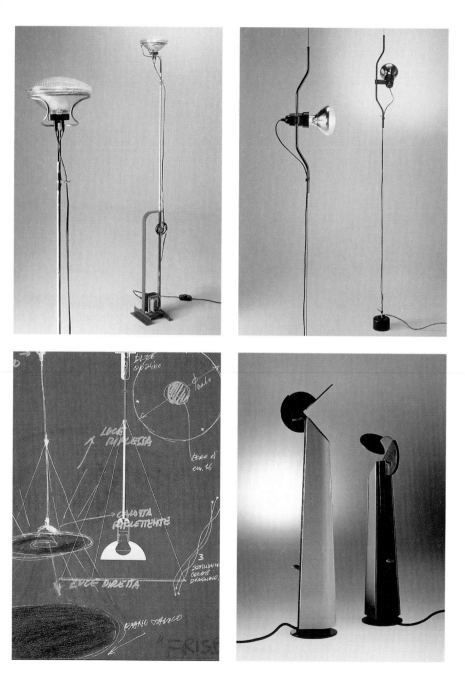

Benjamin Holt, first "Caterpillar" tractor, patented in 1893

Thirty Caterpillar tractor, c. 1930

Early steam tractor,
c. 1890

CATERPILLAR

FOUNDED 1925
PEORIA, ILLINOIS, USA

During the 1890s Benjamin Holt (1849–1920) and Daniel Best independently researched and experimented with steam-driven farm tractors. Both Holt and Best pioneered track-type and gasoline-powered tractors, and during the First World War Holt's tracked "Caterpillar" tractor was used extensively by the Allied Forces to pull artillery and supply wagons. In 1925 the Holt and Best companies finally merged to become Caterpillar, based in Peoria, Illinois. After years of researching diesel as an efficient source of power for track-type tractors, the company launched the landmark *Diesel Sixty Tractor* in 1931. During the 1940s, Caterpillar diversified its product line to include motor graders and electric generators, which were in great demand during the Second World War, as were its tractors. Caterpillar also produced a special engine for the *M-4 Sherman* tank. After the war, Caterpillar introduced numerous new products and in 1950 established its first overseas operation in Britain. Since the 1950s Caterpillar has supplied its powerful *Cat* engines to manufacturers of trucks, trains, boats and construction equipment. Through electric generation systems, *Cat* engines also supply power to areas inaccessible to utility power grids, such as offshore oil rigs and developing nations, and provide emergency power to hospitals, schools etc. During the 1970s Caterpillar developed several innovative products, including what was then the world's largest and most advanced track-type tractor, the *D10*, as well as the 225 hydraulic excavator and the 3400 family of engines. The company was hit badly by the recession in the early 1980s but bounced back with revolutionary in-

D11R CAT tractor,
late 1990s

novations such as the elevated rear sprocket design of its track-type tractors, which increases productivity, simplifies maintenance and extends drive train life. By 1997 Caterpillar had become the world's leading manufacturer of construction and mining equipment, diesel and natural gas engines and industrial gas turbines.

CHRYSLER

FOUNDED 1925
DELAWARE, USA

Walter Percy Chrysler
with a model of Carl
Breer's *Airflow*,
1930s

Walter Percy Chrysler (1875–1940) began his working life as a machinist's apprentice in the railroad industry, but by the age of 33 had already become the superintendent of motive power for the Chicago Great Western Railway. In 1910 he moved to the American Locomotive Company to manage its Pittsburgh works and around this time purchased his first car, a Locomobile Phaeton. A couple of years later, Chrysler joined the Buick Motor Car Company in Flint, Michigan. In 1916 Buick became **General Motors**' first automotive division, and in 1917 Chrysler was appointed its president and general manager. He was promoted to vice-president of General Motors in 1919 but retired from the company a year later. He next worked for Willys-Overland and then the ailing Maxwell Motor Car Company, which he revitalized with the development of the Chrysler *Six* (1924) – America's first high-styled, medium-priced automobile, which set an industry sales record by selling 32,000 units. The Chrysler Corporation was established in Delaware in 1925 as the successor to the Maxwell Motor Car Company and Walter Chrysler was named its chairman. The same year saw the introduction of the highly successful Chrysler *Four*, with a top-speed of 58 mph, and by 1926 the corporation had risen from 57th to 5th place in industry sales. In 1928

Walter Percy Chrysler
with *Four Series 58*,
1925

Dodge truck, 1940s

Chrysler manufactured the first Plymouth and De Soto models and the company also acquired Dodge Brothers, which was highly regarded for its utility vehicles. Although hit by the Great Depression, Chrysler did not cut back on its research and development programme and pioneered several design and engineering innovations, including the "Floating Power" engine mounting system, which reduced vibration to produce a smoother ride. Although the company hired its first in-house stylist, Ray Dietrich (1894–1980), in 1932, design and **styling** at Chrysler essentially took a back seat to engineering until 1950. A notable exception was the Chrysler *Airflow* of 1934. Designed by the company's chief engineer, **Carl Breer**, and his staff engineers Fred Zeder and Owen Skelton, the *Airflow* introduced leading-edge aerodynamic styling, innovative weight distribution, unitized body construction and the industry's first one-piece curved windscreen. While the *Airflow* was way ahead of its time and consequently a commercial failure, it was immensely influential upon later automotive design.

During the Second World War, Chrysler produced 18,000 32-ton *M-4 Sherman* tanks and around 500,000 Dodge trucks for military use. By 1945 the company had supplied over US$3.4 billion worth of equipment to the Allied Forces. After the war, Chrysler responded to the need for increased car and truck production by building or buying 11 plants between 1947 and 1950. In 1949 the company also appointed its first director of advanced styling, Virgil Max Exner (1909–1973), who had previously worked in **Raymond Loewy**'s

Dodge *Viper*, 1992

Chrysler 392 Hemi
Firepower V-8
engine, 1951

design office. Exner introduced European-like styling to the company and developed the "Forward Look", a style that emphasized movement and speed through the use of curved side windows and tail fins. The Forward Look is perhaps best exemplified by Exner's 1957 Plymouth *Belvedere*. During the 1950s Chrysler also introduced several innovations including power steering, key-operated ignition, electric windows and cushioned dashboards for improved safety. In 1951 it also developed the legendary "Hemi" V-8 engine. With deep hemispherical combustion chambers, the Hemi featured large valves that provided high-volume efficiency and produced 20 % more horsepower than standard V-8 engines. The Hemi was discontinued in the mid-1950s but reappeared in 1964 as the ferocious 425 horsepower 426 Hemi.

While it was in production, the 426 Hemi was the perfect power plant for Chrysler's emerging "muscle" cars. Elwood P. Engel (1917–1986) was the designer chiefly responsible for the creation of such beefy and brutishly purposeful cars as the Dodge *Challenger*, the Dodge *Charger*, the Plymouth *Road Runner* and the Plymouth *Barracuda*. From the mid-1960s to the early 1970s, these cars dominated the muscle-car street scene and became

Chrysler *Voyager*,
1999 model

legends in their own time. The oil crisis in the early 1970s put an end to this
period of excess in the American car industry, however, and prompted a demand for smaller, more fuel-efficient vehicles. Chrysler found itself facing a
financial crisis, and in 1979 the celebrated auto executive Lee Iacocca was
brought in to turn the company around. With the well-publicized challenge,
"If you can find a better car, buy it", Iacocca managed to save the company
with the introduction of the "K-cars" in 1981 – the Dodge *Aries* and the Plymouth *Reliant*. This was followed in 1983 by Chrysler's introduction of the
minivan in the form of the Dodge *Caravan* and Plymouth *Voyager*. These vehicles not only created a whole new market – people carriers – but also became the company's best sellers. In 1987 Chrysler re-entered the European
market with export vehicles and also purchased the exotic Italian sports car
manufacturer, **Lamborghini**. During the 1990s Chrysler launched several notable and financially successful models, including the awesome 700 hp
Dodge *Viper* (1992), the **Jeep** *Grand Cherokee* (1993) sports-utility vehicle
and a striking evolution of the *Voyager* MPV (1995). By adapting vehicle design to meet changes in living patterns and customer expectations, Chrysler
has managed to survive and prosper. In 1998 the company merged with the
prestigious German automotive manufacturer, Daimler-Benz AG, and its
product range has consequently adapted to the European market.

CITROËN

FOUNDED 1913
PARIS, FRANCE

Flaminio Bertoni,
Citroën *DS19*, 1962

→Pierre Boulanger,
prototype of the
Citroën *2CV*, 1936

↘Citroën *2CV Cl*,
1959

Citroën poster
showing the *Type A*,
1919

André-Gustave Citroën (c. 1878–1935) trained at the École Polytechnique and then worked as an engineer and industrial designer. From 1905 he began manufacturing components for cars and introduced double helical gears to France. In 1913 he founded the Société des Engrenages Citroën in Paris and two years later began manufacturing ordinance – up to 55,000 shells per day – for the French army. At the end of the war, with arms sales falling, Citroën converted his factory to the production of small and inexpensive cars and hired Jules Saloman as his first automotive designer. In 1919 the company launched the first mass-produced car in Europe, the *Type A*. This car significantly advanced the standards of automotive design with its disc wheels, electric lights and on-board starter. It was, however, the slightly later, more compact and much better-selling *5CV* that firmly established Citroën's reputation in the mass market. In 1934 André Citroën almost bankrupted his business by over-investing in tooling for the *Traction Avant* (later launched as the *7A*) and the company was then taken over by Michelin. The *Traction Avant* was an advanced prototypical design that was at least two decades ahead of its time, with front wheel drive and a long wheel base. Citroën's best-ever seller, the *2CV*, was designed by Pierre Boulanger (1886–1950) and first appeared in 1939. Launched in 1948 as a functionalist, no-frills work-horse, the *2CV* was intended to rival **Volkswagen**'s *Beetle*. Between 1948 and 1990, when production ceased, nearly 3,870,000 *2CVs* were built. Citroën's famous and more elegant *DS19*, designed by Flaminio Bertoni (1903–1964) in 1955, was the evolutionary successor of the *Traction Avant*. With its low-slung forward-looking aesthetic, the *DS19* became known as the "Goddess" in France and was, like the *Traction Avant*, an extremely advanced vehicle for its time. Citroën was eventually taken over by Peugeot in 1975, and although it continues to produce very worthy vehicles, the level of design innovation it achieved in the past has yet to be matched.

Christopher Dresser,
Coalbrookdale Co.
cast-iron grate, 1879

←Christopher Dresser,
Coalbrookdale Co.
cast-iron umbrella
stand, c. 1880

Abraham Darby (c. 1678–1717), a British ironmaster, pioneered the smelting of iron ore with coke in the Coalbrookdale area of Shropshire during the early 18th century. Coalbrookdale's close proximity to supplies of natural resources, low-sulphur coal in particular, helped revolutionize the production of iron. In 1709 Darby was the first to produce marketable iron in a coke-fired furnace and went on to manufacture the first iron rails, iron boat and steam locomotive in Britain. Such was the quality of Darby's iron that it could be cast in thin sheets and could thus compete successfully with brass in the manufacture of pots and other hollow ware. The famous cast-iron bridge near Coalbrookdale was also one of the first of its kind. Designed by Thomas Pritchard, the bridge spans 43 metres across the River Severn and was erected between 1777 and 1779 by John Wilkinson and Darby's grandson, Abraham Darby III (1750–1791), partly to advertise the family iron foundry. As a major centre of iron production for over 100 years, Coalbrookdale was renowned in the 19th century for its cast-iron fireplaces and hall furniture, some designed by **Christopher Dresser**. As the birthplace of the **Industrial Revolution**, Coalbrookdale is now a world heritage site.

Philippe Jacques
de Loutherbourg,
*Coalbrookdale by
Night*, 1801

WELLS COATES

BORN 1895 TOKYO, JAPAN
DIED 1958 VANCOUVER, CANADA

Wells Coates studied engineering at the University of British Columbia, Vancouver, and graduated in 1921. He subsequently moved to Britain and from 1922 to 1924 studied for a doctorate in engineering at London University. From 1923 to 1926 he worked as a journalist for the *Daily Express* and for a brief period was one of their correspondents in Paris. During this time he wrote from a humanist perspective, viewing design as a catalyst for social change. In London in 1928, he designed fabrics for the Crysede Textile Company and interiors for the firm's factory in Welwyn Garden City, in which he used plywood elements. From 1931 Coates was a consultant to Jack Pritchard's **plywood** products company, **Isokon** – a firm pioneering Modernism in Britain. Coates was also commissioned by Pritchard in 1931 to design the Lawn Road Flats, Hampstead, which are seminal examples of British Modern Movement architecture. In 1933 Coates co-founded MARS (Modern Architecture Research Group) and established design partnerships with Patrick Gwynne in 1932 and David Pleydell-Bouverie in 1933. From 1932

Ekco *AD65* radio for
E. K. Cole, 1934

he designed a series of **Bakelite** radios for the Ekco Radio Company, including his famous circular Ekco *AD 65* (1934), which were conceived for industrial production and were among the first Modern products available to British consumers. After the Second World War Coates worked in Vancouver, designing interiors for **de Havilland** and BOAC aircraft, and in the 1950s producing designs for television cabinets. Just as Wells Coates believed that "the social characteristics of the age determine its art", so his designs epitomize the flowering of British Modernism in the 1930s.

The concept of an air-cushion vehicle was first proposed in the 1870s by the British naval architect and engineer, Sir John Thornycroft (1843–1928). Although Thornycroft built and patented several models, he failed to solve the problem of the air cushion becoming "detached" from the craft. It was not until the 1950s, when Christopher Cockerell designed a "skirt" that could contain the air cushion, that the air-cushion principle became viable. In 1959 Cockerell went on to design the world's first hovercraft, the Westland *SRN1*. This amphibious vehicle was capable of carrying three passengers, but required calm water or flat ground to operate properly. The later, much improved and much larger *SRN2* was designed around a spacious central cabin and, depending on the layout, could accommodate 56 to 76 passengers and their luggage. In June 1962 American Navy experts observed trials of the *SRN2* hovercraft in the Solent, while British service chiefs were investigating the military potential of this revolutionary design. Cockerell was knighted in the late 1970s in recognition of his contribution to vehicle design. Today, enormous hovercraft that can accommodate over 400 passengers and some 60 cars frequently cross the English Channel at speeds of up to 60 knots.

Demonstration of Westland *SRN2* Hovercraft at Cowes, 1962

Westland *SRN1* Hovercraft, 1959

Henry Cole, teapot
for Summerly Art-
Manufactures,
c. 1846

SIR HENRY COLE

BORN 1808 BATH, ENGLAND
DIED 1882 LONDON, ENGLAND

Sir Henry Cole worked for several years as an official
in the Public Records Office in London, where he campaigned for better conservation of archival material.
He also devised a scheme for postal reform and studied watercolour painting under David Cox (1783–1859), his works being exhibited at the Royal
Academy. He published a series of children's books entitled *Felix Summerly's
Home Treasury* in 1841, and two years later published the first Christmas card
to a design by J. C. Horsley (1817–1903). In 1846 he became a member of
the Society of Arts; that same year, the Society awarded him a medal for a
tea service which he had created under the pseudonym of Felix Summerly.
Notable for its simplicity of form, this design was manufactured by Herbert
Minton. As one of the greatest champions of the applied arts, in 1847 Cole
founded Summerly's Art-Manufactures, which commissioned designs suitable for industrial production from artists such as John Bell (1811–1895),
Daniel Maclise (1806–1870) and Richard Redgrave (1804–1888). Although

Contemporary
caricature of Henry
Cole, c. 1850

Cole severed his connection with this venture in 1848,
he believed it had "established the compound word
'art-manufactures' in our language". Between 1847 and
1849 Cole organized annual Society of Art exhibitions
to promote art-manufactures and in 1851 became chairman of the Society. In this capacity, Cole was largely responsible for the staging of **The Great Exhibition** of 1851
in London and the assembly of a permanent study collection which would eventually form the basis of the
Victoria & Albert Museum. He also mounted an exhibition of "Examples of False Principles in Decoration" in
an effort to demonstrate the differences between "good
design" and "bad design". Another of Cole's contributions to the advancement of industrial design was his
reform of design teaching, which placed great emphasis
on the creation of simple, functional forms for industrial
production.

GINO COLOMBINI

BORN 1915
MILAN, ITALY

4660 wastebin for
Kartell, 1965

4660 wastebins
and *4650* umbrella
stands for Kartell,
1965

From 1933 to 1952 Gino Colombini worked in the architectural office of Franco Albini (1905–1977), where he helped design commercial and domestic buildings as well as furniture. In 1949 Colombini was appointed technical director of **Kartell**, a recently formed **plastics** manufacturing company. He designed many of the firm's first products, including a kitchen bucket that was awarded the Compasso d'Oro award in 1955. Colombini also designed many other plastic products intended for everyday domestic use, such as a carpet-beater (1957), a milking pail (1958), a lemon squeezer (1958), a children's lunch-box (1958), a washing tub (1957) and various buckets and dustpans (1956–1957). These innovative and highly functional products were among the first to exploit the potential of injection-moulded plastics as materials for high-volume mass production. Together with Leonardo Fioro, Colombini also designed a modular cupboard system for Kartell, which had a **tubular metal** frame supporting doors and shelves made of shock-resistant polystyrene. This forward-looking design of 1956 predicted the later storage-system furniture of the 1960s and 1970s. Colombini's products, which in most cases were based on the principle of structural, material and functional unity of design, helped Kartell to become one of the leading proponents of Italian design in the 1950s. His product designs revolutionized not only the intrinsic quality but the aesthetic form of everyday objects, and in so doing were extremely influential upon succeeding generations of industrial designers.

JOE COLOMBO

BORN 1930 MILAN, ITALY
DIED 1971 MILAN, ITALY

Cesare "Joe" Colombo was active as a painter before turning his attention to industrial design around 1958. He began experimenting with new materials, including reinforced **plastics**, as well as novel construction techniques and manufacturing methods. In 1962 Colombo established his own design office in Milan and worked chiefly on architectural and interior design projects. In 1964 he was awarded the IN-Arch prize for his interiors for a hotel in Sardinia (1962–1964), which included ceiling fixtures constructed of Perspex prisms that diffracted light. With his brother Gianni, Colombo developed this idea in his design for the *Acrilica* lamp (1962) for O-Luce. His first design for **Kartell** was the *No. 4801* chair (1963–1964), which comprised three interlocking **plywood** elements. The fluidity of this design anticipated his later work in plastics, such as the *Universale No. 4860* chair (1965–1967), which was the first adult-sized chair to be manufactured in injection-moulded plastic (ABS). Colombo produced other innovative designs for furniture, including the multi-functional *Boby* trolley system (1968), lighting, glassware, door handles, pipes, alarm clocks and wristwatches. From the beginning of his career, Colombo was interested in domestic systems products, as his early *Combi-Centre* container unit (1963) demonstrates. This interest in systems furnishings gave rise to his *Additional Living System* (1967–1968), *Tube* chair (1969–1970) and *Multi* chair (1970), all of which could be

4860 Universale
chair for Kartell,
1965–1967

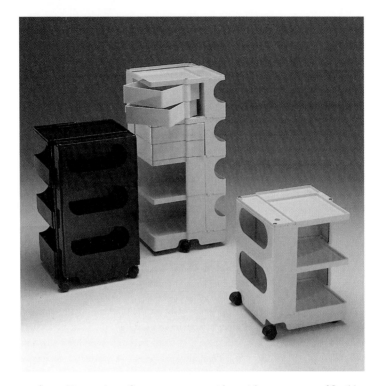

configured in a variety of ways so as to provide a wide assortment of flexible
sitting positions, thus reflecting his primary goal in design – adaptability.
His most forward-looking designs, however, were his integrated micro-envi-
ronments. These included his *Visiona 1* habitat of the future shown at Bayer's
Visiona exhibition in 1969, which comprised a space age "Barbarella-like" interior where furnishings transmuted into structural elements and vice versa.
Traditional furniture items were replaced with functional units, such as the
Night Cell and *Central Living* blocks and the *Kitchen Box*, so as to create a
multi-functional living environment. For his own apartment, Colombo de-
signed the *Roto-living* and *Cabriolet Bed* units (both 1969), and these were
followed by his *Total Furnishing Unit* (1971), which was a highly influential ex-
ample of "Uniblock" design. Shown at the 1972 "Italy: The New Domestic
Landscape" exhibition at the Museum of Modern Art, New York, the *Total
Furnishing Unit* was proposed as a completely integrated machine for living.
Colombo also designed products for O-Luce, Kartell, Bieffeplast, **Alessi**,
Flexform and Boffi.

SAMUEL COLT

BORN 1814 HARTFORD, CONNECTICUT, USA
DIED 1862 HARTFORD, CONNECTICUT, USA

As a young seaman bound for Calcutta, Samuel Colt carved a wooden model that later became the most famous handgun ever designed. It took Colt several years, however, to develop his idea into a working design. His revolutionary multi-shot revolver, which featured a cartridge cylinder that revolved when the hammer was cocked, was also the first handgun to successfully use a percussion action. In 1835 he patented the design and the revolver was put into production by the Patent Arms Manufacturing Company which he established in Paterson in 1836. Colt subsequently designed three types of revolver – pocket, belt and holster – as well as two rifles. In 1842 his factory closed because of insufficient orders, but production resumed five years later when the US Government ordered 1,000 revolvers for use in the Mexican War. In 1855 Colt opened a factory in Hartford which became the world's largest privately-owned armoury. Influenced by **Eli Whitney**, who had earlier pioneered the mass production of firearms, Colt developed a production-line system of manufacturing based on the **standardization** of components for interchangeability – 80% of which were turned out by machine. Colt reputedly declared that "there is nothing that can't be produced by machine", and industrialized methods of production certainly enabled him to produce a remarkable 150 weapons a day by 1856. His revolvers were widely used during the American Civil War, and at the time of his death Colt had manufactured over 400,000 firearms. In the 1880s the Colt *Peacemaker* became legendary in the Wild West, while the Colt .45 semi-automatic pistol was American standard issue during the First and Second World Wars and remained the official sidearm of the American armed forces until 1984. Together with the National Institute of Justice, Colt is currently developing "smart gun" technology which will allow guns to be fired only by their registered keepers.

Samuel Colt's patent firearms factory, Hartford

→Revolver, c. 1849

↘Advertisements for the Colt automatic pistol

Advertisement for Colt's *Official Police* .38 *calibre* revolver

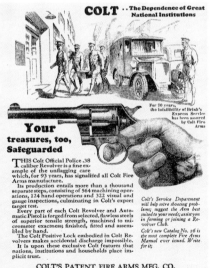

COLT .. The Dependence of Great National Institutions

For 70 years, the infallibility of Brink's Express Service has been insured by Colt Fire Arms

Your treasures, too, Safeguarded

THIS Colt Official Police .38 caliber Revolver is a fine example of the unflagging care which, for 93 years, has signalized all Colt Fire Arms manufacture.

Its production entails more than a thousand separate steps, consisting of 564 machining operations, 124 hand operations and 322 visual and gauge inspections, culminating in Colt's expert target test.

Every part of each Colt Revolver and Automatic Pistol is forged from selected, flawless steels of superior tensile strength, machined to micrometer exactness; finished, fitted and assembled by hand.

The Colt Positive Lock embodied in Colt Revolvers makes accidental discharge impossible. It is upon these exclusive Colt features that nations, institutions and households place implicit trust.

Colt's Service Department will help solve shooting problems; suggest the Arm best suited to your needs; assist you in forming or joining a Revolver Club.

Colt's new Catalog No. 26 is the most complete Fire Arms Manual ever issued. Write for it;

COLT'S PATENT FIRE ARMS MFG. CO.
SMALL ARMS DIVISION
HARTFORD, CONN., U.S.A.
Phil. B. Bekeart Co., Pacific Coast Representative, 717 Market St., San Francisco

.. The ARM of LAW and ORDER

Now Completely Insured!

Of the several forms of insurance not the least valuable is the protection rendered by a Colt because it *prevents* loss. Other forms can only partly restore *after* loss has occurred. Make *your* insurance complete with a Colt Revolver or Automatic Pistol.

Send for interesting booklet, "Romance of a Colt"

COLT'S PATENT FIRE ARMS MFG. CO.
Hartford, Connecticut, U. S. A.
Pacific Coast Representative
Phil. B. Bekeart Co., 717 Market St., San Francisco, Calif.

THE ARM OF LAW AND ORDER

Amory Houghton I

In 1864 Amory Houghton I (1813–1882) bought the Brooklyn Flint Glass Company, which he moved in 1868 to Corning, New York. Specializing in pressed glass products, the company was incorporated as Corning Glass Works in 1875. Two years later, Corning scientists attempted to improve the quality of optical lenses, which marked the company's first foray into scientific research. In 1880 Corning began manufacturing blanks for **Thomas Alva Edison**'s incandescent lamps, and in 1908 the Corning research laboratory was established.

The company subsequently developed heat-resistant borosilicate glass, which it named **Pyrex,** and in 1915 it added Pyrex-branded cookware to its product line. Owing to the popularity of its Pyrex wares, Corning expanded and in 1918 purchased the Steuben Glass Works. Corning continued to develop complex laboratory and scientific apparatus, including large observatory telescope lenses, and in 1935 began commercial production of glass fibres. In 1942 the company started producing cathode ray tubes for radar systems, and during the Second World War completed 174 research projects, including one for "ribbon glass" used in electrical components. Corning commenced automatic production of television bulbs in 1947, followed two years later by the centrifugal casting of television funnels. Colour television bulbs were produced from 1953, and in 1956 the highly successful *Corningware* range was launched – marking the first commercial application of *Pyroceram* glass **ceramics**. In 1980 a range of glass saucepans was introduced under the name of *Visions Cookware*, and since then Corning has pioneered glassware designs for both domestic and laboratory use, while conducting research into other glass applications. In 1994 Corning was awarded the United States National Medal for Technology for its development of six "life-

Television panel and funnel, 1947

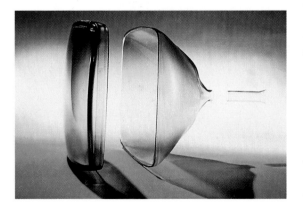

Cornflower casserole dish from Corning-ware range, 1958

"changing" products, including the glass bulb for Thomas Alva Edison's electric lamp, the first mass-produced television tubes and its invention of optical fibres in the 1970s.

Silver Streak iron, 1942–1946 (Corning manufactured the glass shroud element only)

THOMAS CRAPPER

BORN 1836 THORNE, YORKSHIRE, ENGLAND
DIED 1910 LONDON, ENGLAND

Shower fitting,
c. 1902 (detail)

The development of modern sanitation ranks amongst the greatest triumphs of the Industrial Age and one of the areas of design that has benefited humankind the most. One of the most famous figures in this often-overlooked field was Thomas Crapper. Having served an apprenticeship with a plumber in Chelsea, Crapper founded his own plumbing business there in 1861 and began researching and developing new sanitary products. Such was his eventual reputation that in the 1880s his firm was employed to install around 30 lavatories at Sandringham House, the residence of the royal family. While Crapper filed a total of six patents relating to ventilated drains and lavatory mechanisms, he did not invent the flush WC (water closet). He did, however, manufacture a toilet with a syphonic flush known as *Crapper's Valveless Waste Water Preventer*, and his designs for sanitary engineering products were among the most advanced in their time. His manhole covers can still be found all over the south of England; there are even several in Westminster Abbey.

Pages from the
Thomas Crapper &
Company catalogue,
c. 1902

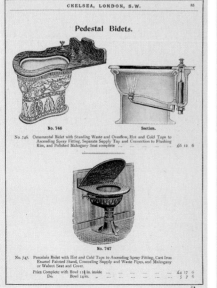

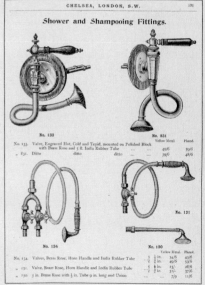

GUSTAV DALÉN

BORN 1869 STENSTROP, SWEDEN
DIED 1937 STOCKHOLM, SWEDEN

In 1906 Dr. Gustav Dalén became chief engineer at the Svenska Aktiebolaget Gas Accumulator Company, a supplier of acetylene gas. Here Dalén undertook much research on gases and turbines, which resulted in the improvement of hot-air turbine engines and the de Laval steam turbine. Dalén became managing director of the company in 1909 and subsequently invented Agamassan – a substance that absorbs acetylene allowing the gas to be concentrated without fear of explosion. In 1912 he won the Nobel Prize for Physics for his invention of the solventil, an automatic sun valve which uses sunlight to regulate a gas light-source so that it is turned off during daylight hours. This innovative design was later used internationally for the lighting of unmanned lighthouses and buoys. Tragically, Dalén was blinded in 1913 by an explosion that occurred during one of his experiments. While he was recuperating, he became aware of the problems his wife was encoun-

Advertisement for the Aga cooker, first introduced in 1924

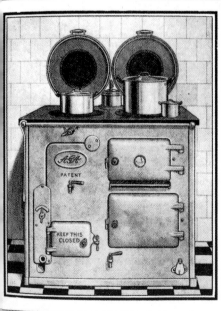

The

AGA COOKER

sets a new standard

in ECONOMY

Burning day and night the normal fuel consumption is 1¼ TONS per ANNUM. The stove requires stoking only once in 24 hours, but is always ready for cooking. The hot spots and ovens are automatically regulated to remain at the correct cooking temperatures.

Reliability is absolute

BELL'S GUARANTEE THE ACCURACY OF THESE CLAIMS and confidently invite your personal investigation at the Ideal Home Exhibition, Stand No. 105, Ground Floor.

Bell's
asbestos
and Engineering
Supplies Limited

Bestobell Works,
Slough, Bucks.

The two-oven *Aga* cooker, 1990s

tering with her antiquated and highly temperamental cast-iron cooking range, which needed constant supervision to keep the temperatures from fluctuating. By applying the concept of heat storage, in 1922 Dalén invented a cooker that combined a small and efficient heat source, two large hotplates and two generous ovens into one robust and compact unit. Known as the *Aga* (taking its name from the initials of his company), Dalén's design achieved a high accumulation of heat that could be delivered in precisely controlled quantities, with a minimum of fuel, within a well-insulated environment. The heat generated was radiant, so that every cubic inch within each oven was utilized. The heat was also thermostatically controlled, which meant the elimination of all knobs and dials. Since the ovens and hotplates maintained different temperatures depending on their distance from the heat source, the *Aga* could roast, bake, steam, simmer, fry, stew, grill, boil and toast – all at once. *Agas* were first imported into the UK in 1929 and are now solely manufactured by Aga-Rayburn in England. All the principal castings for Aga cookers are made at the historic **Coalbrookdale** Foundry in Shropshire – birthplace of the **Industrial Revolution**. The remarkable Aga revolutionized the design of cookers; even today it has many loyal users who consider it the ultimate oven.

DANESE

FOUNDED 1957
MILAN, ITALY

Enzo Mari, *Timor calendar*, 1967

The design-led Milanese manufacturing company Produzioni Danese was founded by Bruno Danese and Jacqueline Vodoz in 1957 to produce tabletop consumer products manufactured in **plastics**. Seeking "perfect" objects, they commissioned freelance designers to create products with a high design integrity. These included the minimalist melamine and anodized **aluminium** *Cubo* ashtray (1957) by Bruno Munari (1907–1998) and the *In Attesa* wastebin by **Enzo Mari**, which had an inclined form to increase the probability of scoring a "goal" when throwing rubbish into it. In 1959 Edizioni Danese issued the results of its research into the "phenomena of visual perception" and later in the 1960s received considerable recognition for its combination of functionalism with sleek Italian **styling**, as demonstrated in Mari's *Formosa* (1963) and *Timor* (1967) everlasting calendars. The majority of Danese products have been designed by Enzo Mari, but other designers who have contributed to the company's range of objects for the home and office include Bruno Munari,

Enzo Mari, *Formosa perpetual calendar*, 1963

Kuno Prey (b. 1958), Angelo Mangiarotti (b. 1921) and Marco Ferreri (b. 1958). In 1994 Danese was purchased by the prestigious Italian furniture manufacturer Alias, and since then has continued its non-conformist tradition of design research in co-operation with the Domus Academy Research Centre, Milan. Among the Danese range are "objects of beauty and spirit, aware of being the guardians of a noble history and tradition in Italian design", which testify to the company's long-held belief that design can be the "bearer of lasting value and innovation".

HUMPHREY DAVY

BORN 1778 PENZANCE, ENGLAND
DIED 1829 GENEVA, SWITZERLAND

In 1796 Humphrey Davy was apprenticed to an apothecary and surgeon wit a view to pursuing a career in medicine. He began his scientific studies in earnest in 1797, establishing his own laboratory, where he undertook experiments with nitrous oxide (laughing gas). A year later, he became the chemical superintendent of the Pneumatic Institution in Clifton, where he researched the therapeutic nature of various gases, especially compositions c nitrogen. He also made several major scientific breakthroughs, including

Various examples
of the Davy Lamp,
1815

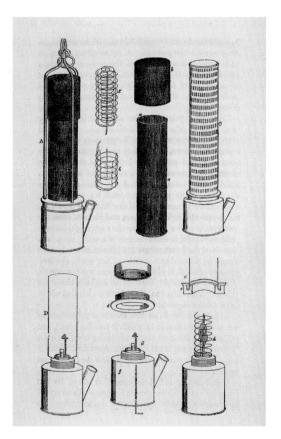

the isolation of both sodium and potassium and the discoveries that the production of electricity in simple electrolytic cells results from chemical action and that chemical combination occurs between substances of opposite charge. It was, however, his design of a miner's safety lamp that brought Davy the greatest public recognition. Miners had traditionally worked by candlelight or by light from other types of naked flame that could ignite the gas known as "firedamp" in deep mines, with disastrous results. Almost 100 lives were lost in this way in the early 19th century, prompting the Society for Preventing Accident in Coal Mines to approach Davy to design a safer lamp. His lamp comprised a wick connected to an oil reservoir and a two-layer metal gauze chimney to surround and confine the flame and conduct the heat of the flame away. Davy was knighted in 1818 and awarded the Royal Society's Royal Medal in 1826

ROBIN DAY

BORN 1915
HIGH WYCOMBE, BUCKINGHAMSHIRE, ENGLAND

In 1935 Robin Day won a scholarship to the Royal College of Art, London, and graduated in 1938. In 1948 he opened a joint design office with his wife, the textile designer Lucienne Conradi (b. 1917). Day worked as a freelance exhibition, graphic and industrial designer, and, at the 1948 "International Competition for Low-Cost Furniture Design" held at the Museum of Modern Art, New York, won first prize with Clive Latimer (b. 1915) for the design of wooden and **tubular metal** storage units. Shortly afterwards, he was commissioned by Hille International to design Modern furniture for the 1949 "British Industry Fair". In 1950 Day designed Hille's **corporate identity** and became its chief designer, while also continuing to consult to other companies, most notably John Lewis. Between 1962 and 1963, he developed an injection-moulded stacking chair, known as the *Polyprop*. This revolutionary design was among the first pieces of furniture to fully exploit the mass-manufacturing potential of thermo**plastics**. Obviously inspired by the earlier fibreglass-shell range of chairs by **Charles and Ray Eames**, the *Polyprop* was more successful in terms of sales because its higher volume method of production meant that it could be manufactured at a lower unit cost. Developed as a seating system with two standard shell types (armchair and side chair) and a wide variety of interchangeable bases, the *Polyprop* has multiple applications, from schools and church halls to football stadiums and airports. Since 1963 over 14 million *Polyprop* chairs have been sold, making it one of the most democratic products of the 20th century. Day believes that "the quality of design [is] important to the well-being of humanity".

Polyprop armchair for Hille International, 1962–1963

DHC-6 Twin Otter,
1965

Geoffrey de Havilland (1882–1965) was one of the great aviation pioneers. In 1920 he founded the de Havilland Aircraft Company, which manufactured the *Moth* (1924–1925) and later the famous twin-engine *Mosquito* fighter-bomber (1939–1940) De Havilland also manufactured the *Vampire* jet fighter (1942–1943) and developed the first civil jet transport with the *Comet*, which first flew in 1949 and began regular transatlantic services in 1958, the same year as the **Boeing** 707. In 1947 a de Havilland subsidiary in Canada, originally founded in 1928, began building the *DHC-2 Beaver*. This legendary bush plane played ed a vital role in the development of remote regions in Northern Canada as well as other inaccessible areas, from Antarctica to equatorial Africa. In 1987 the *Beaver*, which was frequently configured with twin floats, was selected as one of Canada's top ten engineering achievements. Of the 1,692 models built between 1947 and 1967, approximately 1,000 are still in operation – a remarkable tribute to the strength of its design. In 1965 de Havilland introduced the *DHC-6 Twin Otter*, a larger utility bush plane that also went on to carve out a distinct niche in aviation with its exceptional performance and reliability. Since 1997 de Havilland has been a subsidiary of **Bombardier** Inc.

DHC-2 Beaver, 1947

MICHELE DE LUCCHI

BORN 1951
FERRARA, ITALY

Michele De Lucchi studied at the Liceo Scientifico Enrico Fermi in Padua and later trained as an architect under Adolfo Natalini (b. 1941) at Florence University, graduating in 1975. In 1973, together with Piero Brombini, Pier Paola Bortolami, Boris Pastrovicchio and Valerio Tridenti, he founded the architecture and design group Cavart, which promoted a radical design agenda through happenings, publications and seminars. He also collaborated with Superstudio, **Ettore Sottsass** and Gaetano Pesce (b. 1939) and taught architecture at Florence University from 1975 to 1977. He then moved to Milan and began working as a design consultant to the **Kartell** in-house design studio, Centrokappa. De Lucchi later assisted Ettore Sottsass with the planning of the first Memphis exhibition. In 1979 he designed several Post-Modern prototypes of domestic electrical appliances for Studio Alchimia and became a consultant to **Olivetti**. He co-founded Memphis in 1981 and was responsible for the introduction of geometric motifs on the plastic laminates used by the co-operative. In 1986 he founded Solid, a Milan-based design group, and also began teaching at the Domus Academy, Milan. In the early 1990s he established the De Lucchi Group and has since worked widely in

Tolomeo wall light for Artemide, 1995

Japan and Germany. Increasingly, his designs have become more suited to industrial production and his clients have included **Artemide**, Kartell, Bieffeplast, Mandarina Duck and Pelikan. He has won numerous design prizes, including Good Design awards (Japan), Die Gute Form and Deutsche Auswahl awards (Germany) and a Compasso d'Oro (Italy). De Lucchi regards design as a means of communication, a sentiment that has informed his work throughout his career – from contesting young radical to established international industrial designer.

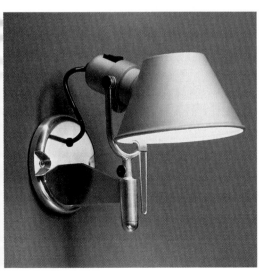

JC DECAUX

FOUNDED 1964
LYONS, FRANCE

Jean-Claude Decaux

In 1955 Jean-Claude Decaux (b. 1937) started up a company specializing in motorway billboard advertising, but after the French government slapped high taxes on this form of publicity in a law passed in 1964, he turned his attentions to urban street advertising. His innovative concept was to provide bus shelters, which would be wholly funded by high-quality advertising, free-of-charge to city councils.

His first bus shelter design was tested in Lyons with the mayor's approval and soon found its way into other French cities, such as Grenoble, Angers and Poitiers. Decaux initially found it difficult to convince advertisers of the benefits of this smaller-format publicity medium, but in the early 1970s his idea finally caught on and the company expanded rapidly, with the *Abribus Standard* shelter (1970) becoming a common sight in French towns and cities.

Jean-Claude Decaux,
SPEA public lavatory,
1980

Decaux also designed street information panels known as *MUPI* (Mobilier urbain pour l'information) in 1972 and *PISA* (Point d'information service

animé) in 1976. He offered local authorities free municipal information displays in return for permission to erect *MUPI*'s and bus shelters. In 1980 Jean-Claude Decaux designed an automatic entry public lavatory known as the *SPEA* (Sanitaire public à entretien automatique) and, a year later, introduced the first illuminated news boards. Using the highest quality materials and instituting a rigorous maintenance program, JC Decaux became a leader in the design and manufacture of street furniture, exporting its products worldwide. During the 1980s and 1990s, the company commissioned leading architects and designers such as **Mario Bellini**, **Philippe Starck**, **Norman Foster**, **Knud Holscher**, Martin Székély (b. 1956) and **Ferdinand Alexander Porsche** to design bus shelters, public toilets, information boards, signage and street kiosks. Impressively, the company manages to progress from the initial design stage of a product's development to final installation in just twelve months. Mindful of the aesthetic

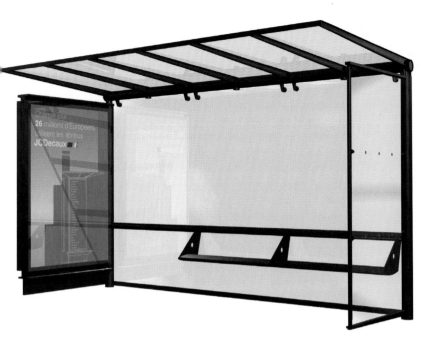

impact of street furniture on our shared environment, Decaux is particularly adroit at producing large ranges of designs that are stylistically suited to different cultural settings.

Jean-Claude Decaux,
Abribus Standard
shelter, 1970

JOHN DEERE

BORN 1804 RUTLAND, VERMONT, USA
DIED 1886 MOLINE, ILLINOIS, USA

→ *D tractor, 1923, the
first Waterloo tractor
to bear the name of
John Deere (pro-
duced 1924 to 1953)*

John Deere served a four-year apprenticeship as a blacksmith in Vermont, before working as a journeyman. He soon gained considerable fame for his careful workmanship and ingenuity. His highly polished hay forks and shovels were in especially high demand throughout Western Vermont. Tempted by tales of a better life in the West, in 1836 Deere moved to Grand Detour, Illinois, where he entered into a business partnership with Major Leonard Andrus. Deere quickly discovered that the cast-iron ploughs originally intended for the light sandy soils of New England could not cope with the clod-like earth of the Mid-West that the pioneers were trying to farm. Having to frequently repair their damaged cast-iron ploughs, Deere became convinced that the problem could be overcome by the addition of a highly polished mouldboard that, if shaped properly, would scour itself clean while turning the furrow slice. In 1837 Deere designed the first commercially successful self-scouring steel plough and began manufacturing the "self-polishers" to sell to local farms. This landmark design, which transformed farming in the West, was followed by another 50 improved models over the next two years. In 1843 Deere managed to arrange the importation of steel from England, and by 1846 his business was manufacturing around 1,000 ploughs per annum. He subsequently established a factory in Moline, Illinois, which by 1857 was producing 10,000 ploughs a year. Deere constantly changed and

*Self-scouring steel
plough, 1837*

improved the design of his ploughs, declaring: "If we don't improve our product, somebody else will." The company was incorporated as Deere & Company in 1868 and the following year John Deere's son, Charles, succeeded him as president. In 1918 the company purchased the Waterloo Gasoline Traction Engine Company in Waterloo, Iowa, and tractors became an important part of the John Deere line. Deere & Company introduced its first tractor, the *Model D*, in 1923, and over the succeeding decades the company diversified its product-line to include cultivators, harvesters, balers and other engine-powered agricultural machinery, as well as construction equipment such as backhoes and loaders. Deere & Company is now the largest farm equipment manufacturer in North America.

↑ *GP* tractor, 1928 (produced until 1935)

MR tractor, 1992

CHRISTIAN DELL

BORN 1893 HANAU, GERMANY
DIED 1974 WIESBADEN, GERMANY

During his apprenticeship as a silversmith at Schleißner & Söhne in Hanau, Christian Dell also studied at the Königliche Preußische Zeichenakademie. From 1911 to 1912 he worked as a silversmith in Dresden before training under Henry van de Velde (1863–1957) at the Kunstgewerbeschule in Weimar from 1912 to 1913. After the First World War he worked as a journeyman and later as a master silversmith for Hestermann & Ernst in Munich. In 1920 he joined the silver workshop of Emil Lettré (1876–1954) in Berlin, establishing his own silver studio a year later in Hanau. From 1922 to 1925 he headed the metal workshop at the Weimar **Bauhaus**, and then taught for seven years at the Frankfurter Kunstschule, where he designed silverware that was made in the school's workshop. During this period, Dell began producing designs that were suitable for industrial production. His lights, such as the adjustable

↘ *Das Frankfurter Register* 1 brochure, 1928

Rondella desk lamp for Rondella, 1927–1928

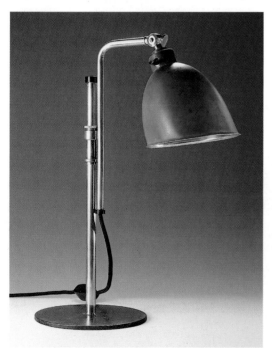

Rondella desk lamp (1927–1928) and his *Idell* range, which was later copied by Helo, were mass-produced by Rondella and Kaiser respectively. Dell designed over 500 lights in total and his *Idell* range remained in production for over 60 years. In the 1930s Dell began experimenting with **plastics** and in 1939 established his own jewellery business in Wiesbaden.

The first issue of
Design magazine,
1949

DESIGN COUNCIL

FOUNDED 1944
LONDON, ENGLAND

In 1944 the British Board of Trade announced the founding of the Council of Industrial Design "to promote by all practicable means the improvement of design in products of British industry". Two years later, the CoID organized the "Britain Can Make It" exhibition to re-invigorate British design after the war so as to boost foreign exports. This successful exhibition attracted over one and half million visitors and did much to raise the profile of design in Britain. In 1949 the CoID launched *Design* magazine to promote "good design" by addressing professional designers, manufacturers, company buyers and managers. The council was later responsible for the selection and arrangement of the 10,000 exhibits at the Festival of Britain in 1951. In 1956 the CoID's Design Centre was opened in the Haymarket, London, to provide "a shopping guide to well-designed British goods for those who want to buy well and save time". This idea was advanced a year later by the appearance of the Design Council Awards and, in 1958, by the introduction of the well-known black and white kitemark. In 1972 the CoID was re-named the Design Council and in 1976 was granted a royal charter. The Design Council was restructured in 1994 and subsequently helped the government modify the national curriculum, so as to place more emphasis on the teaching of design and technology in schools.

Design Council
kitemark, first
introduced in 1958

DESIGN RESEARCH UNIT

FOUNDED 1943
LONDON, ENGLAND

Corporate identity
for British Rail, 1965

Exhibition sales
caravan for Olivetti,
c. 1956

In 1922 Milner Gray and Charles and Henry Bassett founded the Bassett-Gray Group of Artists and Writers, one of the first professional design practices in Britain. Collaborating with artists such as Graham Sutherland (1903–1980), Bassett-Gray represented "a body of artists who design for industrial and commercial purposes". In 1934 Bassett-Gray was joined by **Misha Black**. The following year it was reorganized and renamed the Industrial Design Partnership, even though at this time the practice was still primarily focused on graphic, **packaging** and exhibition design. Following the outbreak of the Second World War, the IDP was shut down and both Gray and Black joined the Ministry of Information. Realizing that there would be a huge demand for consumer goods once wartime restrictions had been lifted, in 1942 Marcus Brumwell (1901–1983) and the design theorist Herbert Read (1893–1968) proposed the idea of forming a collective

of designers and architects who could produce designs for large-scale industrial production after the war. Milner Gray, who was by then head of exhibition design at the Ministry of Information, was consulted on the organization of such a group. The Design Research Unit was subsequently established in January 1943 and from the outset emphasized industrial product design. Both Black and Cray joined the DRU, the former originating the majority of its product designs – from televisions and electric heaters to cameras and trains – while Gray concentrated chiefly on packaging and **signage**. Dorothy Goslett, who helped administrate the office, published a book in 1961 entitled *Professional Practice for Designers*, which was in essence a guide to running a design consultancy. "Designers," she wrote, "have to deal directly with business men who are seldom likely to be sentimental about creative work. ... The relationship between designer and client is therefore strictly a business one. The designer's codes of professional conduct are the foundations on which this relationship is built." As the highest-profile design office in post-war Britain, the DRU did much to elevate the professionalism of industrial design practice and provided "an effective design service for industry" by giving concrete expression to Modern design theory.

Ilford *Pixie* camera,
c. 1960s

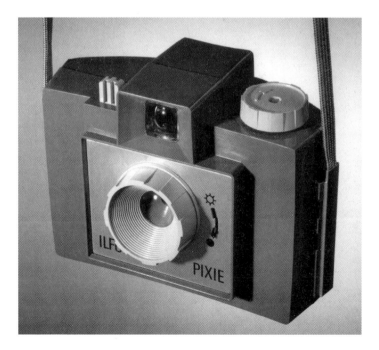

DONALD DESKEY

BORN 1894 BLUE EARTH, MINNESOTA, USA
DIED 1989 VERO BEACH, FLORIDA, USA

Packaging design for
Procter & Gamble,
c. 1951

Donald Deskey studied architecture at the University of California, Berkeley, as well as fine art at the Arts' Student League, New York, and the Institute of Chicago's School of Art. He later trained in Paris at the École de la Grande Chaumière, the Académie Colarossi and the Atelier Léger. After the First World War he worked as a graphic designer for a Chicago advertising agency, but a visit to the 1925 Paris "Exposition Internationale des Arts Décoratifs et Industriels" prompted him to turn his attention to three-dimensional design. He subsequently produced a number of screens for Saks Fifth Avenue in 1926 and a year later designed window displays for this New York department store as well as for Franklin Simon. Also in 1927, he established the Deskey-Vollmer partnership with Philip Vollmer, which concentrated on exclusive metal furniture and lighting, but this venture was dissolved in the early 1930s. During the late 1920s, he developed a stained-wood laminate known as Weldtex, while in 1931 he designed Moderne interiors for the Radio City Music Hall – his most prestigious and influential commission,

Filing cabinet for
Globe-Wernicke,
c. 1951

which exemplified the American Art Deco style. During the 1930s he executed a variety of industrial designs, such as washing machines, vending machines and even a rack for bowling balls, which were exhibited at the Metropolitan Museum of Art in 1934. From the late 1930s until 1975, he was the principal of Donald Deskey Associates and designed everything from graphics and **packaging** for Procter & Gamble to printing presses for American Type Founders. His design work was characterized by streamlined forms and innovative planning. His sleekly styled filing cabinet for the Globe-Wernicke Company, for instance, had several novel features including a combined drawer pull/identification panel and drawer fronts with contoured edges. While chiefly remembered as one of the greatest proponents of the Art Deco style in America, Deskey was also an important pioneer of industrial design consulting.

Deutscher Werk-
bund emblem

Founded in October 1907, the Deutscher Werkbund attempted from its outset to reconcile artistic endeavour
with industrial mass production. Its founding members
not only included designers such as Richard Riemerschmid (1868–1957),
Bruno Paul (1874–1968), **Peter Behrens** and Josef Maria Olbrich (1867–
1908), but also a dozen established manufacturers, including Peter Bruckmann & Söhne and Poeschel & Trepte, and design workshops such as the
Wiener Werkstätte and the Munich-based Vereinigte Werkstätten für Kunst
im Handwerk. Peter Bruckmann (1865–1927) was appointed the association's first president and within a year its membership had risen to around
500. From 1912 the Werkbund began publishing its own yearbook, which included illustrations of and articles on its members' designs, such as factories by **Walter Gropius** and Peter Behrens and cars by Ernst Naumann. The
yearbook also listed members' addresses and areas of specialization in an
attempt to promote collaboration between art and industry. In 1914 the
Werkbund organized a landmark exhibition in Cologne, which included within its grounds Walter Gropius' steel and glass model factory, Bruno Taut's
Glass Pavilion and Henry van de Velde's Werkbund Theatre. A year later, the
Werkbund's membership had swollen to almost 2,000. The increasing diver-

Fritz Hellmut Ehmcke,
tobacco box for the
Deutscher Werkbund
exhibition in Cologne,
1912–1914

gence between handcraftsmanship
and industrial production fuelled
heated debate within the Werkbund,
however, with members such as
Hermann Muthesius (1861–1927)
and Naumann urging for **standardization**, while others such as van de
Velde, Gropius and Taut argued for
individualism. This conflict almost
led to the disbanding of the association. The widespread need for consumer products after the devastation of the First World War, however,
led Gropius to accept the necessity
for standardization and industrial

AEG stand in the main hall of the 1927 Deutscher Werkbund "Die Wohnung" exhibition in Stuttgart

production, although other members, such as Hans Poelzig (1869–1939), continued to resist change. From 1921 to 1926 Riemerschmid was president of the Deutscher Werkbund and during his tenure the functionalists' approach to design was advanced. In 1924 the Werkbund organized the exhibition *Form ohne Ornament (Form without Ornament)*. The catalogue illustrated industrially-produced designs and promoted throughout its text the virtues of plain undecorated surfaces and, ultimately, functionalism. In 1927 the Werkbund staged a unique exhibition in Stuttgart, entitled "Die Wohnung" (The Dwelling), which was organized by Ludwig Mies van der Rohe (1886–1969). Although the focus of the exhibition was the Weißenhofsiedlung, a housing estate project for which the most progressive architects throughout Europe were invited to design buildings, "Die Wohnung" also acted as an important showcase for industrial design, with stands exhibiting the latest products from companies such as **AEG**. The modern **tubular metal** furniture designed by Mies van der Rohe, Mart Stam (1899–1986), **Marcel Breuer**, Le Corbusier (1887–1965) and others, which was used to furnish the interiors of the specially commissioned houses, was also widely publicized and revealed the increasing internationalism of the Modern Movement. Through this landmark exhibition, the Deutscher Werkbund succeeded in achieving a greater acceptance of Modernism both at home and abroad. The Werkbund was eventually disbanded in 1934, and though re-established in 1947, was by then a spent force. Through its activities and in particular its development of the concept of "types" – standardized designs that could be easily assembled from industrially produced components – the Deutscher Werkbund heralded the advent of large-scale industrial production and in so doing had an enormous impact on the evolution of German industrial design.

NIELS DIFFRIENT

BORN 1928
STAR, MISSISSIPPI, USA

Niels Diffrient studied aeronautical engineering at the Cass Technical High School, Detroit, and later trained at the Cranbrook Academy of Art, Bloomfield Hills, Michigan and at Wayne State University, Detroit. He worked in **Marco Zanuso**'s Milan-based design studio from 1954 to 1955 and from 1946 to 1951 in the office of Eero Saarinen (1910–1961). In 1952 he joined **Henry Dreyfuss** Associates, becoming a partner in 1956. While there, he helped to compile data on **anthropometrics** which was eventually published in the three influential volumes entitled *Humanscale*. He also designed aircraft interiors as well as computers and X-ray equipment. In 1981 he founded Niels Diffrient Product Design in Ridgefield, Connecticut and has since designed ergonomically-conceived office systems furniture for Knoll and for Sunar-Hausmann. With his mastery of **ergonomics**, Diffrient is able to promote a highly resolved physical interaction between object and user. Regarding design as a "practical art", Diffrient attempts to provide superlative physical function in the most aesthetically pleasing way possible.

Diffrient chair, office seating system for Knoll International, 1979–1980

CHRISTOPHER DRESSER

BORN 1834 GLASGOW, SCOTLAND
DIED 1904 MULHOUSE/ALSACE, FRANCE

→Electroplated bowl for Elkington & Co., 1885 (this design was roughly sketched in 1864 and illustrated in C. Dresser's *Principles of Design* published in 1873)

Christopher Dresser studied under the botanist John Lindley at the Government School of Design in London from 1847 to 1854 and subsequently lectured there for 14 years. Dresser's advocacy of "Art Botany" – the stylized yet scientifically based depiction of nature – helped replace the overblown and false naturalism common to the High Victorian style with a more formalized type of ornamentation. In 1857 he was appointed professor of "Botany Applied to the Fine Arts" at the School of Design, South Kensington. He also published various articles in the *Art Journal* and three books, which helped establish his reputation and contributed to his receiving an honorary doctorate from the University of Jena.

In 1860 he abandoned botany and embarked on a career in design. He established his own studio, which from the beginning was enormously successful. Dresser supplied designs for metalware, **ceramics**, glass, tiles, textiles, wallpapers and cast-iron furnishings to at least 30 of the most eminent British manufacturers, including Elkington & Co., **Minton** & Co., **Coalbrookdale** Co., Benham & Froud, James Dixon & Sons, Hukin & Heath and Linthorpe Art Pottery. His electroplated metalware designs are especially noteworthy for their geometric simplicity, which prefigured the formal vocabulary of the Modern Movement. These wares demonstrate what Dresser described as the "breadth of treatment, simplicity of execution, and boldness of design" of Japanese applied art. While Dresser was one of the first Europeans to promote the qualities of the art of Japan, his forward-looking designs also reflected his belief in the efficacy of industrial production and his pursuit of "Truth, Beauty, Power". As a leading design reformer and theorist of the 19th century, Dresser was one of the first professional practitioners of industrial design.

→Tea service for Watcombe Terracotta Company, c. 1870

Big Ben alarm clock
for Westclox, 1939

Vacuum cleaner for
Hoover, c. 1950 –
with new stream-
lined form

HENRY DREYFUSS

BORN 1904 NEW YORK, USA
DIED 1972 PASADENA, CALIFORNIA, USA

Henry Dreyfuss trained at the Ethical Culture School in New York prior to apprenticing with the industrial designer **Norman Bel Geddes**. While at Geddes' office from 1923 to 1929, he concentrated chiefly on theatrical work and designed costumes, sets and lighting for the Strand Theater, New York and for R. K. O.'s vaudeville theatres. Dreyfuss also worked as a consultant to Macy's before establishing his own design office in New York in 1929. From 1930 onwards he designed telephones for **Bell Telephone** Laboratories, including the *Model 300* (1937), the *Model 500* (1949) and the *Trimline* telephone (1964). Dreyfuss also created a model of "The City of Tomorrow" for **General Electric**, which was displayed at the 1939 **New York World's Fair**. Between 1938 and 1940, Dreyfuss designed two trains for the New York Central Railroad, including the streamlined *20th Century Limited* (1938). Later, in 1947, he executed a curious hybrid prototype, the *Convair* flying car for Vultee. Dreyfuss' straightforward business-like approach to the design process, which included working closely with engineers, contributed to the success of

Model 300 telephone
for Bell Laboratories,
1937

his office. His large corporate clientele included AT&T, American Airlines, **Polaroid**,
Hoover and **RCA**. His designs were characterized by the use of sweeping sculptural forms,
and as such exemplified **streamlining** in American design. Like **Raymond Loewy**, Norman Bel
Geddes and **Walter Dorwin Teague**, Dreyfuss
re-styled many products for manufacturers so
as to increase consumer demand through
stylistic rather that technical innovation. Some
of his designs bore a facsimile of his signature
– an early example of designer labelling. Dreyfuss was a founder member of the Society of
Industrial Design and the first president of the
Industrial Designers Society of America. He
was also a long-term faculty member of the engineering department at the California Institute
of Technology. Dreyfuss' greatest contribution
to industrial design practice, however, was
his research into **anthropometrics** – the find-

–Convair car/plane
or Consolidated
Vultee Aircraft, 1947

–Convair car/plane
or Consolidated
Vultee Aircraft, 1947

ngs of which were published in his influential books, *Designing for People* (1955) and *The Measure of Man* (1960). He also published a sourcebook of international symbols, which acknowledged the communicative power of symbols over words.

20th Century Limited locomotive for New York Central Railroad, 1938

DUCATI

FOUNDED 1926
BOLOGNA, ITALY

Ducati bicycle with
Cucciolo motor,
c. 1946

→Ducati *MY 916
Biposto* motorbike,
1998

→*M900 Monster*
motorbike, 1993

↘*750 GT* motorbike,
1971 – the first hy-
persport production
motorbike

Ducati *Mariana 125* –
first design by
Taglioni, c. 1955

The Società Radio Brevetti Ducati initially manufactured radio components that were based on patents filed
by Adriano Ducati (1903–1991). The company grew and
in 1935 built its first factory. After the Second World War,
the Ducati brothers developed a small auxiliary bicycle
motor known as the *Cucciolo* (1946), and when it later
acquired a frame, the first Ducati motorcycle was born.
In the early 1950s Ducati began producing motorbikes
such as the futuristic 175 cc *Cruiser* (1952). In 1955 the
company was joined by the brilliant engineer-designer Fabio Taglioni (b.
1920), who had already designed several motorbikes that were notable for
their originality. At Taglioni's insistence, the high-performance designs he
produced for Ducati were tested in long-distance races such as the Milan-
Taranto and the Giro d'Italia. Taglioni began developing the desmodromic
system in 1955, but it was not used in a production bike until 1968, when the
450 Mark 3D appeared. Ducati's *Scramblers* found much success in America,
and with the late 1960s maxibike boom, Taglioni designed the twin-cylinder
desmodromic *750*, which won the 1972 Imola 200 Miglia. Developed from
this winning design, the *750 Super Sport* was later taken up to a 900 cc dis-
placement and won the 1978 Formula 1 TT. During the 1980s – the Superbike era – Ducati expanded its market by introducing new high-performance
models with distinctive **styling**. In 1993 Miguel Galluzzi (b. 1959) created the
legendary *Monster* and a year later the *916* was born. Combining "functionality and harmony of appearance, logic and emotion", it was subsequently
nominated "Motorcycle of the Year".

JAMES DYSON

BORN 1947
NORFOLK, ENGLAND

While training in furniture and interior design at the Royal College of Art, London (graduation 1970), James Dyson designed the *Sea Truck* for the inventor Jeremy Fry, which won a **Design Council** Award and the Duke of Edinburgh's special prize in 1975. After working as a designer for the Marine Division of Rotork for four years, Dyson left to develop the *Ballbarrow*. This innovative and better-performing barrow, which used a ball instead of a wheel, became a highly successful product. Dyson did not profit greatly

DC01 cleaner, 1993

from the *Ballbarrow*, however, because he sold his interest in it to fund the development of a bagless vacuum cleaner. The concept of a bagless vacuum cleaner first captivated Dyson in 1978 when he realized that, when in use, even new bags quickly become clogged with a fine layer of dust, which dramatically and increasingly reduces suction. Between 1979 and 1984, Dyson built 5,127 prototypes of his revolutionary vacuum cleaner, which used centrifugal force to lift dust and separate it from the air, just like the much larger cyclone towers used in saw-mills and spray-paint shops to remove hazardous particles from the atmosphere. By using a cyclonic method of suction and eliminating the bag, Dyson was able to create a vacuum cleaner that provided a high level of constant suction. In 1983 his first prototype, the pink and lilac *G-Force*, was featured on the cover of *Design* magazine and in an article entitled "Form Follows Fluff". When attempting to commercialize his cleaner, however, Dyson discovered that the established vacuum cleaner manufacturers did not want to invest in his invention as it spelt the end of the lucrative **planned obsolescence** of the replacement bag market. In 1985 Dyson took the *G-Force* to Japan and found a licensee to make it. Six years later, the cleaner won the International Design Fair prize in Japan. Having also eventually managed to find licensees in North America, Dyson was able to establish his own research centre and

factory in Chippenham in 1993 with the royalties generated from overseas sales. Launched the same year, the famous yellow and grey Dyson *DC01* (*Dual Cyclone*) became the best-selling vacuum cleaner in Great Britain within two years, despite its relatively high price. Every aspect of Dyson's design was driven by logic, such as the transparent collecting container, which allows one to see when it is full. The *DC01*'s aesthetic also distinguished it from other cleaners on the market. As Dyson put it: "we wanted it to look like a piece of **NASA** technology. Its superior performance had to be visible. It had to look the business." For its marketing, Dyson adopted a "less is more" approach with bold no-nonsense advertisements that directly communicated the design's benefits with simple slogans such as "say goodbye to the bag" and "100% suction, 100% of the time". In 1995 Dyson launched a cylinder model – the Dyson *DC02* – and opened a larger manufacturing facility in Malmesbury. The following year, the *DC01* was selling at the phenomenal rate of 32,000 units per month, five times that of its nearest competitor, demonstrating clearly that the public were more than prepared to pay a premium price for a better-performing design. In 1996 the exhibition "Doing a Dyson" was staged by the Design Museum in London, tracing the development of the Dyson cleaner, and the following year Dyson became the first British company to be awarded a European Design Award.

CHARLES & RAY EAMES

BORN 1907 ST LOUIS, MISSOURI, USA
DIED 1978 ST LOUIS, MISSOURI, USA

BORN 1912 SACRAMENTO, CALIFORNIA, USA
DIED 1988 LOS ANGELES, CALIFORNIA, USA

Charles and Ray Eames did more to change the public perception of Modern design than just about anyone else in the 20th century. With their mission "To get the most of the best to the greatest number of people for the least", the Eameses had an enormous impact on industrial design practice, producing throughout their careers highly successful and extraordinarily innovative work that exemplified the timeless ideals of good design. The achievements of Charles and Ray Eames continue to be revered today, as they are increasingly recognized as having single-handedly changed the course of design. After training and teaching at the Cranbrook Academy of Art in Bloomfield Hills, Michigan, Charles Eames and Eero Saarinen (1910–1961)

PAW swivel chair for
Herman Miller, 1950

– assisted by Ray Kaiser – submitted entries to the "Organic Design in Home Furnishings" competition held at the Museum of Modern Art, New York, in 1940. Their prize-winning seating designs incorporated two state-of-the-art manufacturing techniques – the compound moulding of **plywood** into complex curves and the bonding of wood to metal by means of electronic cycle-welding, a process that had recently been pioneered by **Chrysler.** Through the foam-lined monocoque seat shells of these chairs, Eames and Saarinen also advanced the revolutionary idea of continuous contact and support, thus heralding a totally new direction in Modern furniture design. Although wartime restrictions on materials and processes meant this highly innovative seating programme was not manufactured, it informed all of the Eameses subsequent chair designs. In 1942 the Eameses were commissioned by the US Navy to design plywood leg and arm splints and a plywood litter. They established their own company to manufacture these products, which were their first to be mass-produced. In 1945 the Eameses designed a series of compound-moulded plywood children's furniture, which was produced by the Evans Products Company. This was soon followed by their famous series of plywood chairs (1945–1946), which were initially manufactured by Evans and then later by **Herman Miller**. Employing a system of standardized interchangeable components, these revolutionary chairs were the first pieces of furniture with compound-moulded elements to be truly mass-produced. In 1948 the couple proposed a series of moulded fibreglass chairs, which won second prize at the Museum of Modern Art's "International Competition for Low-Cost Furniture Design". Their designs led on to the innovative seating programme which they developed with Herman Miller, Zenith Plastics of Gardena, California, and the engineering department of the University of California, Los Angeles, between 1948 and 1950, and which was some of the earliest unlined plastic seat furniture to be mass-produced. Based on the concept of a universal seat shell that could be used in conjunction with a variety of interchangeable bases to provide numerous variations, this product range, which was manufactured by Herman Miller, was extremely influential. The Eameses slightly later *ESU* (Eames Storage Units) series of 1950 – a modular system of standardized steel frames with either plastic-coated plywood or lacquered Masonite panels, sliding door and drawer options – offered similar functional flexibility and could be configured into room dividers, cupboards, shelves and desks. The *ESU* series bears a passing resemblance to the Eameses' own house (1949), which was largely constructed of pre-fabricated components. The Eames house was designed as part of the Case Study House program sponsored by the journal *Arts & Architecture*, and as such speculated on the idea of **product architecture** – buildings as industrially produced consumer products. Through their work, which also included film-making, photography and exhibition design, the Eameses skilfully and eloquently communicated the values of appropriateness, social morality, egalitarianism, optimism, informality and **essentialism**. While balancing the poetic with the pragmatic, the Eameses above all demonstrated in practice how cutting-edge materials and technology can be harnessed to create high-value yet accessible Modern products.

Harley Earl, Cadillac
62 Sedan de Ville,
1956, photographed
with a Douglas *A4
Skyhawk* which was
put into production
the same year

Harley Earl's father established the Earl Carriage Works in 1908, but two years later renamed the venture the Earl Automobile Works to reflect the transition from horse-drawn to motorized vehicles. His business involved customizing cars for film stars and building chariots for use in films. In 1914 Harley Earl began studying at Stanford University, but this was soon abandoned so that he could train as a designer at his father's auto works. After a period of time he began designing luxurious cars for the Hollywood glitterati, including the soon-to-be-disgraced actor Roscoe "Fatty" Arbuckle and the legendary director Cecil B. De Mille. In 1919 the business was purchased by Cadillac's West Coast distributor, Don Lee, and the works were subsequently devoted entirely to car customization. Around this time, Harvey Earl devised an innovative modelling technique using clay that promoted greater sculptural expressiveness, which he used as a **styling** aid and which later became standard practice throughout the car industry. Whereas, in the early years of the car industry, manufacturers only had to worry about their ability to meet demand, by the early 1920s, with productivity no longer an issue, the market was growing more competitive. The chairman of **General Motors**, Alfred

La Salle, with Harley
Earl (seated) and
L. T. Fisher, 1927

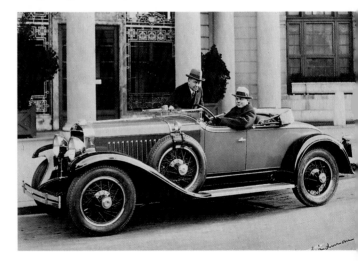

Sloan, realised that aesthetics would play an increasingly important role, and in 1925 he invited Earl to Detroit. Earl's first brief for GM was to bring "something" to the new La Salle brand, which was pitched in-between the well-heeled Buick and luxurious Cadillac models. The resulting 1927 La Salle was the first mass-produced car to be designed by a stylist rather than by an engineer. Sloan realized that the company would profit if it could produce new cars each year that differed from the previous year's model. This idea of annual cosmetic changes to promote stylistic obsolescence led to the creation of General Motors' Art & Color Section in 1928. Earl became supervisor of this unprecedented styling department, which was re-named the Style Section in 1937. He oversaw the development of the very first concept car, the 1938 Buick *Y-Job*, which had the first two-tone paintwork, wrap-round windscreen and "fishtail" rear fender. Earl was also responsible for convincing GM to build a sports car, which resulted in the classic 1953 Chevrolet *Corvette*. Earl's styling, however, became increasingly extreme – with cars becoming lower, longer and more chrome-laden – culminating in the extraordinary 1959 Cadillac *Eldorado* with rocket-ship tail fins. Earl's brand of exaggerated styling was highly influential upon the automotive industry, and his contribution is still remembered at GM in the immortal lines: "Our father who art in styling ... Harley be thy name!"

ECCO DESIGN

FOUNDED 1989
NEW YORK, USA

Beepwear watch/
pager for Motorola/
Timex, 1999

For Women Only
hairbrushes for
Goody, 1997

Ecco Design was founded by Eric Chan in 1989 and has since expanded into a creative team that specializes in the design of consumer products, furniture and computers as well as interactive multimedia, telecommunications, sporting and medical equipment. It has won numerous prizes, including five Good Design awards. The companies it has worked for include **Herman Miller**, Colgate, Bausch & Lomb, Timex and Motorola. Ecco Design states that: "As designers, we challenge ourselves to utilize today's technology to create products that bring harmony between man, nature and society in an increasingly global and multicultural market. We are most interested in translating complicated technology into products that go beyond mere function, that are friendly, understandable, and sensi-

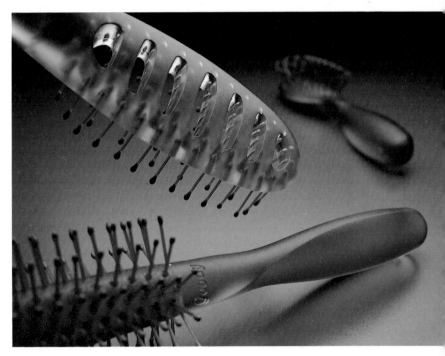

LS550 numeric pager
for Motorola, 1998

tive to people's needs." Indeed, Ecco Design's products are often distinguished by a tactility that expresses the inner qualities of the materials it has used, for example, the over-sized rubber handles of the *Grip 'Ems* children's toothbrushes for Colgate (1999) and the softly formed translucent plastic handles of the *For Women Only* hairbrushes for Goody (1997), winner of *ID Magazine*'s Design Distinction award. In 1993 Ecco Design developed the *SolarBlind*, a prototypical venetian blind that absorbs solar energy during the day and emits light in the evening. Ecco Design believes that "the designer's task is to mediate between people and object, poetry and logic, technology and nature".

THOMAS ALVA EDISON

BORN 1847 MILAN, OHIO, USA
DIED 1931 WEST ORANGE, NEW JERSEY, USA

Edison's original
carbon filament
lamp manufactured
by the Corning Glass
Works, 1879

→Advertising display
light-bulb box made
by the Edison Lamp
Works, General
Electric, 1923

Drawing showing
the elements of a
light bulb, c. 1880

Thomas Alva Edison pioneered some of the greatest and widest-ranging inventions of the 20th century, yet as a child his education was hampered by dyslexia and severe hearing problems. He was eventually removed from school and taught instead by his mother, who was a former schoolteacher. She gave him an elementary science book which outlined experiments that could be done at home. At the age of ten, Edison was allowed to set up his own science laboratory in the basement of his home. At the age of twelve, he began working as a trainboy for the Grand Trunk Railway and moved his laboratory into a baggage car so that he could do experiments during the layover periods in Detroit. The laboratory soon caught fire, however, and Edison lost his job. After this, Edison sold newspapers along the railroad and reputedly saved the life of a station official's son. The father was so grateful to Edison that he taught him how to use a telegraph. He later worked as a telegrapher in Toronto and, having to send a signal every hour, Edison devised a transmitter and receiver that could automatically telegraph a message even when he was asleep. He later moved to New York and began working on other inventions including a telegraph machine that could send multiple messages, an electric vote recorder and an improved stock market ticker, the rights to which he sold to the Gold & Stock Telegraph Company for a massive $40,000 in 1870. By now wealthy, Edison established a workshop in Newark, New Jersey, to manufacture stock tickers, high-speed printing telegraphs and an improved version of Christopher Sholes' typewriter that was the first model suitable for commercial use.

There's a Right Lamp
For Every Purpose

In 1876 he built an "invention factory" in Menlo Park, New Jersey, and subsequently spent the majority of his time experimenting in this new science laboratory that had 60 employees. He worked at the laboratory for a decade and often had as many as 40 projects running at once. During this period, Edison was applying for up to 400 patents per annum. The numerous inventions conceived at Menlo Park and subsequently transformed into landmark products included the wireless telegraph (1875), the memograph (1876), an improved carbon-button telephone transmitter (1877), the phonograph (1877–1878), the incandescent electric light bulb (1878–1880) and the wireless induction telegraph (1885). Of these, the phonograph and the electric light bulb must be considered his greatest triumphs. The phonograph evolved from his research into telegraphy and was the first-ever recording machine. By developing a device that indented a strip of paper to record Morse Code, Edison observed that when the strip was run quickly,

Thomas Alva Edison

it emitted distinctive noises. The concept was adapted to sound recording by using a stylus that vibrated when exposed to sound, which in turn indented tin foil wrapped around a rotating drum. His incandescent electric light bulb was the result of many years of research and some 1,200 experiments. Using carbonized filaments made from burnt cotton thread, Edison's light bulbs could burn for up to 48 hours and were first used on the steamship *Columbia*. In 1878 he founded the Edison Electric Light Company to develop and eventually market this life-changing invention – the firm was incorporated as the Edison General Electric Company in 1892 and later became known as **General Electric**. In 1882 Edison established the first "electric light-power station", which enabled New York to become the first city to be illuminated by electricity. Another far-reaching discovery made by one of Edison's engineers in 1883 led to the development of the electron tube, which helped establish electronics as a completely new branch of science. Having founded an enormous new laboratory for the "business of inventing" in West Orange in 1887, which employed over 5,000 workers, Edison

Poster advertising the Edison phonograph

Edison phonograph, c. 1880

went on to design improved versions of his phonograph, the first-ever motion picture camera (1891), the kinetograph and projecting kinetoscope for still and moving films (1897), the bipolar dynamo (1899) and the reversible galvanic battery (1900). Over his remarkable life, a total of 1093 patents were issued to Edison inventions. "All progress, all success, springs from thinking", he declared – but his phenomenal achievements were also the result of dogged perseverance and his ability to learn from his innumerable failures. As Edison put it so succinctly: "Genius is 1 % inspiration, and 99 % perspiration."

Edison bipolar dynamo generator manufactured by the Edison Machine Works, 1899

Peter Müller-Munk, *Ekcoware* saucepans, c. 1950

EKCO PRODUCTS COMPANY

FOUNDED 1888
CHICAGO, ILLINOIS, USA

From the early years of the 20th century, the manufacture of kitchen tools became increasingly mechanized and competitive, with quality often becoming the first casualty in the battle to make and sell goods more cheaply. From the 1920s Ekco mass-produced inexpensive kitchen tools of limited durability that differed little from those of their competitors. In 1935, however, the company brought out some kitchen spoons with holes in their handles that enabled them to be hung up – an innovative feature that helped Ekco gain market share but that was soon copied widely. Around 1945 Ekco's president decided to develop a line of high-quality kitchen tools with superior durability. Since there was no company "designer", the firm's tool-room foreman and product development engineers were given the responsibility of designing this new product range. While the resulting stainless steel and plastic *Flint* range introduced in 1946 was significantly more expensive than competitor products, it sold very well. The success of the *Flint* range demonstrated the public's willingness to pay a premium price for "good design in everyday objects" and established Ekco as a market leader in the 1950s. Richard Latham of **Raymond Loewy** Associates, who designed the packaging for *Flint*, assisted Ekco with project planning from the 1950s.

Flint kitchen tools, first introduced in 1946

KENJI EKUAN

BORN 1929
TOKYO, JAPAN

→Soy sauce
dispenser for
Kikkoman, 1961

Dubbed "the **Raymond Loewy** of Japan", Kenji Ekuan is Japan's most prominent industrial designer. He initially trained as a Buddhist priest before studying design at the National University of Fine Arts & Music, graduating in 1955. With fellow students he founded the **GK** (Groupe Koike) **Design Group** in 1953, which initially concentrated on industrial design. A scholarship from the Japan External Trade Organization (JETRO) enabled him to study industrial design at the Art Center College of Design in Pasadena, California, for one year. On his return to Tokyo in 1957 he became president of GK Industrial Associates. As well as designing motorcycles and audio equipment for **Yamaha** for over 40 years, GK has also created many other products, including packaging for Kikkoman, bicycles for Maruishi and cameras for **Olympus**. GK later expanded its remit to include graphic design, **signage** and urban planning. With his many publications on design practice, Ekuan has emphasized the need to humanize technology, democratize design and "seek the soul in material things". As both a world-class designer and a Buddhist monk, Ekuan has always pursued a harmonious connection between the material and spiritual worlds, and has been fascinated by the "communicative quality of design as a tangible medium".

Narita Express train
for the East Japan
Railways Company,
1991

Alvar Lenning,
Assistant food
processor for
Electrolux, 1940

ELECTROLUX

FOUNDED 1919
STOCKHOLM, SWEDEN

The Electrolux Group is the world's largest producer of household appliances, with over 55 million products being sold worldwide each year. The origins of the company can be traced back to Axel Wenner-Gren's invention of a very early household vacuum cleaner, the *Lux 1*, which was launched by AB Lux in 1912. In 1919 Lux merged with another Stockholm-based company, Elektromekanista, to form Electrolux, and Wenner-Gren was appointed general manager of this new entity. Three years later, Swedish engineering students Baltzar von Platen and Carl Munters invented a cooling machine which converted heat to cold by applying the absorption process, and which could be powered by electricity, kerosene or gas. This highly innovative refrigerator was subsequently manufactured in limited quantities by their company, AB Arctic. In 1925 Electrolux purchased the company, increased production of the refrigerator and over the next decade grew rapidly, establishing manufacturing plants in France, Britain and America. In the late 1930s Electrolux recognized the competitive advantage offered by high-style design and com-

Ralph Lysell,
Design for the *L460*
refrigerator for
Electrolux, 1948

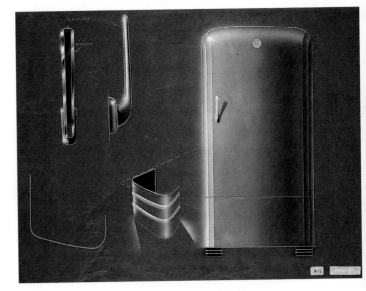

missioned **Raymond Loewy** to
design several streamlined products. 1940 saw the launch of Alvar
Lenning's revolutionary labour-
saving streamlined *Assistant* food
processor, which could mix, grind,
grate, squeeze and chop. Later in
the 1940s and early 1950s, Swedish
designers **Sixten Sason** and Ralph
Lysell (b. 1907) produced other
streamlined designs for Electrolux,
including Sason's famous bullet-
shaped *Model 248* vacuum cleaner
(1943). In 1963 Electrolux established its own in-house design department under the leadership of
Hugo Lindström. Over the following
decades, Electrolux increased its
share of the white goods market
and household appliance sector
with various company acquisitions,
including Eureka, Husqvarna, **Zanussi** and White Consolidated Inc.,
which owned brands such as Frigidaire, Kelvinator and White Westinghouse. One of Electrolux's
design directors, Michael Green,
notes: "In today's marketplace,
appliance manufacturers that
don't make a serious commitment
to design very quickly turn into
commodity producers ... this company includes design in the Values
and Vision statement."

ERCO

FOUNDED 1934
LÜDENSCHEID, GERMANY

Heidi & Dieter Witte,
7310 spotlight,
1969–1970

Erco's origins can be traced back to a lighting factory
established by Arnold Reininghaus in 1934. In 1968 the
factory's executive director, Klaus Jürgen Maack, came
up with the concept of "Light, not luminaries" that defined the future direction of the company. Viewing light as the fourth dimension of architecture –
in that light interprets and defines space and brings rooms to life – the firm
pioneered revolutionary track light systems in the late 1960s. Erco designs,
such as the 7310 spotlight (1969–1970) by Heidi and Dieter Witte, are typically distinguished by functionalism, **standardization** and suitability for
mass production. Initially mainly used in contract settings, Erco track systems, spotlights and flush-mounted downlights had a strong sense of technical structuralism and were highly influential upon the designs of other
lighting manufacturers. Many well-known designers and consultancies
have designed lighting for Erco, including **Mario Bellini**, **Knud Holscher**,
Frogdesign, **Norman Foster** and **Franco Clivio** (b. 1942). Erco is a leader
in the field of architectural lighting with its state-of-the-art products, which
are installed in such buildings as the Reichstag in Berlin and the Guggen-
heim Museum, Bilbao.

Mario Bellini, *Eclipse*
spotlight, 1986

Speedglas welding mask for Hörnells, 1982

In 1979 two Swedish design groups both founded in the mid-1960s, Designgruppen and Ergonomi Design, amalgamated to form the Ergonomi Design Gruppen, based in Bromma. The 14-member group is dedicated to the research and development of safe, reliable and efficient designs based on ergonomic principles. It studies user-system problems by building full-scale models to evaluate designs in experimental situations. Ergonomi's most notable design is the *Eat and Drink* combination cutlery, drinking vessels and plates manufactured by RSFU Rehab for use by the disabled. Other award-winning products include the *Genotropin Pen* for Pharmacia & Upjohn, an ergonomic coffee pot for SAS, the *Ergo* range of hand tools for Sandvik, a baby carrier for BabyBjörn and an ergonomically resolved stapler for Esco. In 1972 two of the group's founders, Maria Benktzon (b. 1946) and Sven-Eric Juhlin (b. 1940), undertook research into muscular ability and its relationship to the actions of gripping and holding, and have since specialized in **design for disability**. Ergonomi Design Gruppen is guided by the belief that: "Design is not only appearance. It is form, function and economy. Conscientious product development can increase sales, reduce production costs, open new markets and improve quality profile. Design becomes a strategic tool for success." The office is also renowned for its design of machines which reduce the risk of repetitive strain injury and accidents.

E Series knives for Tupperware, 1995–1996

ERICSSON
4
1956

Review

L. M. ERICSSON

FOUNDED 1876
STOCKHOLM, SWEDEN

Cover of the
Ericsson Review,
1956

Model 88 telephone,
1909

Lars Magnus Ericsson (1846–1926) founded the L. M. Ericsson Company in Stockholm in 1876. The business initially concentrated on the repair of telegraph equipment, but by 1878 it was producing telephones that were based on an earlier design by **Alexander Graham Bell**. Soon after, L. M. Ericsson began manufacturing telephones of its own design, which exported throughout Europe, and which were the first to feature a combined transmitting/receiving handset. In 1909 Ericsson launched a cradle telephone, which was enormously influential throughout Europe. Seeking to update this landmark design, in 1930 the company commissioned the artist Jean Heiberg (1884–1976) and the Norwegian engineer Johan Christian Bjerknes to design a telephone with a **Bakelite** casing. Although not the first telephone to use **plastics** in its construction, this sculptural design was highly influential and inspired **Henry Dreyfuss'** later *Bell 300* model (1930–1933). Between 1940 and 1954, Hugo Blomberg

(1897–1994), Ralph Lysell (1907–1987) and Gösta Thames (b. 1916) designed and developed another groundbreaking model for Ericsson, the *Ericofon* (launched in 1956). This unusual and forward-looking telephone had a completely unified form that integrated the earpiece, mouthpiece and dial. The design of the *Ericofon*, which incorporated new lightweight materials such as plastic, **rubber** and nylon, was made possible by the increasing **miniaturization** of technology. Fun yet stylish, it was the most popular one-piece telephone for over three decades. Today, L. M. Ericsson continues its commitment to innovative

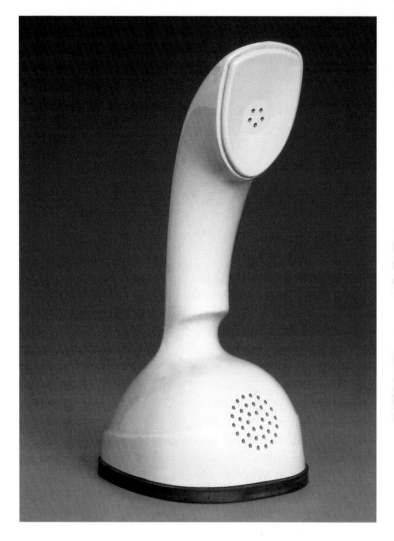

design and cutting-edge technology and remains one of the leading players
in the telecommunications industry, with over 100,000 employees. The company is the world leader in mobile telephone systems, connecting nearly
40% of the world's mobile callers. Ericsson is at the forefront of the WAP
(Wireless Application Protocol) revolution – the bridge between mobile communication and the Internet – with landmark products such as the *R320*,
the company's first mobile phone with a WAP browser.

HARTMUT ESSLINGER

BORN 1944
ALTENSTEIG, GERMANY

Hartmut Esslinger trained as an electrical engineer at the University of Stuttgart before studying industrial design at the Fachhochschule in Schwäbisch Gmünd. In 1969 he established his own industrial design consultancy, Esslinger Design, in Altensteig. His first client was the electronics company Wega Radio. Wega's subsequent acquisition by **Sony** in 1975 introduced Esslinger to the lucrative consumer electronic market that was booming in Japan. His sleek designs, such his *Concept 51K* hi-fi system for Wega (1975), brought him widespread recognition and were remarkable for their Modern dematerialist aesthetic. His work initially appeared to be influenced by the functionalist approach to design promoted by **Max Bill** and **Hans Gugelot** at the Hochschule für Gestaltung in Ulm. It was, however, also driven by a rejection of "the prevailing Modern sentiments of 60s' Germany" and sought instead a "vision of what design could still yet become as opposed to what

Esslinger Design,
Tri-Bel showerhead
for Hansgrohe,
1973

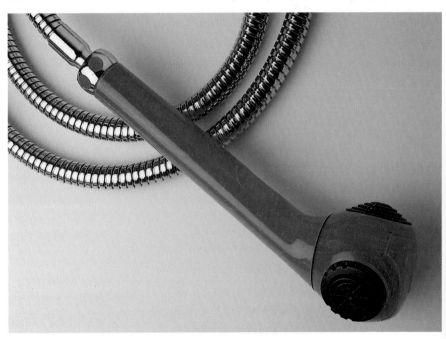

it has already settled for being". In 1982 Esslinger renamed his consultancy **Frogdesign** (frog = Federal Republic of Germany) and opened an office in Campbell, California, so as to meet the design needs of the burgeoning computer industry in Silicon Valley. In 1984 the consultancy designed the off-white *Apple Macintosh* for **Apple Computer** and in so doing redefined the aesthetic parameters of personal computers. Two years later Esslinger opened a sister office in Tokyo, and during the 1990s established branches in technology hot-spots such as the "Silicon Prairie" of Austin, Texas, "Silicon Alley" in New York, "Media Gulch" in San Francisco and "Silicon River" in Düsseldorf. Frogdesign has also designed cameras, synthesizers, binoculars, communications equipment and office seating for such companies as **RCA**, **Kodak**, **Polaroid**, Motorola, Seiko, **Sony**, **Olympus**, AT&T, **AEG**, König und Neurath, **Erco**, Villeroy & Boch, **Rosenthal** and **Yamaha**. Esslinger attempts to humanize technology with sculptural forms or visual references to create more user-friendly products. He has always had an international outlook and recognizes the "benefit of bringing Asian teamwork, European discipline, German precision, American optimism and Californian craziness together" to produce a very unique and culturally melding creative enterprise.

EUCLID

FOUNDED 1931
EUCLID, OHIO, USA

First large-capacity
self-powered hauler
for off-road use,
1933

In 1933 Euclid Road Machinery introduced its first hauling unit that could undertake more work at a lower cost than conventionally designed trucks. The reputation of Euclid's equipment was significantly enhanced when it was used in the construction of the Hoover Dam (1930–1936). In 1934 Euclid introduced the first-ever 20-ton coal hauler and by the 1940s the name "EUC" had become a generic term for all off-road dump trucks. During the Second World War, Euclid produced equipment for the US military and afterwards began assembly-line production of its off-highway equipment, which could now manage up to 40-ton loads. Realizing that speed and capacity was limited by the horsepower of available engines, in 1949 Euclid devised the "Twin Power" concept using two diesel engines mounted side by side. This system permitted the hauling of much heavier loads on steeper gradients at faster speeds. In 1968 Euclid introduced the *R105* – the first ever off-highway hauler with a load capacity of more than 100 tons. 1996 saw the introduction of the *R260* mining hauler, which can carry an astonishing 260 tons.

R260 mining hauler,
1996

Ferrari logo

FERRARI

FOUNDED 1929
MODENA, ITALY

In 1920 Enzo Ferrari (1898–1988) joined **Alfa Romeo** and over the next 20 years went from test-driver to racing driver to director of the Alfa Racing Division. In 1929 he founded the Scuderia Ferrari in Modena with the purpose of helping members compete in motor races. In 1940 the Scuderia became an independent company, the Auto Avio Costruzioni Ferrari. The workshop moved to Maranello in 1943 and at the end of the war changed its name to Ferrari. The first Ferrari racing car was the *125 GT Sport*, which won the Rome Grand Prix in 1947. In 1969 Enzo Ferrari sold 50% of the company to **Fiat** (this investment increased to 90% in 1988), but it was nevertheless able to maintain a strong autonomy. In 1972 Ferrari built the Fiorano track close to its factory so as to test its Formula 1 and GT cars. The experience gained on the racetrack has always had a direct influence on Ferrari road cars. By combining extraordinary performance with beautiful, visually seductive lines, Enzo Ferrari began a tradition of creating, as he put it, "expressions of superlative engineering" and "cars that make themselves desired". Over a period of 40 years, models such as the legendary *250 GT* (1960), the *Testarossa* (1984) and the *550 Maranello* (1996) have been exquisitely styled by **Pininfarina** and have come to epitomize the best of Italian design.

Pininfarina, Ferrari
Testarossa, 1984

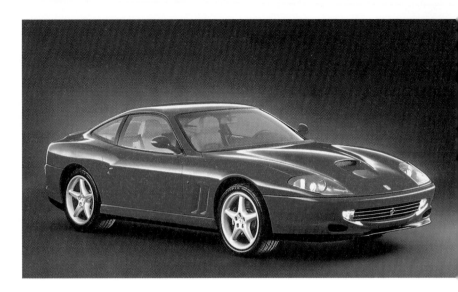

Pininfarina, Ferrari
550 Maranello, 1996

Pininfarina, Ferrari *Mythos*,
1989 (concept car)

FIAT

FOUNDED 1899
TURIN, ITALY

Fiat is Italy's largest automotive company and one of the world's largest industrial groups, with diversified production ranging from buses and trains to farm tractors and aircraft. Founded in 1899 as the Fabbrica Italiana Automobili Torino by a group of investors including Giovanni Agnelli (1866–1945), Fiat received instant recognition for its luxury cars as well as its racing triumphs. The company grew rapidly in terms of both sales and production, and built its first American factory in 1909 in Poughkeepsie, New York. Later, in an attempt to increase production, Agnelli decided to build the largest automotive plant in Europe, which opened in Lingotto in 1922. Implementing assembly-line methods of production, Fiat used this factory to transform the car from a relatively exclusive item into a much more accessible product, a philosophical approach to manufacturing that the company has adhered to throughout its history. Its most successful pre-war car was the compact Fiat 500, or "Topolino" (Little Mouse) as it became known, which was designed by **Dante Giacosa**. Launched in 1936, the 500 was the smallest mass-produced car in the world and over 519,000 models were produced between 1936 and 1955. Its rounded lines broke with the tradition of box-shaped forms, while its light metal body enabled extremely fluid production.

Dante Giacosa,
Fiat 500, 1936

→ Fiat *Nuova 500A*, 1957

During the post-war period, Fiat's objective was to produce the sort of cars that the large American manufacturers were not making – cars with smaller engines that everyone could afford. This led directly to models such as the Fiat 600 (1955) and the Fiat *Nuova 500* (1957). These inexpensive and diminutive vehicles, together with their variations, were manufactured in their millions and did much to democratize car ownership in Italy. They also became potent symbols of Italy's economic miracle. In 1978 Fiat introduced the world's first flexible robotic assembly lines in its plants in Rivalta and Cassino, and a year later merged with Lancia, **Ferrari**, Autobianchi and Abarth to form Fiat Auto SpA. During the 1980s and 1990s, the company continued producing "people's cars" such as the Fiat *Panda* (1980), the Fiat *Uno* (1983) which was styled by **Giorgetto Giugiaro** and became Fiat's best-ever selling car, and the intermediate sized Fiat *Punto* (1993). In 1999 Fiat launched the versatile 6-seater *Multipla*, a groundbreaking mid-range MPV with non-conformist **styling** that was voted Car of the Year 2000 by the BBC. Highly innovative designs such as these are developed in Fiat's Advanced Design centre, which employs a youthful team devoted specifically to generating new car concepts.

↘ *Fiat Uno*, 1983, styled by Giorgetto Guigiaro,

Fiat *Multipla*, 1999

FISKARS

FOUNDED 1649
FISKARS, FINLAND

Olavi Lindén,
Clippers secateurs,
1996

As the oldest industrial company in Finland, Fiskars can trace its origins to an ironworks established by Peter Thorwöste in southern Finland in 1649. From the 1820s, Fiskars was forging high-quality tableware, kitchen knives and scissors, but it was not until 1967 that the company came to the world's attention when it launched its famous orange-handled *O-series* scissors (1963). Developed by the engineer and wood-carver **Olof Bäckström**, the prototype for the ergonomically conceived ABS handles of these scissors was carved from wood. Licensed worldwide, the scissors were also manufactured as pinking shears and in a left-handed version. Over the succeeding years this benchmark ergonomic design, now known as the *Classic*, has been developed into a family of 18 scissors. More recently, under the guidance of Olavi Lindén (b. 1946), the Fiskars design team has developed a range of garden tools that has been awarded numerous design prizes. Known as *Clippers* (1996), these tools reflect Fiskars' "commitment to quality and systematic application of ergonomic principle, which result in products that are easy, safe and pleasurable to use".

Olavi Lindén, *Handy
axe and Clippers
secateurs*, 1994 &
1996

Philippe Starck,
Miss Sissi table lamp,
1991

Achille & Pier Giacomo
Castiglione, *Luminator*
floor lamp, 1962

FLOS

FOUNDED 1962
MERANO, ITALY

In the late 1950s a spray-on self-skinning plastic coating was developed in the United States. This new material, known as Cocoon, was imported into Italy by Arturo Eisenkeil, who began seeking applications for it. In 1962 he founded the lighting company Flos in partnership with the avant-garde furniture manufacturers Dino Gavina (b. 1932) and Cesare Cassina (1908–1981). That same year, he launched the company's first range of lamps, which were designed by **Achille** and **Pier Giacomo Castiglioni**. Soon afterwards Flos produced its first Cocoon lamps, designed by the Castiglioni brothers and by Tobia Scarpa (b. 1935). From the outset, Flos distinguished itself from its competitors not only by commissioning avant-garde designers but through its strong **corporate identity**. To this end, the graphic designer Max Huber (b. 1919) designed its first catalogues, while the Castiglioni brothers designed interiors for its Milan showroom in 1968. In 1974 Flos acquired the lighting company Arteluce, which had been established by Gino Sarfatti (1912–1985). Since then, the two companies have produced lighting products by an international array of designers, including **Mario Bellini**, Matteo Thun (b. 1952), Marc Newson (b. 1963), **Jasper Morrison**, Antonio Citterio (b. 1950) and Konstantin Grcic (b. 1965). In 1988 Flos launched its first lamp designed by **Philippe Starck**, the *Arà*, which was followed by *Miss Sissi* (1991). In 1994 the innovative rubber-like *Drop* wall lamp (1993) designed by Marc Sadler (b. 1946) was awarded a Compasso d'Oro, as was the company for its long-commitment to design innovation. Many of the lamps, especially those by Achille Castiglione, are classic examples of Italian industrial design. The fact that they continue to sell well is a tribute to Flos's commitment to technical and aesthetic excellence.

Henry Ford I

While Henry Ford (1863–1947) did not invent the automobile, it was his vision that made the car accessible to literally millions of people. His primary goal was to "build a motor car for the great multitude ... it will be so low in price that no man ... will be unable to own one." In the 1860s, two events occurred that enabled his dream to be fulfilled – the invention of the open-hearth process in 1864, which heralded the birth of modern steel manufacture, and the oil industry's laying of a pipeline in the Allegheny River valley in 1865, which signalled the beginning of an enormous network of petrol stations that would eventually fuel Ford cars. Funded by twelve investors, the Ford Motor Company was established in June 1903 in a converted Detroit wagon factory. The first car was sold only a month later and was described as "the most perfect machine on the market.... so simple that a boy of fifteen can run it". Over the following five years, Henry Ford directed a research and development programme that resulted in a plethora of models.

Model T, 1925

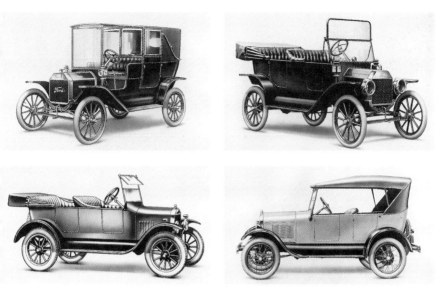

These vehicles, which were coded alphabetically, included two-, four- and six-cylinder cars, some of which were chain driven while others were shaft driven. Not every model made it into production, and of those that did, not all were successful: while the *Model N*, which retailed for a modest $500, sold well, for example, the expensive *Model K* limousine did poorly. Following the failure of the *Model K*, Henry Ford insisted that the company's destiny lay in the manufacture of inexpensive cars. At this time, George Selden held a patent for "road locomotives" powered by internal combustion engines, and most car manufacturers paid him royalties. Ford, however, believed Seldon's claim of exclusivity to automobile technology was invalid and refused to pay any royalties. In 1911, after many years of costly litigation, Ford finally won the suit brought against him by Selden and in so doing freed the entire automobile industry from what was, clearly, an impediment to its progress. Despite this legal distraction, the Ford Motor Company continued to expand and in October 1908 the pivotal *Model T* came into being. Described by Henry Ford as the "universal car", the low-cost and reliable *Model T* became an instant success. The *Model T* was so much in demand that Henry Ford instigated a system of compartmentalized production and came up with the concept of a moving assembly line. This approach to manufacturing, which became known as **Fordism**, dramatically reduced assembly time and resulted in over 15 million *Model T*s being built between 1908 and 1927. The succeeding *Model A* (named after Ford's first car) placed

Ford *Escort GT*,
1967

Ford *Fiesta 1.6*,
1976

Ford *Focus*,
1999

greater emphasis on safety and comfort than its predecessor and became known as "the baby Lincoln" because of its softer contours. Although sales were impacted by the Great Depression, some 5 million *Model A*s were produced before it was succeeded in 1932 by the *V-8*. In 1948 Ford launched the *F-Series*, a pick-up truck that was "built stronger to last longer". Becoming the best-selling vehicle in North America (a title it still holds) – over 27 million *F-Series* trucks have been manufactured to date. In Europe, Ford became better known for its practical "everyman" cars, such as the *Escort* and *Fiesta*. During the late 1990s, the company evolved a new language of design known as "New Edge Design", which combines "smooth, sculpted surfaces with clear, crisp intersections". This fresh approach led to such innovative vehicles as the *Ka*, *Puma*, *Cougar* and *Focus*, which was voted Car of the Year in Europe in 1999 and in America in 2000. Ford's commitment to design innovation resulted in it commissioning the product designer Marc Newson (b. 1963) to develop a concept car for the 21st century, the Ford *012C*, which features a number of imaginative details such as a slide-out luggage tray and a single LCD headlight. Ford also acknowledges its heritage and its new **retro design** *Thunderbird* is an intriguing reinterpretation of the legendary 1950s original. Like all motor manufacturers, Ford has become increasingly mindful of environmental issues and is committed to developing a new generation of vehicles that radically reduce CO_2 emissions.

ord *Ka*, 1996

NORMAN FOSTER

BORN 1935
MANCHESTER, ENGLAND

*Enercon E66 Wind
Energy Converter,*
c. 1999

Lord Foster studied architecture and planning at the University of Manchester, graduating in 1961. He continued his studies at Yale University in New Haven, Connecticut, graduating in 1963. The same Year, Foster established the Team 4 partnership in London with fellow Yale scholarship student Richard Rogers (b. 1933), Georgie Wolton and Wendy Cheesman. In 1967 Foster independently founded Foster Associates in London and from 1968 to 1983 collaborated on several projects with **Richard Buckminster Fuller**. In 1977 Foster received the Royal Institute of British Architects Award for his black glass-clad office building for Willis Faber & Dumas in Ipswich (1973–1975), and has since then become internationally celebrated for his numerous architectural projects, including the new Reichstag (German Parliament in Berlin (1995–1999). Like his buildings, Foster's industrial designs have a High Tech aesthetic, as evidenced by his *Nomos* furniture system (1986–1988) for Tecno and his lighting systems for **Erco** (1986) Other lighting designs by Foster, which have strong architectural qualities, include the *RA* system (1998) for **Artemide** and the *Saturn* down-lighter (1999) for Guzzini. Foster's interest in energy-efficient design led to the development of the elegant *Enercon E66 Wind Energy Converter*, which was selected by the **Design Council** as a Millennium Product in 1999. By combining state-of-the-art materials and technologies within a strictly Modern formal vocabulary, Foster creates inherently logical and beautifully engineered industrial designs.

Remote control for
Sky Digital Satellite
ervice, 1998

Stephen Frazer (b. 1948) studied industrial design at the Central School of Arts & Crafts and later at the Royal College of Art in London. In 1978 he founded Frazer Designers, which specialized in the design of leading-edge electronic products. From 1983 he began designing products for Psion, including the original Psion organizer. Around 1987 the consultancy began producing designs for the early cell-phones manufactured by NEC – an area that it is still actively involved in. Frazer views cell-phones and other similar products as tools that allow entry to much larger services. The actual production costs of electronic products such as these are today very low – the majority being manufactured in the Far East – and they have become extremely accessible. Over 3.8 million remote controls for Sky's digital satellite service, for example, were distributed within twelve months of its launch in Britain in 1998. Frazer describes the remote control as a "visible and tangible key" to Sky's entertainment service, and believes that design will play an increasingly important role within the electronic products market; as the technology used in one product differs little from that used in another, design will be used more and more as the primary means of product differentiation. Frazer Designers approaches the design process holistically, while considering both the technical and emotional requirements of new products. Its aim is to "stimulate that moment of personal desire by creating visual product identities that will attract and intrigue".

Concept microwave
oven (prototype),
1993

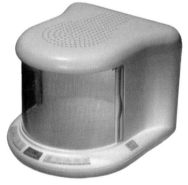

Solido headset
for IMAX, 1994

Frogdesign's origins lie in the industrial design consultancy founded by **Hartmut Esslinger** in 1969 in Altensteig, Germany. A leader in the design of consumer products, Esslinger Design, as it was originally called, attracted a clientele which included **Sony**, **AEG**, **Zeiss**, **Olympus** and **Apple Computer**. In 1982 the name Frogdesign (frog = Federal Republic of Germany) was adopted when the consultancy opened an office in Campbell, California, to cater to the design requirements of Silicon Valley. Unlike other design offices, Frogdesign actively promoted itself during the 1970s and 1980s with a strong and eye-catching advertising campaign in specialist design-related journals, which raised its profile enormously. In 1983 it became the first consultancy to win a "million dollar contract" (with Apple Computer) in Silicon Valley. Significantly, the office re-invested in computer-aided machining tools, taking it to the forefront of design technology. Three years later, another international office was opened in Tokyo to meet the demands of the burgeoning Japanese consumer electronics market. Frogdesign developed a house-style that combined Modern Movement functionalism – as pioneered at the **Bauhaus** and later at the Hochschule für Gestaltung in Ulm – with organic forms. This rational vocabulary, however, was occasionally tempered by the use of visually

Z-Lite laptop
computer for Zenith
Data Systems,
1992

expressive and sometimes whimsical references to a product's function – as in the case of the *Personal Communicator* (1992) for AT&T, which has elements that resemble ears. Esslinger sees Frog (as it is now called) as more than just an industrial design firm and believes that design must be both creative and enabling. In Esslinger's words: "The purpose of design is to make our artificial environment more human. My goal is and always was to design mainstream products as art."

Philippe Starck,
*S1 door handle
for FSB, 1991*

FSB

FOUNDED 1881
BRAKEL, GERMANY

Founded by Franz Schneider in 1881 to produce devotional objects and brass cabinet-fittings, FSB became renowned for its architectural ironmongery. In the early years of the 20th century, Franz Schneider Brakel GmbH diversified its product line to include door and window fittings. In the 1950s the company introduced a range of "moulded-to-the-hand" products that were created by its in-house designer, Johannes Potente. The first "named" designer to be commissioned by FSB was **Dieter Rams**, who developed a series of twelve fittings that adapted to windows and doors of every size and shape. Reflecting Rams' belief that "unassuming is better than showy, colourless better than gaudy, harmony better than divergence, balanced better than exalted", the design of his **aluminium** and thermoplastic *rgs 1, 2, 3* Series (1986) was driven entirely by logic. FSB has subsequently produced designs by **Jasper Morrison**, **Philippe Starck**, Franco Clivio, Hartmut Weise, Nicholas Grimshaw (b. 1939) and Erik Magnussen (b. 1940), which range from the sculptural to the functional so as to suit a wide range of tastes. As Otl Aicher (1922–1991), the co-founder of the Hochschule für Gestaltung in Ulm, observed, the uniting feature of all FSB products is the human hand. With his assistance, FSB published the *Good Grip Guide*, which asserted that "any object designed to be held has got to make allowance for the specific needs of the thumb, the forefinger, the ball of the thumb and the palm."

Hartmut Weise, *1025*
door handle for FSB,
. 1991

NAOTO FUKASAWA

BORN 1956
KOFU, JAPAN

→17" monitor
for NEC, 1998 –
co-designed with
Sam Hecht

Naoto Fukasawa trained as a product designer at Tama Art University and later worked as the chief designer of the R&D Design Group at the Seiko Epson Corporation in Japan. Since 1989 Fukasawa has worked for ID Two (later **IDEO**), leading a design team that has formulated a coherent visual language for NEC products ranging from computer notebooks to LCD projectors, and which have received several G Mark and Hanover iF awards. He has also assisted **Apple Computer** in developing a new vocabulary of design in which the ergonomic form of the product conforms to the user's actions. Fukasawa's *Left Ventricular Heart Assist* (1990), a computer that externally controls an artificial heart for a recovering patient, is not as he describes "a clean, health-goods type of design" but a no-nonsense and robust design that ensures the continuous beating of the heart. Fukasawa, who is now design director of IDEO Japan, is fascinated by the interaction between man and technology. He has collaborated with Sam Hecht (b. 1969) on a number of products that attempt to humanize computer technology, and which are born out of his belief that "perfection is admitting to oneself the existence of imperfection."

*Left Ventricular
Heart Assist* for
Baxter, 1990 –
co-designed with
Sam Hecht

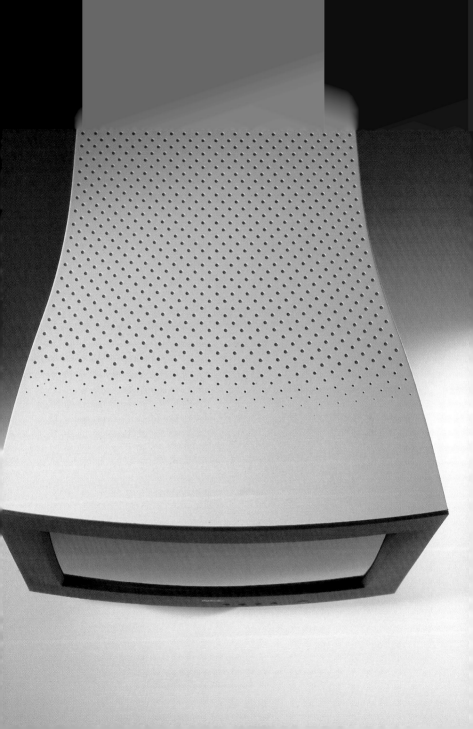

RICHARD BUCKMINSTER FULLER

BORN 1895 MILTON, MASSACHUSETTS, USA
DIED 1983 LOS ANGELES, CALIFORNIA, USA

Richard Buckminster Fuller studied mathematics at Harvard University from 1913 to 1915 and in 1917 enrolled at the US Naval Academy in Annapolis. While at the Academy he began his "theoretical conceptioning", which included a proposal for "flying jet-stilts porpoise" transport that was eventually published in 1932. After leaving the military, he founded the Stockade Building Company in 1922, which was a financial failure and led to his personal bankruptcy. He committed himself to finding global solutions to social problems after the death of his four-year-old daughter in 1922, which he believed was due to inadequate housing. In 1933/34 he founded the 4D Company in New York to develop his design concepts, which were driven by his ambition to evolve a "design science" that would bring about the best solutions with the minimum use of energy and materials. He based this concept on the Modern Movement principle of getting the most with the least and named it "Dymaxion", which was derived from "dynamic" and "maximum

Dymaxion car, 1932

efficiency". In 1929 he launched the magazine *Shelter* and was its publisher and editor from 1930 to 1932. From 1932 to 1938 he was director and chief engineer of the Dymaxion Corporation, which he set up to develop and produce three streamlined prototype cars that were based on his dymaxion principles and were inspired by aircraft design. It was claimed by Fuller that the Dymaxion car of 1934 could accelerate from 0 to 60 mph in three seconds and provide 30 mpg fuel consumption. The prototype car, however, was not progressed due to several serious design flaws. From 1927, Fuller also developed the Dymaxion House concept and between 1944

and 1947 a prefabricated metal dwelling known as the Wichita House. Although the company he set up to manufacture this industrially produced architecture received an astonishing 38,000 orders after its press launch, he was not prepared to commence manufacture of the Wichita House until its design had been perfected. His backers subsequently grew disheartened with the time it was taking and this **product architecture** project was shelved. Fuller's most famous invention was the geodesic dome of 1949, which had a wide range of applications from industrial to military to exhibitions. The geodesic dome remains the only large dome that can be set directly on the ground as a complete structure. It is also the most economic space-filling structure and the only practical type of building that has no limiting dimensions. The outcome of Fuller's "more for less" approach, the geodesic dome employed a minimum of materials and could be easily transported and assembled. This remarkable design offered a means of producing ecologically efficient housing for the mass-market, and Geodesics Inc. was set up in 1949 to develop the concept. Fuller was also the first person to coin the expression "spaceship Earth". Prolific communicator, humanist, polymath and futurist, Buckminster Fuller believed that the creative abilities of humankind were unlimited, and that technology and comprehensive, anticipatory design-led solutions could eliminate all barriers to humanity's expansion into a positive future.

ACHILLE GAGGIA

BORN 1895 ITALY
DIED 1961 ITALY

Ambrogio Fumagalli,
*International Two
Group* coffee
machine, 1954

Espresso maker,
c. 1948

Working as a bartender in Milan, Achille Gaggia grew increasingly dissatisfied with the traditional machines that used steam to brew coffee. He began experimenting in his attic and developed new models that used a piston system to drive boiling water at a high pressure through coffee grounds, so as to make a better-tasting drink. Around 1937 he installed his first machines in several cafés in Milan, and in 1944 he patented his innovative espresso-making process. Three years later, he established a manufacturing company, Brevetti Gaggia, to produce his machines. His first design, known as the *Classico*, was powered by either gas or electricity and was intended only for café use. Around 1948, however, a two-handled espresso maker was developed for domestic use. By pumping the hinged arms, pressure was created in the heated water canister, which in turn forced the water through the grounds. This model was very popular throughout the 1950s, as was the later *Gilda* (1952–1954), named in tribute to a film starring Rita Hayworth. The coffee machine which Ambrogio Fumagalli designed for Gaggia in 1954 was the first to produce two separate quantities of coffee, thus ensuring that a constant supply could be on hand in cafés. By using expressive forms inspired by automotive and aeronautical styling, Gaggia's innovative machines gave the convention of coffee making and coffee drinking a previously unknown glamour.

NORMAN BEL GEDDES

BORN 1893 ADRIAN, MICHIGAN, USA
DIED 1958 NEW YORK, USA

Norman Bel Geddes studied art at the Cleveland Institute of Art and later
trained briefly at the Art Institute of Chicago. In 1913 he worked as a draftsman in the advertising industry in Detroit, where he designed posters for
Packard and **General Motors**. In 1916 he wrote a stage play and subsequently worked as a theatre-designer for six productions in Los Angeles. In 1918
he became a set designer for the Metropolitan Opera Company in New York,
before moving to Hollywood in 1925, where he designed lavish film sets. Influenced by Frank Lloyd Wright (1867–1959) and the German Expressionist
architect Erich Mendelsohn (1887–1953), in 1927 Geddes turned to architecture and product **styling**. While he was a highly successful industrial design
consultant – notably working for the Toledo Scale Company in 1929 and the

*Soda King Syphon
for the Walter Kidde
Sales Company, 1932*

Standard Gas Equipment Corporation in 1931 – he is
best remembered as a design propagandist. In 1932 he
published a book entitled *Horizons*, in which he outlined
his approach to industrial design and his belief in the
supremacy of the teardrop shape. Geddes also designed
futuristic cars for the Graham-Paige automobile company (1928) and streamlined products such as radios
for Philco (1931), radio casings for **RCA**, metal bedroom
furnishings for Simmons (1929) and the well-known
Soda King Syphon for the Walter Kidde Sales Co. (1932).
One of his most notable achievements was his **standardization** of kitchen equipment. The design of his
modular and streamlined *Oriole* stove (1931) for the
Standard Gas Equipment Company was inspired by the
construction of skyscrapers and, as such, had a steel
frame onto which white vitreous enamel panels were
clipped. Geddes also designed the General Motors'
"Futurama" display for the 1939 **New York World's Fair**,
which speculated on the future (as it was believed it
would be in 1960) and correctly predicted the freeway
system. Geddes was one of the greatest exponents of
streamlining and was also an important pioneer of
industrial design consulting.

Complete line of
heating devices,
1907

→WD-560S combination washer-drier,
1954

Model AW1 washing
machine, 1933

The origins of this phenomenally successful company can be traced to the inventor **Thomas Alva Edison**, who established the Edison Electric Light Company in 1878 and a year later invented the first practical electric light bulb. This venture, renamed the Edison General Electric Company in 1889, eventually merged with the Thomson-Houston Electric Company in 1892 to form the General Electric Company. Following Edison's earlier idea of establishing "invention factories", the company founded the GE General Engineering Laboratory in 1895 to conduct research into advanced engineering and instrumentation. In 1900 the company established the GE Research Laboratory, one of the first industrial laboratories in America to undertake basic scientific research. These laboratories were crucial to the development of many inventions and discoveries, including the high-frequency alternator (1906) devised by Dr. Ernest Alexanderson (1878–1975) which revolutionized radio broadcasting, the gas-filled incandescent lamp (1913) invented by Dr. Irving Langmuir (1881–1957) that was the forerunner of the light bulbs still used today, and the first practical X-ray tube (1913) invented by Dr. William Coolidge (1873–1975), which was the predecessor of the modern medical X-ray tube. During the late 1920s General Electric also pioneered television broadcasting in America, using technology developed by Alexanderson. In addition, the company was renowned for its labour-saving domestic appliances, from electric heaters and washing machines to vacuum cleaners and refrigerators – all of which literally transformed the

SHES AND DRIES IN ONE COMPLETELY AUTOMATIC OPERATION

Combination Washer-Dryer

UNDERCOUNTER MODEL WD-560S

The first home television reception at the Schenectady residence of Ernest Alexanderson, c. 1928

Arthur BecVar, redesign of cylinder vacuum cleaner for General Electric, c. 1950

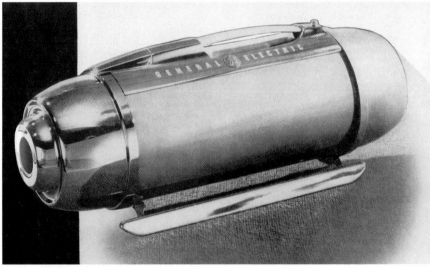

American way of life. In the 1940s, GE produced engines that powered the first American jet-propelled aeroplane, the **Bell** *P-59 Airacomet* (1942) and the world's then fastest aeroplane, the **Lockheed** *P-80 Shooting Star* (1945). The company later produced the *J-79*, the first jet engine to propel an aircraft

GE satellite station in California, c. 1992

at twice the speed of sound. During the 1960s, GE contributed its technical expertise to **NASA**'s Apollo moon-landing mission (1969) and developed weather satellites. In 1975 its research labs designed an essential component for the development of the CAT (computerized axial tomography) scanner and during the 1980s acquired the medical diagnostic imaging company, CGR. Continuing its commitment to television broadcasting, GE purchased **RCA** in 1986, which included the NBC television network, and subsequently launched two new networks, CNBC (1989) and MSNBC (1996). For over a century, General Electric has been driven by technological innovation and its life-changing and life-enhancing inventions have touched us all. Central to its continuing success are the corporation's twelve multi-disciplinary laboratories that reflect the diversity of its operations: from lighting systems to satellite systems, from aircraft engines to washing machines, from state-of-the-art **plastics** to medical diagnostic equipment. The research and development currently being undertaken in General Electric's laboratories will almost certainly shape the future of industrial design, just as it has done over the past 100 years.

Oldsmobile, 1903

GENERAL MOTORS

FOUNDED 1908
HUDSON, NEW JERSEY, USA

General Motors is the world's largest automotive manufacturer and throughout most of the 20th century was the world's largest industrial corporation. For this reason, it has historically been regarded as a barometer of the American economy. Seven years before its incorporation in 1908, the company manufactured the first car to be produced in quantity in America. The year of its official founding, GM launched the first electric headlamp, and in 1910 introduced "closed bodies' as standard equipment. In 1912 the company pioneered the first all-steel car body, which provided much greater strength and safety than earlier automobiles. In 1924 GM established the first automotive proving-ground test facility in Milford, Michigan. During the early 1920s, the company's chairman predicted the increasing importance of **styling** in the automotive industry and subsequently commissioned **Harley Earl** to work on the 1927 La Salle, the first mass-produced car to be developed by an automobile stylist. In 1928 GM established its own styling de-

La Salle, 1934

Pontiac, 1935

Cadillac, 1941

partment, which was headed by Earl and was the first of its kind. This influential in-house facility devised annual stylistic changes in an effort to accelerate the aesthetic life cycle of automobile models and, as a result, increase sales. As well as instigating this programme of annual cosmetic changes, GM continued to produce real innovations, such as the first built-in trunk (1933), which revolutionized the overall form of the car. Under Earl's guidance, GM also developed its first concept car, the Buick *Y-Job* (1938). By 1940 GM had produced a massive 25,000,000 vehicles, which sheds some light on the extent of pre-war car ownership in America. After the war GM introduced several innovations to its product line, including the "airfoil' fender (1942–1948), the curved windscreen (1948) and the modern, high-compression overhead valve V-8 engine, the first of which were used on the company's Cadillac and Oldsmobile models. The famous Chevrolet *Corvette* launched in 1953 became the first GM concept car to have evolved unaltered from a Motorama auto show to a full-scale production model. Although GM had set up the first safety-testing laboratory in 1955, the fateful Chevrolet *Corvair* which it produced in 1960 was famously attacked for its lack of safety and tendency to roll over by the young lawyer, Ralph Nader, in his book *Unsafe at any Speed* (1965). Nader's legal action against GM and his subsequent victory led the US Congress to pass 25 pieces of consumer

Corvette, 1963

legislation between 1966 and 1973, which opened the floodgates for product-liability law suits in America. Learning from this difficult experience, GM introduced the first energy-absorbing steering column (1966), the first side-guard door beam (1969) and the first airbags in production vehicles (1974). In 1975 it also began fitting the first-ever catalytic converters on all its cars sold in America. In more recent times, in 1996 GM launched the first modern from-the-ground-up electric car, the *EV1*. This remarkable zero-emission vehicle, with its sophisticated lightweight **aluminium** body, has the most aerodynamic shape of any production car ever. It is powered by a 137 hp, 3-phase AC induction motor and is offered with two battery technologies: an advanced high-capacity lead acid and an optional Nickel Metal Hydride, which doubles the *EV1*'s range from 55–95 miles to 75–130 miles. Today General Motors boasts many major automobile brands, including Chevrolet, Chevy Trucks, Pontiac, Oldsmobile, Buick, Cadillac, GMC and Saturn, and employs over 397,000 people with operations in 73 countries. As a truly global corporate giant and with the will to innovate such revolutionary vehicles as the *EV1*, General Motors seems likely to continue to dominate the automotive industry as it has done for nearly 100 years.

Giacinto Ghia

GHIA

FOUNDED 1915
TURIN, ITALY

An Italian legend, Ghia SpA is one of the most famous *carrozzeria* or coachbuilders in Italy. Founded in Turin in 1915 by Giacinto Ghia (1887–1944), who had previously learned about car production in Diatto's workshop, the company initially manufactured designs for **Bertone**. It later became acknowledged for its own designs for luxury cars and racing cars. Among its competition car designs, the aerodynamic "torpedo" form came to characterize Ghia's work. After the death of Giacinto Ghia, young designers, most notably Mario Boana, revived the company's outlook. During the 1950s, the Ghia works created specially designed bodies for Bentley, Delahaye and Talbot. The prototype of the Ghia *Gilda* (1955) was one of the first to undergo aerodynamic wind-tunnel testing, and the data collected resulted in the addition of fins for increased stability – a feature that was later "borrowed' by American manufacturers for stylistic purposes. During this same period Ghia also styled many American cars, and in so doing influenced the car designs of Detroit-based manufacturers in general. Throughout the 1960s, the company continued **styling** cars for mainstream volume-manufacturers such as **Fiat**, **Volkswagen**, **Alfa Romeo**, **Renault**, **Chrysler** and

→De Tomaso *Mangusta*, 1966

↘Maserati *Ghibli*, 1968

VW *Karman Ghia*, 1961

Ford. While Ghia designs such as the VW *Karman Ghia* (1961), the De Tomaso *Mangusta* (1966) and the Maserati *Ghibli* (1968) had a recognizable character, they did not match the elegance of cars styled by some other Italian *carrozzerie*. In 1972 Ghia became a wholly-owned subsiduary of the Ford Motor Company and subsequently styled the first Ford *Fiesta* (1973). Today, as part of Ford's Advanced Design group, Ghia is a proponent of "New Edge Design", which blends "the smooth, sculpted surfaces of traditional aero design with clean, crisp intersections to produce a sharper, more defined image."

DANTE GIACOSA

BORN 1905
ROME, ITALY

Drawing of Fiat 500A

Dante Giacosa began working for **Fiat** in 1926 and around 1932 was asked to design a small yet stylish car. His work resulted in the 500A, which was launched in 1936. Also known as the "Topolino" (Little Mouse), this compact and streamlined vehicle marked one of the greatest advances in small car design. The 500A's rounded form broke with the convention of the box shape, while its lightweight metal body was highly suited to mass production. Between 1938 and 1939, it was Italy's biggest-selling vehicle export to America. The 500A was the world's smallest mass-produced car, with 519,000 being built between 1936 and 1955. Giacosa went on to design modified versions of the Topolino – the 500B saloon and station wagon (1948) and the 500C (1949), which incorporated a number of new features such as interior heating. These models were succeeded by Giacosa's diminutive *Nuova 500* (1957), of which 3.6 million were built. He also designed the Fiat *124*, the Fiat *128* and the Fiat *130*. While the 500A helped democratize car ownership in Italy, Giacosa's *Nuova 500* became a potent symbol of Italy's post-war economic miracle.

Fiat 500A, 1936

GILLETTE

FOUNDED 1901
BOSTON, MASSACHUSETTS, USA

King Camp Gillette

Reared in Chicago, King Camp Gillette (1855–1932) became the star sales-man at the Crown Cork & Seal Company, which mainly manufactured cork bottle caps. His mentor at the company, who noticed his predilection for mechanical tinkering, suggested that he "try to invent something like the Crown Cork product which, when used once, is thrown away" so that the customer would keep coming back. Gillette became obsessed with this idea and one morning in 1895, while struggling to hone his straight razor, he had a vision of a razor with a separate handle that clamped a thin double-edged disposable steel blade between two plates. After six difficult years, in 1901 Gillette eventually founded the American Safety Razor Company, which in 1902 became the Gillette Safety Razor Company. While not the first "safety razor", Gillette's innovative refillable razor and blade system offered much greater convenience. It was first manufactured in 1903 and patented a year later. During the First World War, the company supplied 3.5 million razors and 36 million blades to the US armed forces. In 1932 the company intro-duced the famous double-edged *Blue Blade*, which dominated the blade market for many years. Gillette diversified its interests with its acquisition of the Paper Mate Pen Company in 1955 and **Braun** AG in 1967. As a leading

Early Gillette safety razor, c. 1904

Patent drawing of King Camp Gillette's first safety razor, 1901 – the final production model varied slightly

Trac II, 1971

↗ *Sensor*, 1990

→ *Mach 3*, 1998

manufacturer of electric razors, Braun was renowned for its product design team, which was headed by **Dieter Rams**, and which now began developing products for Gillette. In 1971 the company introduced the world's first twin-bladed cartridge system, the *Trac II*, which became an immediate best-seller. Later, after ten years' development, Gillette launched the *Sensor* system (designed by the Braun design team) in 1990. This revolutionary product featured thin spring-mounted blades set in a pivoting plastic cartridge. In the year of its introduction, 24 million razors and an astonishing 350 million cartridges were sold. As the most successful product launch in Gillette's history, the lessons learned from *Sensor* were clear: it reaffirmed the company's commitment to "spend whatever it takes to gain technological supremacy in a category, and then produce innovative products that will capture consumers, even at a premium price". This philosophy led directly to Gillette's latest flagship product, the remarkable triple-bladed *Mach 3*. By far the most expensive wet razor systems product today, the ultra-thin *Mach 3* blades were the first ever to be coated with a diamond-like film. Throughout the 1980s and 1990s, Gillette continued diversifying by purchasing Oral-B, **Waterman**, **Parker Pen** and Duracell, with the result that, by the year 2000, it had become the world's second most profitable brand after Coca-Cola.

WILLEM HENDRIK GISPEN

BORN 1890 AMSTERDAM, NETHERLANDS
DIED 1981 THE HAGUE, NETHERLANDS

Having studied architecture and worked as an architectural draftsman and designer of ornamental ironwork, in 1911 Willem Gispen travelled to England, where he was exposed to the Arts & Crafts ideal of "art-manufactures." He subsequently became a member of the Netherlands Union of Handicraft & Industrial Art and in 1915 opened a factory to manufacture wooden furniture. A year later, he founded the Gispen Fabriek voor Metaalbewerking in Rotterdam to produce chiefly architectural fittings, often using bent **tubular metal**. As part of the Dutch avant-garde, Gispen received commissions from his peers and was involved in the great functionalism versus formalism debate. In 1924 the factory embraced mechanized production and three years later launched its GISO range of lighting. Combining functionalism with geometric purity, the lights were advertised using bold eye-catching Modern graphics. Bolstered by the amount of orders he received in the wake of the **Deutscher Werkbund** "Die Wohnung" exhibition of 1927, in 1929 Gispen opened a large factory in Culemborg to mass-produce his tubular metal furniture and lighting. The remarkable success of designs such as these lay in Gispen's ability to reconcile industrial production with artistic expression.

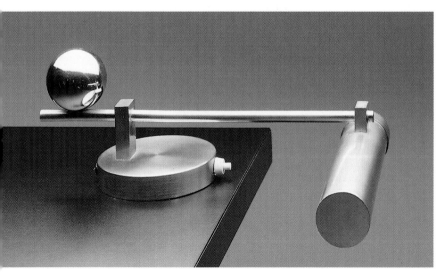

GIORGETTO GIUGIARO

BORN 1938
GARESSIO, ITALY

→ *Macchina Sportiva* sports watch for Seiko, 1995

After studying technical drawing and graphic design at the Accademia di Belle Arti in Turin, Giorgetto Giugiaro initially worked in **Fiat**'s design department. While at **Fiat**, he also trained at an engineering technical college in Turin. From 1959 he headed the **styling** department at **Bertone** in Turin, where he collaborated with Nuccio Bertone (b. 1914) and designed several cars, including the **BMW** *3200 CS* (1961), the **Alfa Romeo** *Giulia GT* (1963) and the Fiat *850 Spider* (1965). In 1965 Giugiaro became director of the **Ghia** automotive-styling studio, where he designed the Fiat *Dino Coupé* (1967). Three years later, he founded Ital Styling (later re-named ItalDesign) with Aldo Mantovani (b. 1927). This enormous design consultancy, which now employs over 750 people, offers car manufacturers a complete design service from pre-production studies to the construction of working prototypes. As well as designing sleek sports cars, such as the Maserati *3200 GT Coupé* (1998), Giugiaro also designed more utilitarian vehicles such as the **Volkswagen** *Golf* (1974), the Fiat *Panda* (1980) and the Fiat *Uno* (1983). In 1981, ItalDesign's industrial design section was established as an independent office, Giugiaro Design. It has since designed numerous products and vehicles, including the *Logica* sewing machine for Necchi (1982), the **Nikon** *F4* and *F5* cameras (1988 & 1996), the *High Speed Ferry* catamaran for SEC (1992), the *ETR460 Pendolino* locomotive for the Italian Railway Company (1995), the *Macchina Sportiva* watch for Seiko (1996), and *Grand Prix* ski boots for Nordica (1997). Giugiaro Design also works in the areas of **signage** and **packaging**. Driven by the "pursuit of function", Giugiaro skilfully combines physical requirements with those of production methods, while being acutely aware of marketing considerations. He does not attempt to impose a formal design language on his products, but instead works out "the aesthetic and functional answers on the basis of need".

Logica sewing machine for Necchi, 1982

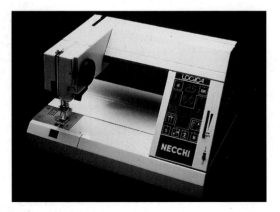

GK DESIGN GROUP

FOUNDED 1957
TOKYO, JAPAN

*FM350FWX
all-terrain vehicle
for the Yahama
Motor Co., 1995

Signage for Expo
'70, Osaka, 1970

The GK Design Group is one of Japan's foremost multi-disciplinary design consultancies. Founded in 1957 by **Kenji Ekuan**, the GK (Groupe Koike) Design Group initially specialized in industrial design. Over the succeeding years, the office widened its remit and became involved in communication, graphic and environmental design. Believing in an integrated approach to design, sometimes referred to as "total design", the studio has built up an impressive network of subsidiaries so as to offer a comprehensive range of services from research (market surveying and product planning) to promotion (**corporate identity** and event planning). Over the last 40 years, GK Design has designed motorcycles, audio equipment, a snowmobile, a personal watercraft and an all-terrain vehicle for **Yamaha**; locomotives for the East Japan Railway Company, **packaging** for Kikkoman, Konica and the Kirin Brewery; and public toilets for Restatio. It has also designed a diversity of products for other manufacturers, including a parking meter, a sewing machine, saucepans, watches, dental equipment and cosmetic packaging. GK Design is additionally noted for its design of **signage** and urban planning, especially its work for Expo '70 in Osaka and Expo '85 in Tsukuba. The work of the GK Design Group is distinguished by exquisite engineering and a functionalist formal vocabulary.

Street furniture
and signage for Expo
'85, Tsukuba, 1985

omachi train for the
East Japan Railway
Company, 1997

KENNETH GRANGE

BORN 1929
LONDON, ENGLAND

Kenneth Grange trained at the Willesden School of Arts & Crafts in London between 1944 and 1947. Later, while completing his national service, he was trained as a technical illustrator by the Royal Engineers. He subsequently worked as an assistant for various architecture and design practices before founding his own London-based design office in 1958. Two years later, he redesigned a food mixer for Kenwood that had been originally launched in 1950. His resulting *Chef* food mixer (1960), which was restyled in just four days, was inspired by earlier kitchen appliances designed by Gerd Alfred Müller (b. 1932) for **Braun**. Grange worked as a design consultant to Kenwood for over 40 years and regularly updated this classic British design. In 1972 Grange, together with Theo Crosby, Alan Fletcher, Colin Forbes and Mervyn Kurlansky formed the design partnership **Pentagram**. Among Grange's many notable designs as part of Pentagram are the **Kodak** *Pocket Instamatic* camera (1975), the **Parker** 25 range of pens (1979), the *Royale* and *Protector* wet razors for Wilkinson Sword (1979 & 1992) and the *ST50* travel steam iron for Kenwood (1995). Grange also designed the exterior body of British Rail's *125 Intercity* high-speed train (1971–1973), Adshel bus shelters for **London Transport** (1990) and the restyled classic London black cab for London Taxis International (1997). Grange has received numerous awards for his esign work, including ten **Design Council** awards and the Duke of Edinburgh's Prize for Elegant Design in 1963. From 1985 to 1987 he served as

↘*Royale* razors for Wilkinson Sword, 1979

Kenwood *Chef A701* food mixer, 1960

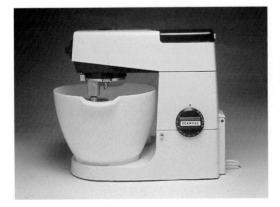

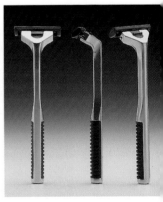

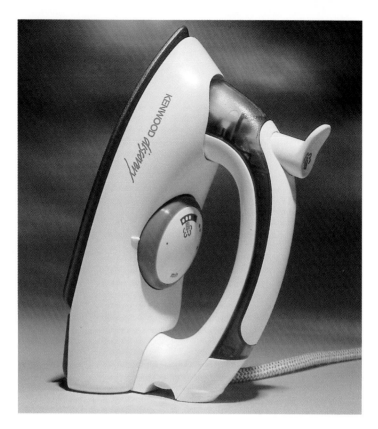

Master of the Faculty of Royal Designers for Industry and was appointed
president of the Chartered Society of Designers in 1987. His work was also
the subject of a one-man show at the Boilerhouse at the Victoria and Albert
Museum, London, in 1983. In 1997 he retired from Pentagram and established his own independent design office. Grange regards design as a means
of innovation and believes that design should be an integral part of the manufacturing process. Grange combines the traditional attributes of British design – honesty, integrity and appropriateness – to create high-quality, me-
ticulously-detailed products that meet the criteria for "good design". Throughout his long career Grange has shown a remarkable consistency – his gently
elegant designs have all had an underlying robustness – which reflects his
skill in balancing functional requirements with those of aesthetics. In many
ways, Grange's work eloquently defined British design during the last half
of the 20th century.

"The Exhibition as a Lesson in Taste"

Southern Entrance to the Transept

The *Report on Arts and Manufactures* drawn up by a Select Committee in 1836 concluded that, so far as industrial design was concerned, Great Britain was being surpassed by its European rivals. During the 1840s Queen Victoria's husband, Prince Albert of Saxe-Coburg Gotha, and **Henry Cole**, an official in the Public Records Office who had founded the short-lived but highly influential Summerley's Art-Manufactures, attempted to address this situation. They campaigned tirelessly for design reform that would align art more closely with industrial production, and endeavoured to revitalize the Society of Arts with a series of exhibitions dedicated to industrial design. When these efforts proved unsuccessful, they decided to organize an exhibition to promote their cause themselves. In 1848 they began planning an international fair that had a three-fold purpose: "exhibition, competition and encouragement". Drawing enthusiastic support both nationally and internationally, the organizers arranged an open competition for the design of the great hall that would house all the exhibits. In 1850, after a cumbersome design had been officially selected, an inspired plan by Joseph Paxton (1801–1865) was substituted. Paxton, who was the superintendent of the gardens at Chatsworth, the Derbyshire home of the Duke of Devonshire, ingeniously met all the rigorous specifications, from material cost to construction time, with an enormous glass house which was based on an earlier design of his for a conservatory. This elegant structure, which became known as the Crystal Palace, comprised a remarkable construction of industrially prefabricated iron components and provided 33 million cubic feet of space. Sited in Hyde Park, the building enclosed an area of 19 acres and easily incor-

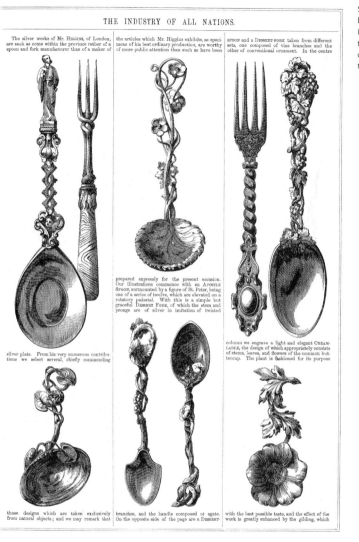

The silver works of Mr. HIGGINS, of London, are such as come within the province rather of a spoon and fork manufacturer than of a maker of the articles which Mr. Higgins exhibits, as specimens of his best ordinary production, are worthy of more public attention than such as have been SPOON and a DESSERT-FORK taken from different sets, one composed of vine branches and the other of conventional ornament. In the centre prepared expressly for the present occasion. Our illustrations commence with an APOSTLE SPOON, surmounted by a figure of St. Peter, being one of a series of twelve, which are elevated on a rotatory pedestal. With this is a simple but graceful DESSERT FORK, of which the stem and prongs are of silver in imitation of twisted column we engrave a light and elegant CREAM-LADLE, the design of which appropriately consists of stems, leaves, and flowers of the common buttercup. The plant is fashioned for its purpose silver plate. From his very numerous contributions we select several, chiefly commending those designs which are taken exclusively from natural objects; and we may remark that branches, and the handle composed of agate. On the opposite side of the page are a DESSERT-with the best possible taste, and the effect of the work is greatly enhanced by the gilding, which

Silverware exhibited by Mr Higgins of London – typical of the highly decorated designs shown at the exhibition

orated the park's mature trees. It could be argued that Paxton's exceptionally innovative building was the highlight of the event, which was named the "Great Exhibition of the Industry of All Nations" and was held in 1851. Over 4,000 exhibitors participated in the exhibition, nearly half of whom were non-British. While some of the most popular exhibits were industrial machines, such as hydraulic presses, steam engines, pumps and automated

Whitworth's stand
of machine tools at
the Great Exhibition

cotton mules (spinning machines), many household items, such as furniture, glass, **ceramics** and metalware, also attracted much attention. The domestic exhibits, however, were for the most part notable for their profusion of superfluous decoration, which was drawn from an eclectic range of styles – Greek, Egyptian, Renaissance, Rococo, Baroque, Celtic – and the "vegetable world". Everything from scissors to carriages was encrusted with ornament, and what was meant to be a "lesson in taste" had become, much to the horror of the organizers, a celebration of decorative excess. Although the exhibition was extremely popular, attracting over 6 million visitors, and was a political and economic success, it was to a large extent seen by its organizers as a vindication of their concerns. Afterwards, some of the exhibits were used to form a study collection for the teaching of design that eventually became the basis of the Victoria & Albert Museum. Between 1852 and 1854 the Crystal Palace was moved to Sydenham Hill in Upper Norwood, London, where it was re-erected in a somewhat different form. It remained there until it was destroyed by fire in 1936.

WALTER GROPIUS

BORN 1883 BERLIN, GERMANY
DIED 1969 BOSTON, MASSACHUSETTS, USA

Walter Gropius studied architecture at the Technische Hochschule, Munich, and the Technische Hochschule, Berlin. Between 1908 and 1910 he worked in **Peter Behrens**' office, where he designed offices and furniture for the Lehmann department store in Cologne. In 1910 Gropius established an architectural partnership with Adolf Meyer (1881–1929) and became a member of the **Deutscher Werkbund**. He was an active member of the Werkbund and initially opposed Hermann Muthesius' (1861–1927) advocacy of **standardization** and design "types". In 1911 Gropius designed his well-known Fagus factory, which incorporated an innovative curtain wall suspended from the building's vertical elements. He also designed a steel and glass model factory for the 1914 Deutscher Werkbund exhibition in Cologne. After the appalling devastation of the First World War and influenced by the success of Henry **Ford**, Gropius came to realise the need for standardization and an industrial approach to design if society was to be modernized. He subsequently merged the two art schools in Weimar to form the Staatliches

Walter Gropius &
Adolf Meyer, Nickel-
plated brass door
furniture, 1928

Bauhaus in an effort to unify the arts and to "provide artistic advisory services to industry, trade and craft". As director of the design school from 19▮
to 1928, Gropius instigated a system of workshops headed by "Masters".
During this period he created several pieces of white-painted furniture, co-
designed metal door furniture with Adolf Meyer, and developed a prefabricated house for the Weißenhofsiedlung estate built as part of the Werkbund
exhibition "Die Wohnung" (The Dwelling) in Stuttgart in 1927. The purpose
built premises which Gropius designed for the Bauhaus following its move
to Dessau in 1925 dramatically embodied the school's shift towards industrial modernity.

In 1934 Gropius moved to London, where he worked in partnership with the
architect E. Maxwell Fry (1899–1987) until 1937. While in London, Gropius
also worked for Jack Pritchard's company, **Isokon**, where he was appointed
controller of design in 1936. In 1937 he emigrated to America and became
professor of architecture at Harvard University. One of the greatest proponents of the Modern Movement, Gropius was also one of the most influential industrial design educationalists of the 20th century.

HANS GUGELOT

BORN 1920 CELEBES, INDONESIA
DIED 1965 ULM, GERMANY

After studying architecture, Hans Gugelot worked with Max Bill until 1954. From 1954 to 1965 he worked in the design department at **Braun**, where he helped develop a house style with a strong visual identity based on functionalism and **essentialism**. Between 1954 and 1965 Gugelot also directed the product design department 2 at the Hochschule für Gestaltung in Ulm, where he promoted the "form follows function" approach to design. Among his best-known products, the *Phonosuper SK 4* radio-phonograph (1956) was co-designed with **Dieter Rams** and nicknamed "Snow White's Coffin" because of its clear acrylic lid and hard-edged geometric formalism. Other notable designs for Braun included the *Sixtant* electric shaver (1962). Gugelot also designed the *S-AV 1000* carousel slide projector for **Kodak** (1962) and built-in furniture for Bofinger (1954). While Gugelot's career was short, his influence – particularly upon the development of German product design – was immense.

Sixtant electric shaver for Braun, 1962 (co-designed with Gerd Alfred Müller)

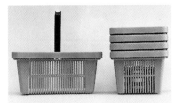

GUSTAVSBERG

FOUNDED 1825
GUSTAVSBERG, SWEDEN

Sven-Eric Juhlin,
shopping basket,
1970

⤵Carl-Arne Breger,
covered bucket, 1959

Wilhelm Kåge, *Soft
Shapes* earthenware,
1937, with *Gra
Ränden* pattern
produced from
1945 to 1969

The Gustavsberg **ceramics** factory is situated some
20 kilometres east of Stockholm. In its early years the factory sourced virtually all of its raw materials from Cornwall and also used imported British
labour skilled in china-making. In the 1820s and 1830s Gustavsberg designs
were hand-painted with simple floral patterns. The factory subsequently introduced transfer printing, a technique imported from Britain, whereby the
patterns it used also often closely followed those of British wares. One of
the factory's owners reputedly possessed a double-walled suitcase in which
he smuggled copperplates between Sweden and England. Nevertheless,
the factory soon became celebrated for its wares and achieved commercial
success when its Parian figurines (also based on British antecedents) were
awarded a gold medal at the 1867 "Exposition Universelle" in Paris. In the
mid-1860s Gustavsberg broadened its production of both "art" wares and
tableware, but during the 1870s it came under increasing attack for the
"bad taste" of its products. The management responded to this criticism
by issuing ceramics that were truly Swedish in nature and by commission-

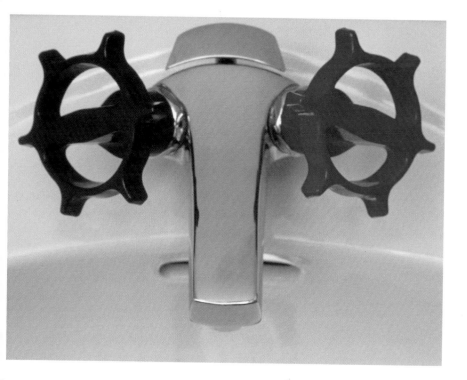

Jan Landqvist, taps, 1979

ing August Malmström and Magnus Isäus (1841–1890) to produce designs in the Viking style. By 1900, under the directorship of Wilhelm Odelberg, the factory had 1,000 employees and was manufacturing ceramics designed by Gunnar Wennerberg (1863–1914) in the Art Nouveau style. The factory's output later came under the influence of the Svensk Slöjdföreningen (Svenska Form), an organization that was inspired by the **Deutscher Werkbund**, which promoted an alliance between art and industry. In 1937 Gustavsberg was purchased by the Swedish Co-Operative Union and Wholesale Society and the factory was modernized. Two years later it began production of sanitaryware and in 1945 began manufacturing products in **plastics**. In 1949 Stig Lindberg (1916–1982) took over from **Wilhelm Kåge** as artistic director. During his tenure, Gustavsberg produced both industrial products and more limited artistic wares. During the 1970s, Sven-Eric Juhlin (b. 1940) – who founded **Ergonomi Design Gruppen** – and Jan Landqvist designed sanitaryware and plastic products mainly for institutional use. In 1987 Gustavsberg was sold to the Finnish company Wärtsilä, and is today divided into three small companies, which continue to produce ceramics in its old factory.

HANSGROHE

FOUNDED 1901
SCHILTACH, GERMANY

Philippe Starck, *Axor Starck* shower head, 1999

→ Phoenix Product Design, *Axor Allegroh Novo* single lever basin mixer, 1997

↘ Phoenix Product Design, *Axor Steel* single lever basin mixer, 1997

Phoenix Product Design, *Axor Azzur* single lever basin mixer, 1997

In 1901 Hans Grohe (1871–1955) established a plumbing and manufacturing business in an abandoned metal-pressing mill in the Black Forest. In 1908 the company, named Hansgrohe, issued its first catalogue and began supplying plumbing wholesalers. In 1928 Hansgrohe diversified its product range to include shower attachments with ceramic handles, and over the succeeding years its export business grew. The company managed to survive the war years and in 1953 introduced the world's first adjustable wallbar for showers. In 1968 it launched the *Selecta*, the first-ever fully-adjustable hand-shower, which became an instant success with over 20 million being sold worldwide. In 1974 Hansgrohe commissioned **Hartmut Esslinger** to design its *Tribel* showerhead, which was the first to feature three spray settings. By pioneering other innovative features, Hansgrohe became a leader in shower technology and by 1998 was the largest producer of showerheads, manufacturing over 2.5 million per annum. Viewing design as the essential connection between the company and its customers, Hansgrohe has in recent years commissioned Phoenix Product Design in Stuttgart and **Philippe Starck** to design sanitaryware products which have gone on to win both Good Design and Rote Punkt awards.

JAMES HARGREAVES

BORN C. 1720 NEAR BLACKBURN, LANCASHIRE, ENGLAND
DIED 1788 NOTTINGHAMSHIRE, ENGLAND

Hargreaves' *Spinning Jenny*, 1764

Around 1760, James Hargreaves was employed by Robert Peel to co-design an improved carding machine for combing wool. Later, while working as a weaver, Hargreaves developed the first practical spinning machine. Like many weavers, he had his own loom and spinning wheel at home. When his daughter Jenny accidentally knocked over the wheel one day, so the story goes, Hargreaves observed how the spindle continued revolving. This gave him the idea of how a whole line of spindles could be worked off a single wheel. He built the first *Spinning Jenny* around 1764, which used eight spindles to spin thread and thus increased output eightfold. Hargreaves' invention, which could be easily operated by children, was initially used only at his home. When he later began selling the machine in 1768, Lancashire hand-spinners, who were fearful of its impact on their livelihoods, marched on his house and destroyed the machinery. He eventually moved to Nottingham and established a small cotton mill. As Hargreaves did not apply for a patent until 1770, the *Spinning Jenny* was copied and improved by others, which resulted in its spinning capacity being increased from eight to 80 threads. At the time of his death, there were over 20,000 *Spinning Jennies* in operation in Great Britain, making it one of the great inventions of the **Industrial Revolution**.

Woman working at the *Spinning Jenny*, c. 1870

HARLEY-
DAVIDSON

HARLEY-DAVIDSON

FOUNDED 1902
MILWAUKEE, WISCONSIN, USA

45-degree V-Twin
engine, 1909

In 1901 William Harley (1880–1943) and Arthur David-son (1881–1950) began experimenting with ways to "take the work out of bicycling". A year later, they found-ed the Harley-Davidson Motor Company in a shed that belonged to Davidson's father. By 1903 the friends had produced their first motorcycle, which was powered by a single 167cc engine, and later the same year they pro-duced another two motorcycles. In 1906 the company was incorporated and the following year its production had risen to 150 motorcycles per annum. The famous 45-degree V-Twin en-gine, which could produce a top speed of 60 mph, was introduced in 1909. By 1913 Harley-Davidson was dominating motorcycle racing in America and its production leapt to 12,904 motorcycles for that year. Having previously been proven effective by the US military in border skirmishes with Pancho Villa, over 20,000 Harley-Davidson motorcycles were deployed during the First World War. Following the war, the company grew rapidly and by 1920 had become the largest manufacturer of motorcycles in the world. The repu-

Early model, c. 1913

tation of Harley-Davidson's motorcycles for speed was enhanced when a works bike became the first to win a race with an average speed of 100 mph in 1921. During the 1920s, the Teardrop gas tank and front brake were introduced and in 1936 the overhead valve Knucklehead engine came into being. During the Second World War, 90,000 motorcycles were produced for Al-

Motorcycle with Panhead engine, produced 1948–1965

lied use, and in 1948 the famous Panhead engine was launched. 1957 saw the introduction of the *Sportster*, described as the "father of superbikes", and a year later the smooth-riding *Duo Glide* with its hydraulic suspension was launched. In 1965 an improved model known as the *Electra Glide* was ushered in, and a year later the Shovelhead engine succeeded the V-Twin. The 1970s saw a motorcycle boom in America and many Harley-Davidson enthusiasts began to customize their bikes, often in the chopper style. By 1975 the company was producing over 75,000 motorcycles a year. While facing increasing competition from Japanese manufacturers, the brand identity of the company continued to strengthen and today it enjoys a 56% share of the US market, with growing demand in Europe, Japan and Australia. For many years Harley-Davidson has powerfully symbolized American individualism, and this has no doubt contributed to its success and its recognition the world over as the "King of the Road".

Road King motorcycle, 2000

JEAN HEIBERG

BORN 1884 OSLO, NORWAY
DIED 1976 OSLO, NORWAY

Plaster prototype
of the *DBH 1001*
telephone for
Ericsson, c. 1930

Jean Heiberg studied painting in Munich and then trained under Henri Matisse (1869–1954) in Paris from 1908 to 1910. He remained in Paris until 1929, becoming a follower of the Fauves and receiving widespread recognition for his own paintings and sculptures. After his return to Oslo, he was commissioned to design the casing of a newly developed **Bakelite** telephone, which was to replace the existing metal types. The internal layout of this early plastic telephone had been devised by the Norwegian engineer, Johan Christian Bjerknes, for a joint venture between the Norsk Elektrisk Bureau and the Swedish telephone manufacturing company, **L. M. Ericsson**. Heiberg's sculptural housing for the *DHB 1001* telephone (1930), which was developed using plaster prototypes, was less angular than earlier models. The coherent relationship between the handset and the main instrument body established the standard for telephone design until the 1950s. The *DHB 1001* was produced from 1932 and was widely distributed throughout Scandinavia, Britain, Italy, Greece and Turkey. The telephone was also manufactured under licence by **Siemens** in England as well as in France and America. Heiberg's design later inspired **Henry Dreyfuss**' *Model 300* telephone (1937) for **Bell** Laboratories.

DBH 1001 telephone
for Ericsson, put into
production in 1932

KEITH HELFET

BORN 1946
CAPE TOWN, SOUTH AFRICA

Keith Helfet trained as a mechanical engineer at the University of Cape Town and later in 1977 completed a Masters degree in automotive design at the Royal College of Art, London. The following year, he joined **Jaguar** Cars and began working as a designer. In 1981, he was promoted to the **Styling** Management Team and was for the first four years given responsibility for the sports car project to be a successor to the legendary *E-Type* known as the Jaguar *XJ 41*. Although this car was not put into production, the design was later used as the basis for other projects. Between 1985 and 1999, Helfet designed both sports cars and luxury saloons, most notably the Jaguar *XJ220* supercar, which was launched as a concept car in 1988 and then as a limited production car in 1991. Working with the Jaguar Design Team, Helfet worked on the *XJ-Series* saloon (1994), the *XK8* sports car (1996) and the *XJ-8* saloon, both of the latter winning the coveted Italian "Most Beautiful Car" award. Like his other car designs, Helfet's latest project, the Jaguar *F-Type* Sports Roadster, combines sculptural form with design originality, while acknowledging the great tradition of the marque. This design, which derives from Helfet's earlier high-performance Jaguar *XK 180* concept car (1998) received universal acclaim when it was unveiled at the 2000 Detroit Motor show, winning the "Car of the Show" award. The elegance and beauty of Helfet's car designs led the international medical equipment company Elscint to commission him to design three patient-friendly MRI (magnetic imaging resonance) body scanners (1994, 1995 & 1996). These products signalled a new approach to the design of medical equipment, which took into account for the first time the aesthetic aspect and its psychological impact on the patient.

Jaguar *XJ220*, 1991

Jaguar *XK180*, 1998

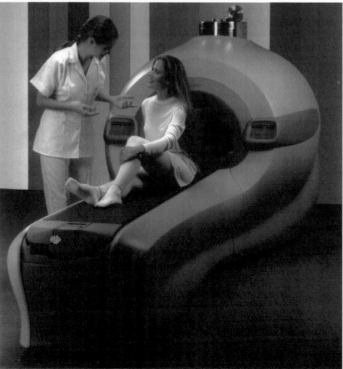

Elscint MRI scanner,
1994–1995

D.J. De Pree

Robert Propst (b. 1921) wrote of Herman Miller: "Our aspiration has always been to design things that are fashion-proof; that have more quiet, enduring qualities; that try to retreat to more elementary qualities". The origins of this remarkable company can be traced back to 1923, when D.J. De Pree and his father-in-law, Herman Miller, acquired the majority holding in the Michigan Star Furniture Company and subsequently renamed it the Herman Miller Furniture Company. Like many furniture manufacturers at the time, Herman Miller produced reproduction furniture that was popular within the main-stream market. In 1930, however, when De Pree met the designer Gilbert Rohde (1894–1944), the company's destiny changed. Rohde proposed a high quality range of Modern furniture that was of the utmost simplicity, so that all the value went into materials and construction rather than decorative sur-face treatment. The subsequent success of Rohde's Modern furniture led Herman Miller to abandon the manufacture of reproduction furniture alto-gether in 1936, and by 1941 the company had opened a New York showroom to display its new Modern designs. In 1946 **George Nelson** succeeded Ro-hde as design director and subsequently brought in other talented design-ers, including **Charles Eames**, Isamu Noguchi (1904–1988) and Alexander

Robert Propst,
Action Office II
system, 1968

Girard (1907–1993), to create high-quality Modern furnishings for the company. Over the following years, Herman Miller produced designs that met
De Pree's criteria for good design – "Durability, Unity, Integrity, Inevitability".
These included Charles and Ray Eames' moulded **plywood** chairs (1945–
1946) and plastic shell series of chairs (1948–1950) and George Nelson's
Comprehensive Storage System (1959) and *Action Office I* (1964–1965). 1968
saw the launch of Robert Propst's landmark *Action Office II*, which literally
transformed the office landscape. Since then, the company has been a
leader in the contract furniture market and has produced a series of pioneering office chairs by Bill Stumpf (b. 1936) and Don Chadwick (b. 1936) – the
Ergon (1976), the *Equa* (1984) and the revolutionary *Aeron* (1992). As one of
America's most admired corporations, Herman Miller continues to flourish
through its quest for design and manufacturing excellence.

GEOFF HOLLINGTON

BORN 1949
ESSEX, ENGLAND

Proficia internet
telephone for
Camelot Inc., 1997

Geoff Hollington trained as an industrial designer at Central School of Arts & Crafts and later studied **environmental design** at the Royal College of Art, London, graduating in 1974. Between 1976 and 1978 he was a member of the design team for the new town of Milton Keynes, and as such was responsible for landscaping and street furniture. In 1980 he founded Hollington Associates with the aim of concentrating on industrial design – from consumer products and medical equipment to personal electronics and office furniture systems. Since then he has worked on numerous projects for such companies as **Herman Miller**, **Parker Pen**, **Gillette**, **Matsushita**, Cable & Wireless, NEC, Filofax, Lloyd Loom, **Ericsson**, Hille, Artifort and **Kodak**. In 1992 his groundbreaking *Relay* furniture group (1990) for Herman Miller received a Gold Award from the Industrial Designers Society of America. Hollington has readily embraced the introduction of **computer-aided design** and high-tech engineering tools, yet still believes in the primary importance of "the front end, innovative, thoughtful part" of the design process. More recently, the office has added "interaction design" to its remit and has designed around 100 interactive exhibits for the Wellcome Wing at the Science Museum in London.

Digital TV remote
control for Cable &
Wireless, 1999

KNUD HOLSCHER

BORN 1930
RØDBY, DENMARK

Knud Holscher studied architecture at the Royal Danish Academy of Art, Copenhagen, and from 1960 to 1964 worked as an assistant to **Arne Jacobsen**. Together with Alan Tye, he subsequently designed the *Modric* range of sanitary fittings, which received a British Design Award and the British **Aluminium** Industry's prize in 1966. After returning to Denmark from Great Britain, he became a partner in the Krohn & Hartvig Rasmussen architectural office in 1967. In 1970 he won another British Design Award for an office **signage** system, and two years later designed his highly successful *d-line* range of stainless-steel door furniture, which has won numerous awards and has grown to become the world's most extensive range of architectural ironmongery and bathroom fittings. In 1975 Holscher designed stainless-steel salad servers and tongs for Georg Jensen that were remarkable for their simplicity. Today, Knud Holscher Industrial Design is one of Scandinavia's leading design consultancies and numbers **J C Decaux**, **Erco**, **Philips**, Fritz Hansen and Facit among its clients. The office has also won numerous competitions for street furniture and signage systems.

d-line staircase
system, 1999

*Dream 246cc
motorbike, 1960*

⌐*CUV-ES Canopy
(Clean Urban
Vehicle), early
1990s*

"You meet the nicest
people on a Honda"
– advertising cam-
paign for the *Super
Cub*, 1962

From an early age, Soichiro Honda (1906–1991) was fascinated with anything mechanical. As a young man, he built and drove racing cars and won several races before retiring from competition after seriously injuring himself. In 1937 he founded a piston ring manufacturing company, which was later acquired by **Toyota**, and in 1946 established the Honda Technical Research Institute, which produced the first Honda prod-uct, the *A-type* bicycle motor, a year later. This venture was re-named the Honda Motor Company in 1948 and in 1949 launched its first motorcycle, the *Dream D-type*. During the 1950s Honda introduced its first *H-type* power engine (1953) and its lightweight 100cc *Super Cub* motorcycle (1958). It subsequently established sister companies in America (1959) and Europe (1960), and in 1962 finished the construction of the famous Suzuka circuit, near its factory. The same year, the company began marketing the *Super Cub* motorcycle with its well-known "You meet the nicest people on a Honda" advertising campaign in America, which helped to alter attitudes towards motorcycle riding. The result was that, by 1967, some 5 million *Super Cubs* had been sold, making it the best-selling motorcycle in the world (some 25 million have been sold to date). In 1963 Honda introduced its first sports car, the *S500*, and the *T360* light truck. As part of its research and develop-

You meet the nicest people on a Honda. And the remarkable thing is the low cost of it all. Prices start about $215? Insurance is painless. Upkeep negligible. Honda's four-stroke engine demands 200 miles from a gallon of gas. And gets it. Plenty of drive. That's how you stay at the top of the class. World's biggest seller. HONDA

Mantis sportsbike, early 1990s

ment programme, Honda entered Formula 1 racing in 1964, winning its first victory in 1966 in Mexico. Since then, F1 racing has remained an important aspect of the company's business. Honda also went on to become the first company to win the Constructor's Championship in all motorcycle classes at the 1966 World Grand Prix. Apart from its racing triumphs, Honda's mini-compact *N360* (1966) became the best-selling car in Japan in 1968. Another highly successful compact car was launched in 1972, the omnipresent fuel-efficient Honda *Civic*, which sold 1 million units within its first two years of production and 10 million units by 1995. The larger Honda *Accord* was introduced in 1976 and achieved a similar sales record. During the 1980s, Honda continued to expand abroad and introduced several new car models, including the mini-compact *Today* (1985) and the luxury-end *Legend* (1985). In 1989 the *Accord* became the best-selling car in America. In 1992 the company established a huge high-tech manufacturing plant in Swindon, England. Ever mindful of environmental concerns, in 1993 Honda became the first manufacturer to produce engines that met the new Californian emissions regulations. Honda's quest for technological innovation continues to produce challenging designs in the areas of motorcycles, cars and power products, and since 1986 the company's engineers have been developing a walking two-legged "humanoid robot" that has a degree of artificial intelligence.

HOOVER

FOUNDED 1908
NORTH CANTON, OHIO, USA

Vacuum cleaner for
the Electric Suction
Sweeper Company,
1908

↘Henry Dreyfuss,
Hoover vacuum
cleaner *Model 150*,
1946

Advertisement for
the Hoover vacuum
cleaner *Model 150*,
1936

The first Hoover vacuum cleaner was based on a design by a department store janitor, James Murray Spangler (1848–1915). He realised that if the sweeping action of a carpet cleaner was combined with the relatively new idea of using suction to remove dirt and dust from carpets, performance would be dramatically improved.

His wood and tin prototype used a pillowcase for the bag and although it was a relatively clumsy device, it worked fairly well. Lacking the funds to commercialize this new invention, Spangler managed to interest his friend, W. H. Hoover (1849–1932), who was a manufacturer of leather goods. In 1908 the Electric Suction Sweeper Company (later re-named the Hoover Suction Sweeper Company in 1910 and then The Hoover Company in 1922) was established and Spangler became its production manager. Frank Mills Case subsequently re-designed the cleaner, giving it an **aluminium** casing. These first electric vacuum cleaners were initially produced at the rate of six to eight machines a day and weighed nearly 40 pounds, but the development in 1909 of a small high-speed universal motor by Hamilton & Beach,

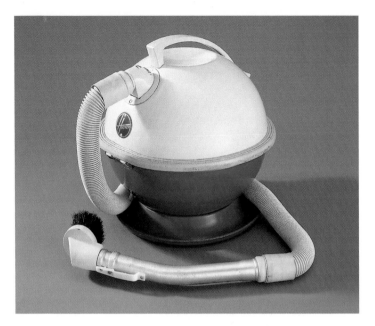

Hoover *Constellation*
vacuum cleaner,
1955

Racine, allowed their weight to be decreased to around five or six pounds.
The same year, Hoover began a research and development programme,
which resulted in the introduction of the "Agitator" to all its upright models in 1926. This innovative feature comprised a rotating bar with spiralled
bristles that helped loosen dirt and dust caught in carpets through a beating action. The famous Hoover slogan of 1919, "It beats, as it sweeps, as
it cleans" was used very effectively to market this improvement in vacuum
cleaner performance. Also in 1926, another feature appeared for the first
time – the power switch was moved to the handle on the *Model 700* cleaners. From the 1930s to the early 1950s, the company commissioned **Henry
Dreyfuss** to design a range of streamlined cleaners, including one that featured an illuminated lighting strip. In 1955 the futuristic looking *Constellation* cleaner was introduced which, when operating, floated on a cushion
of air just like a hovercraft. Extremely popular in America, the design of this
model reflected contemporary interest in future space technology. In testimony to the brand's continuing influence, the Hoover name is still recognized around the world, although the Hoover Company is no longer an international company. Maytag Corporation, which acquired Hoover in 1989, di-
vested its European operations (including UK operations) in the mid-1990s.

ELIAS HOWE

BORN 1819 SPENCER, MASSACHUSETTS, USA
DIED 1867 BROOKLYN, NEW YORK, USA

Elias Howe's sewing machine, c. 1845

Elias Howe was interested in mechanics from childhood, and later worked as a machinist in a factory producing machinery for the cotton industry. While there, it was suggested to him that a fortune could be made by developing a machine that could sew. For the next five years, he undertook experiments that eventually led in 1845 to the construction of the first practical sewing machine. The following year, he patented his handwheel-operated lock-stitch sewing machine, which carried the cloth on a spiked feed-plate past a vertically positioned needle, so that the eye-pointed needle could then pierce the cloth in the same direction as the handwheel was turned. His invention received little attention in America at first, so his brother Amasa took it to England, where the patent rights were sold to William Thomas for £250 (then equivalent to US$1,250).

Detail of Elias Howe's sewing machine

Around this time, Howe also moved to England, where he worked for a pittance while continuing to perfect his invention so that it could be used for the sewing of leather and corsets. In the meantime, sewing machines were being widely manufactured in America that infringed his patents – including those produced by Isaac **Singer**. After three years in Britain, Howe returned to the United States and began litigation. His rights were eventually recognized in 1854 and he subsequently received royalties from all sewing machines manufactured in the United States. Howe's innovative design heralded the birth of the sewing machine industry and quite literally revolutionized clothing manufacture in both homes and factories around the world.

Eliot Noyes, IBM
Model A typewriter,
1948

IBM

FOUNDED 1911
NEW YORK, USA

In 1911 three firms – the Tabulating Machine Company
(producing electrical machines that processed data
using a punched card system), the Computing Scale
Company (holding the patent for Julius Pitrat's computing scale) and the
International Time Recording Company (manufacturing a mechanical time
recorder) – merged and later re-named International Business Machines
(IBM) in 1924. Four years later, the data capacity of punched cards almost
doubled from 45 to 80 columns of information, which heralded the development in the early 1930s of a new series of machines that could not only perform addition and subtraction, but could also undertake full-scale accounting calculations. In 1935 IBM launched its first electric typewriters and, a
year later, provided accounting machines for America's Social Security Program, which was referred to as "the biggest accounting operation of all
time". Developed in collaboration with Harvard University, the ASCC (Automatic Sequence Controlled Calculator) was introduced in 1944 and was the
first machine capable of automatically executing lengthy computations. Four

Eliot Noyes, 705
Electronic Data
Processing Machine,
1954

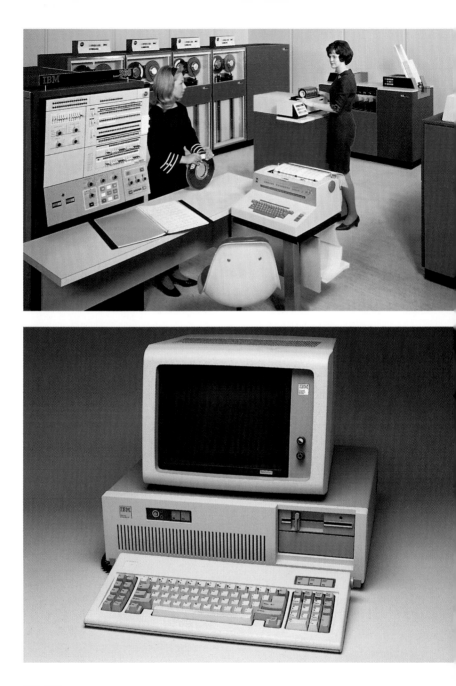

Eliot Noyes, Main-
frame computer
system 360, 1964

128,00-BIT CMOS
RAM microchip,
early 1990s

years later, IBM launched the first computer that combined electronic computation with stored instructions. During the 1940s, **Norman Bel Geddes** advised IBM on the design of its office products. 1952 saw the introduction of the first computer capable of scientific calculations, the IBM *701*, which used magnetic tapes to store data equivalent to 12,500 punched cards. **Eliot Fette Noyes**, who had previously worked in Geddes' office, designed the *705 Electronic Data Processing Machine* in 1954 and two years later was appointed corporate design director of IBM. Noyes was responsible for several revolutionary products, including the *Selectric I* typewriter (1961) with its innovative golf ball typing head. He also developed a strong **corporate identity** for IBM, both through the design integrity of his own products and by commissioning graphics from Paul Rand and buildings from leading architects such as **Marcel Breuer**. During the 1950s and 1960s, IBM continued to develop state-of-the-art computing systems for a range of uses, from accounting and ticketing operations to helping **NASA** with its space programme. Computer processors and peripheral units, such as those of IBM's landmark *System/360* (1964), were still huge in size but were becoming increasingly powerful. In 1968 IBM introduced its *3986* Braille typewriter – an important **design for disability** – and two years later entered the photocopying market with the IBM *Copier*. During the 1970s, IBM developed the *3614 Automatic Banking Machine* and the *3660 Supermarket System* that could read universal product codes. The late 1970s saw the development of computers that used the new semiconductor technology and electronic typewriters with microprocessors. IBM introduced its first personal computer in 1981, but by the mid-1980s was finding it difficult to compete with **Apple Computer's** more user-friendly models. As one of the largest corporations in America, IBM was slow to react to technological change and as a result began losing market share. As a pioneer in the field of information processing, IBM has nevertheless affected all areas of life – from banking and shopping to energy generation and weather forecasting. The character of the company is hallmarked by technological innovation and a strong design ethos that reflects its core values – respect for the individual, customer service and excellence.

IBM personal
computer, c. 1984

IDEO

FOUNDED 1969
LONDON, ENGLAND

Tracy Currer & Nick
Dormon, Digital
Radio for BBC, 1997

→Rick Lewis, *Visor*
communicator for
Handsprung Inc.,
1999

Christopher Loew,
SyncMaster 400 TFT
flat screen monitor
for Samsung, 1996

IDEO's founder, Bill Moggridge (b. 1943), studied industrial design at the Central School of Arts & Crafts, Lon-
don, graduating in 1965. He later undertook a research fellowship in typography and electronic communications at the Hornsey School of Art. In 1969
he established the design office, Moggridge Associates, and ten years later
opened a second office in San Francisco, called I. D.Two, to cater to Silicon
Valley's fledging computer industry. One of his first products was the GRiD
Compass laptop computer (1980). Over the succeeding years, believing that
the design of an interface must be an integral part of a product's development, Moggridge christened and pioneered a new discipline known as
"Interaction Design". In 1991 Moggridge's office merged with David Kelley
Design and was renamed IDEO. By offering a full product development
service, the consultancy's interdisciplinary teams translate ideas into ready-
to-manufacture products. IDEO's non-linear design methodology, which

includes brainstorming sessions, can at times appear
chaotic; as the business guru Tom Peters noted: "IDEO
is a zoo.... Experts of all flavours commingle in 'offices'
that look more like cacophonous kindergarten class-
rooms than home of one of the world's most successful
design firms". This way of working, however, generates
a powerful synergy that harnesses both creativity and
cutting-edge technology to produce design solutions
that are both novel and imaginative. While at the forefront of technological developments, IDEO also realises
that emphasis must be placed on human needs, and
has therefore adopted a "user-focused' approach to design so as to create user-friendly products. IDEO has
designed over 2,000 medical, computer, telecommunications, industrial, furniture and consumer products
for a wide-ranging clientele, including **Apple Computers**
Black & Decker, **Nike**, British Airways, Baxter Healthcare, Amtrak, Deutsche Telecom, Microsoft, NEC,
Nokia, Samsung, **Siemens** and **Steelcase**.

Hyvälysti travel kit for Kai International, 1992

TAKENOBU IGARASHI

BORN 1944
HOKKAIDO, JAPAN

Takenobu Igarashi trained at Tama Art University, graduating in 1968, and went on to further studies at the University of California, Los Angeles in 1969. He later established his own design office specializing in graphic design, **corporate identity** and product design. Igarashi's early poster designs were remarkable for the way in which he was able to translate simple numbers and letters from the Western alphabet into striking eye-catching compositions. He also created sculptures that were almost architectural in their complexity. In 1985 he founded the Takenobu Igarashi Studio and began producing sculptural yet restrained designs, such as his *Zao* stool (1992) and series of iron platters (1990). Interestingly, Igarashi is able to imbue three-dimensional objects with a strong graphic quality, as illustrated by his *YMD* paperknife (1992) for ClassiCon, which has a handle that casts a wave-like shadow on the blade. Igarashi's sensitivity to both volume and line is, however, best expressed in his design of compact personal objects such as the *Hyvälysti* gardening and grooming kits. Igarashi is a professor at the Tama Art University and a visiting professor at the University of California, as well as vice-president of the Alliance Graphique International.

Hyvälysti gardening kit for Kai International, 1990–1991

IITTALA

FOUNDED 1881
IITTALA, FINLAND

Göran Hongell,
Aarne glassware,
1954

Kaj Franck, *Kartio*
glassware, 1958

The Swedish glassblower Per Magnus Abrahamsson founded the Finnish glassworks, Iittala, in 1881. The company's early glassware revealed influences from Europe and mainly consisted of simple undecorated designs for household use. In 1917 Iittala was acquired by the Karhula Glassworks, as part of the Ahlström Group, and numerous improvements were subsequently instigated at the factory. By the 1930s the glassworks was concentrating primarily upon the production of blown-glass rather than pressed glass. Aino Aalto's range of functional glassware (1932) and Alvar Aalto's famous *Savoy* vase (1936) were initially manufactured by Karhula and were only later produced by Iittala. In an effort to develop a range of art glass, in 1946 Iittala organiz-

ed a competition which was won jointly by Tapio Wirkkala (1915–1985) and Kaj Franck (1911–1989). The unconventional work of both designers, such as Wirkkala's *Kantarelli* vases (1946), heralded a new direction for Iittala and a new period in Finnish glass design. Iittala was later joined by Timo Sarpaneva (b. 1926), who produced sculptural art glass and in the mid-1950s also designed the company's first modern utility glassware, the *i-Collection*, which was suited to mass production. "Frosted" designs by Wirkkala and Sarpaneva subsequently exerted much influence on mainstream glassware in the 1960s and 1970s. In 1990 Iittala was purchased by the Hackman Group and has since commissioned designs from Konstanin Grcic (b. 1965), Marc Newson (b. 1963) and Harri Koskinen (b. 1970).

Morten Kjelstrup &
Allan Østgaard,
Mammut child's
chair, 1993

IKEA (

FOUNDED 1943
ÄLMHULT, SWEDEN

With a loan from his father of £150, the 17-year-old
Ingvar Kamprad (b. 1926) founded IKEA in the town
of Älmhult, a few miles from his home village, in 1943.
Situated in the Småland region, historically one of the
poorest parts of Sweden, the company originally sold
a variety of products through mail order, from cattle soap to stockings. In
the early 1950s, IKEA began retailing furniture at factory prices through mail-
order catalogues. This move was met with such resistance by the established furniture trade that IKEA was not allowed to exhibit at the large furniture trade fair held in Stockholm, and suppliers threatened to boycott the
company. Observing that "most nicely designed products were very, very
expensive", Kamprad wanted to manufacture well-designed practical products that the majority of people could afford. With this "social mission" to
democratize design, Kamprad realised that there were three essential prio-
rities – aesthetics, function and suitability for mass production. During
the late 1950s he recruited his first designers, Gillis Lundgren, Bengt Ruda
and Erik Wørts, who, while work-
ing within the Modern idiom, designed furniture for self-assembly –
a revolutionary development in furniture design. Lundgren's *Regal*
bookshelf (1959) was one of the
first IKEA products sold in flat-pack
form. Another concept which IKEA's
designers helped to pioneer was
"modular thinking" – the development of modular components and
interlinking systems rather than individual one-off products. This led
to the company's co-ordination of
its product range and promotion of
an integrated home-furnishing look
– an early and important example
of lifestyle retailing. The advent of

Knut & Marianne
Hagberg, *Puzzle*
children's furniture,
1988

Noburo Nakamura,
Poem chair, 1977

particleboard in the 1960s heralded the arrival of really low-cost furniture and influenced IKEA's product-line immensely. An inexpensive yet relatively hard-wearing material, particleboard continues to be used extensively by IKEA for all kinds of self-assembly furniture. While the 1970s were a period of general insecurity, IKEA prospered and expanded. During this time the company retailed several products in **plastics**, including the *Telegono* lamp (1970) by the avant-garde Italian designer, Vico Magistretti (b. 1920). Throughout its history, however, IKEA has mainly concentrated on producing interpretations of well-known classic designs – from the ubiquitous **Thonet** café chair to the *Polyprop* by **Robin Day**. The 1970s were also the decade of stripped pine and brightly coloured textiles, and IKEA developed this informal look within

its product range specifically for younger consumers. The following decade, IKEA began producing stylish yet practical furniture designed by Niels Gammelgaard (b. 1944) – his *Moment* table (1987) won the Excellent Swedish Design prize. Other award-winning designs included the *Puzzle* range of children's furniture (1988) designed by Knut and Marianne Hagberg, which combined primary colours and simple forms and became a best seller. Throughout the 1990s, IKEA continued to expand its international operations and popularize its version of Scandinavian Modernism. Today the company boasts annual sales of £3.8 billion. Kamprad is not content with just producing home furnishings, however, and plans to diversify his product line into small pre-fabricated flats. As the design theorist Victor Papanek wrote: "One thing is certain: IKEA will continue in the forefront – ecologically, socially and culturally – of making things that work, possess beauty and are affordable."

INDIAN

FOUNDED 1901
SPRINGFIELD, MASSACHUSETTS, USA

Detail of Indian *Chief*
motorbike, 1947

Having built several racing bicycles around 1900, Oscar Hedstrom constructed an engine-powered "pacer" bike, which was used for the training of cyclists. This practical two-wheeler caught the attention of the entrepreneur George Hendee. In 1901 Hendee and Hedstrom founded the Hendee Manufacturing Company (later renamed Indian Motocycle Manu-facturing Company), and the following year the first Indian motorcycles were manufactured. These single-cylinder bikes were the first commercially mar-keted gasoline-powered motorcycles in the United States. In 1903 the com-pany launched a V-Twin engine-powered motorcycle, which was updated over the succeeding years to incorporate two- and three-speed gearboxes and swingarm rear suspensions. The revolutionary *Hendee Special* intro-duced in 1913 was the first motorbike to have an electric starter and a full electrical system. Prior to the First World War, Indian was the largest motor-cycle manufacturer in the world. Later it created such well-known designs as the *101 Scout* (1919), the *Chief* (1920) and the *Sport Scout* (1935). Before the Great Depression there were over 20 American motorcycle manufactur-ers, but only Indian and **Harley-Davidson** survived. Indian later supplied large numbers of motorcycles to the US military, but in 1953 it went out of business, leaving a design legend that has refused to die.

Indian *Chief*
motorbike, 1946

Tom Ahlström &
Hans Ehrich,
toothbrush for
Colgate, 1986

In 1968 Tom Ahlström (b. 1943) and Hans Ehrich (b. 1942) founded **A&E Design** in Stockholm, a design consultancy that became celebrated for its design for disability and won numerous awards for its life-enhancing and ergonomic products. In 1982 Ahlström and Ehrich established InterDesign as a subsidiary of A&E Design, and in 1997 the venture became an independent industrial design consulting company. Although mainly specializing in **plastics**, InterDesign has worked in a wide range of fields. It has designed sanitaryware for **Gustavsberg** (1984), a locomotive for ASEA-Traction (1985), a range of toothbrushes for Colgate (1986), automatic injection devices for SHL Medical (1997) as well as dental floss and tooth pick dispensers for Jordan, electric circuit breakers for Norwesco and lighting fixture components for Belysia. In 1992 the consultancy received the Swedish Environment Design Award for its returnable milk bottle (1992) designed for Arla. In the development of its product designs, InterDesign uses computer-aided design (CAD) systems to rapidly determine product structure, and stereolithography to validate the **ergonomics**, aesthetics and integration of components – all before tooling is engaged. According to InterDesign: "Designing the first product for a company is not without consequences. Designing the second and subsequent products means implementing a design strategy that includes product image in perfect coherence with user ergonomics. The whole must satisfy and attract the direct client (the distributor) as well as the client of the client (the end-user). This is the way to build fidelity, notoriety and competitive differentiation."

Tom Ahlström, Hans
Ehrich & Jochen
Ratjen, returnable
milk bottle for Arla,
1992

Marcel Breuer,
chaise longue for
Isokon, 1935–1936

ISOKON

FOUNDED 1931
LONDON, ENGLAND

Having studied engineering and economics at Cambridge University, Jack Pritchard (1899–1992) began working as a designer in 1925 for Venesta **Plywood**, a leader in the development of this new and exciting material. While there, he commissioned leading architects and designers such as Le Corbusier (1887–1965), László Moholy-Nagy (1895–1946) and **Wells Coates** to design exhibition stands for Venesta. In 1931 Pritchard, together with his wife and Coates, founded Isokon with the aim of bringing "modern functional design to houses, flats, furnishings and fittings". For a while, **Marcel Breuer** acted as the design controller of the firm and produced five plywood furniture designs between 1935 and 1937. **Walter Gropius** also worked for Isokon and was its design director from 1936 to 1937. One of Isokon's best-known products was the small *Penguin Donkey* (1939) bookcase designed by Egon Riss specifically for the storage of paperbacks published by **Penguin Books**. Benefiting from the exodus of **Bauhaus**-associated designers during the mid-1930s, Isokon became a leading exponent of Modernism in Britain.

Cover of Isokon
catalogue, c. 1936

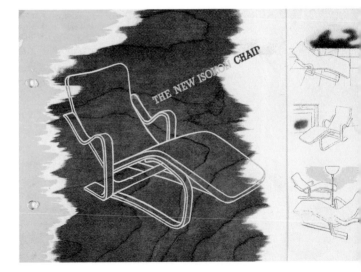

THE NEW ISOKON CHAIR

ALEC ISSIGONIS

BORN 1906 SMYRNA (IZMIR), TURKEY
DIED 1988 BIRMINGHAM, ENGLAND

Born in Turkey as the son of a Greek merchant, Alec Issigonis emigrated to England in 1922. Although he apparently received no formal training as an engineer, he did study for a time at Battersea College, London. He was subsequently employed as a draughtsman by Rootes Motors in Coventry and later moved to Morris Motors in Oxford, where he worked as a suspension designer. He later became Morris's chief engineer. 1948 saw the launch of his revolutionary Morris *Minor*, which was influenced by American automotive **styling** from the 1930s and 1940s. This diminutive yet curvaceous vehicle had a unitary body construction which made it suitable for large-scale mass production. As the first Modern British car, the Morris *Minor* was immensely popular and became the first all-British car to exceed sales of one million. It remained in production until 1971 and to this day has a devoted following of collectors and enthusiasts.

In the 1950s Issigonis briefly worked for other companies before returning to Morris, which had by now been renamed the British Motor Corporation (BMC). In response to petrol shortages resulting from the Suez Crisis and

Advertisement featuring the 1949 model of the Morris *Minor*

the resurgence of interest in the **Volkswagen** *Beetle*, Issigonis designed the small, fuel-efficient and inexpensive Morris *Mini*. Launched in 1959, this front-wheel drive, box-like vehicle was only three metres long. It employed a radical layout that included a space-saving transverse engine, which allowed the compact cab to accommodate four adult passengers in relative comfort. The *Mini* was influential not only in terms of its design but also culturally – it was the first truly "classless" car, as likely to be driven by celebrities as by ordinary citizens. No car came as close to epitomizing the Swinging Sixties as the Mini, although in later years its safety record was seriously questioned. The *Mini* was sold in several versions, including the wooden-framed, van-like *Traveller*. In 1962 Issigonis designed the slightly larger *1100* which was similarly manufactured under Morris, Austin and badges. Like his earlier models, this design was also influential upon later compact cars such as the **Renault** *5* and the Volkswagen *Golf*. In 1967 Issigonis was made a fellow of the Royal Society and two years later was knighted for his immense contribution to the British car industry. He eventually retired in 1971. By 1988 – the year of his death – over five million *Minis* had been sold.

Drawing of the
Morris *Mini Minor*.
c. 1959

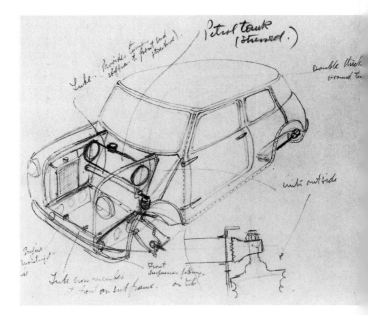

JONATHAN IVE

BORN 1967
LONDON, ENGLAND

Jonathan Ive studied design at Newcastle Polytechnic and later worked for the London-based industrial design consultancy **Tangerine**, where he designed a wide range of products, from televisions and VCRs to sanitaryware and hair combs. At Tangerine he also assisted in the development of the *PowerBook* for **Apple Computer** in 1991. While working on this project, Ive noted that the bland product identity of computers was the result of their arbitrary configurations, and concluded that there was a huge opportunity in creating exciting new products that did not conform to the conventional grey, beige or black boxes. At this stage, the computer industry was mainly concerned with the internal aspects of its machines – processing speed and memory capacity – and little if any thought was being given to their external form. As a consequence, the industry as a whole was suffering from what Ive describes as "creative bankruptcy". Frustrated that, as an external consultant, he could only have a small impact on the future development

Advertisement for
iMac, 1999

of computers, in 1992 Ive joined the design team at Apple, where he became director of design. It was not until Steve Jobs' return to Apple, however, that the design team was given the freedom to concentrate on the "pursuit of nothing other than good design". Jobs realised that Apple needed to recapture its once strong identity, which had been diluted by a "design by focus group" mentality. Significantly, his first day back at the helm marked the beginning of the *iMac* project. The translucent turquoise *iMac* (1998) with its unified curvaceous organic form, broke all conventions. At long last, here was a computer that looked cool and had a strong identifiable character. At its launch, it became obvious that the rather beleaguered Apple (whose market share had shrunk to a miserable 3 % in 1997) had backed a winner that could restore its fortunes – an astonishing 150,000 *iMacs* were sold over the weekend following its introduc-

tion. Helped by a high-profile advertising campaign, which included the in-sightful slogan "Chic, not Geek", *iMac* became the best-selling computer in America – impressively, because of its design rather than its technology. Ive's designs have managed to powerfully differentiate Macs from PCs, and it may well be that the people who buy into Apple Computer products on the strength of their looks will become hooked on the company's excellent user-friendly operating system.

ARNE JACOBSEN

BORN 1902 COPENHAGEN, DENMARK
DIED 1971 COPENHAGEN, DENMARK

← *Series 7* chairs for
Fritz Hansen, 1955

Arne Jacobsen initially trained as a mason before studying architecture at
the Kongelige Danske Kunstakademi, Copenhagen, from where he graduated in 1927. He later worked in the architectural practice of Paul Holsøe
from 1927 to 1929. Around 1930 he established his own design office in
Hellerup and began practising independently as an architect and interior
designer. His early work was influenced by Modernists such as Le Corbusier
(1887–1965), Gunnar Asplund (1885–1940) and Ludwig Mies van der Rohe
(1886–1969). Jacobsen became one of the first designers to introduce Modernism to Denmark through such projects as the "House of the Future",
which he co-designed with Fleming Lassen in 1929. His first important architectural commissions were for the Bellavista housing project in Copenhagen (1930–1934) and the functionalist Rothenborg House, Ordrup (1930),
which was conceived as a *Gesamtkunstwerk*. For his most fully integrated
project, the SAS Air Terminal and Royal Hotel in Copenhagen (1956–1960),
Jacobsen designed every detail from textiles to light fittings. Prior to this,
in 1945 Jacobsen had also begun designing for industrial production and
this led most notably to his *Ant* chairs (1951–1952) and his *Series 7* chairs
(1955) for Fritz Hansen, which remain among the most commercially successful seating programmes ever produced. Jacobsen also designed the *AJ*
range of lighting (1955–1960) for **Louis Poulsen**, the *Cylinda-Line* metalware
collection (1967) for **Stelton**, cutlery for Georg Jensen, bathroom fittings for
I. P. Lunds and textiles for August Millech, Graucob Textilen and C. Olesen.

Cylinda-Line teapot
for Stelton, 1967

From 1956 until 1965, he was a professor at the Skolen for Brugskunst
in Copenhagen. In the 1960s, Jacobsen's most important architectural
scheme was St Catherine's College,
Oxford – a wholly unified project
for which he designed site-specific
furniture. Jacobsen's elegant essentialist industrial designs possess a
timeless appeal, and because of
this they have enjoyed remarkable
success.

C-type on its way to
victory at Le Mans,
1953

JAGUAR

FOUNDED 1922
BLACKPOOL, ENGLAND

As a young motorcycle enthusiast, William Lyons (1901–1985) met William Walmsley, who had constructed a stylish motorcycle sidecar. Realising the sales opportunity of such a design, Lyons founded the Swallow Sidecar Company in Blackpool in 1922. The company's torpedo-shaped **aluminium** sidecars became an instant success and the firm grew rapidly. Lyons spotted another business opportunity in 1927, when Herbert Austin launched his revolutionary Austin *Seven*, the first small and inexpensive British car intended for the masses. Lyons subsequently constructed a chic two-seater body that was mounted on an Austin *Seven* chassis. Known as the Austin *Seven Swallow*, the car provided inexpensive luxury and proved so popular that a saloon version was introduced in 1928. Having moved the Swallow factory to Coventry around 1929, Lyons and Walmsley decided to design their own chassis, which were built by the Standard Motor Company. The resulting *S1* and *SS2* Coupés caused a sensation when launched in 1931. With their long bonnets and low bodies, they looked luxurious and expensive but were actually modestly priced. After Walmsley left the company in 1934, Lyons recruited the design engineer, Harry Weslake, and established an engineering department, which was headed by the youthful William Heynes. Working together, Weslake and Heynes developed a completely new saloon and sports car range,

Jaguar *XK-E* roadster,
1965

JAGUAR

A special kind of motoring which no other car in the world can offer

THE MARK TEN SALOON

Powered by the XK'S' Type 3.8 litre engine with 3 carburettors. Independent suspension front and rear. Disc Brakes on all 4 wheels

THE MARK 2 SALOON

Available as 2.4, 3.4 or 3.8 litre models powered by the world famous XK engine. Disc Brakes on all 4 wheels

THE 'E' TYPE G.T. MODELS

Available as Fixed Head Coupe or Open Two Seater (with detachable hard top as optional extra). 3.8 litre XK'S' Type engine with 3 carburettors. Independent suspension front and rear. Disc Brakes on all 4 wheels.

Jaguar advertisement from *The Motor*, 1963 – showing the *Mark-Ten* saloon, the *Mark 2* and the *E-type GT* model

which was launched at the 1935 London Motor Show. In tribute to their feline grace, agility and power, these cars were named "SS Jaguar". At the outbreak of the Second World War, car production ceased and the company manufactured sidecars for military use and also built aircraft. After the war, car production resumed, the SS name was dropped and the company became known as Jaguar Cars. In 1948 the first post-war model was launched, the rather conservative *Mark V*. Meanwhile, Jaguar had developed a new XK engine that had an impressive output of 160 bhp. For the Motor Show that same year, Lyons designed a suitable body for this new powerful engine and the result was one of the greatest sports cars of all time, the classic *XK120*. Claimed to have a top speed of 120 mph, the *XK120* was marketed as the fastest production car in the world. To prove the claim to the press, Jaguar speed-tested the car in Belgium, where it clocked 126 mph; when the windscreen was removed, it went even faster, achieving 133 mph. Orders soon flooded in, and it was not long before a new five-seater saloon with the XK engine, the *Mark VII*, was introduced at the 1950 Motor Show with an eye to the American market. Following the racing successes of the *XK120*, Jaguar decided to produce a real racing car, the *XK120C*, or *C-type* as it became known. To reduce weight on the car, a triangulated frame was adopted and a sleek aerodynamic body was designed by Malcolm Sayer. Although only 53 *C-types* were built, it was an influential design which took first, second and fourth place at the 1953 Le Mans. The similarly styled *D-type* followed, but from 1955 Sayer began developing a new car which would be both a superlative road car and racing car. The legendary *E-type* was subsequently unveiled at the Geneva Motor Show in 1961. Arguably the most beautiful car ever designed, with its seductive **styling** and top speed of nearly 150 mph, the *E-type* set new standards in car refinement. Costing £1,830, the *E-type* was half the price of an Aston Martin and a third of the

price of a **Ferrari** and stunningly embodied the value-for-money ethos that Lyons had always tried to perpetuate at Jaguar. In 1963 the stately *S-type* saloon was launched, followed five years later by the unmistakable *XJ6* saloon. Around this time Jaguar developed a new high-performance yet "silky smooth" engine, the famous V12, which was used to power later *E-types* and the *XJ12* launched in 1972, which was the world's fastest four-seater production car. As the luxury car market became increasingly competitive during the 1980s, Jaguar realised the absolute necessity of collaborating with one of the automotive giants. In 1990 it was acquired by the **Ford** Motor Company, since when it has undergone a renaissance. In 1991 the beautiful *XJ220* supercar was put into limited production and in 1996 the exquisite *XK8* was launched at the Geneva Motor Show. Both these cars, as well as the *XJ-Series* saloon (1994) and the *F-type* concept roadster (2000), were designed by **Keith Helfet**, who has skilfully evolved Jaguar's "signature" styling while embracing state-of-the-art technology. Jaguar's commitment to design has given the marque a strong visual identity and enabled the company to build rather than rest upon its long and illustrious heritage.

F-type concept
roadster, 2000

PAUL JARAY

BORN 1889 VIENNA, AUSTRIA
DIED 1974 ST. GALLEN, SWITZERLAND

Audi front chassis
with streamlined
body designed by
Paul Jaray, 1930s

Engineer Paul Jaray studied in Vienna and Prague before moving to Germany, where he worked as an aircraft designer for Flugzeugbau Friedrichshafen from 1913 to 1914. Between 1914 and 1923 he worked for Luftschiffbau **Zeppelin**, developing airships. His hydrogen *LZ38* airship was over 163 metres long and had a gas volume of 33,780 m³. His later helium *LZ126* airship was even larger, measuring over 200 m in length. This model became known as the *Los Angeles* and was the first airship to cross the Atlantic during the winter months. As an important pioneer of **aerodynamics**, Jaray designed a wind tunnel for Zeppelin in 1919 and was responsible for the adoption of the teardrop shape in airship design. Jaray also established scientific principles and mathematical models for **streamlining**, and in 1922 filed a patent for streamlined automobile bodies. His revolutionary enclosed car design, with its wrap-around windscreen, integrated headlamps and flush door handles, was at least ten years ahead of its time and was extremely influential upon later designs such as the 1936 **Volkswagen Beetle** and the 1938 Tatra *V-8*. **Audi** also produced several experimental designs based on Jaray's patents. In 1921 Jaray patented an aerodynamic bicycle known as the *J-Rad*, of which 2,000 were produced. The rider sat in a semi-recumbent position and ropes rather than chains drove the bicycle. Jaray's *J-Rad* design was an early antecedent of later fully recumbent high performance vehicles, such as those designed by **Mike Burrows**.

Patent drawing of an
aerodynamic car,
1922

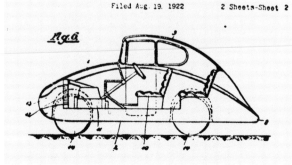

MOTOR CAR

Filed Aug. 19. 1922   2 Sheets-Sheet 2

Joseph Cyril Bamford

JCB

FOUNDED 1945
UTTOXETER, STAFFORDSHIRE, ENGLAND

JCB was founded in 1945 by Joseph Cyril Bamford. His first design, a trailer, was produced in an Uttoxeter lock-up garage out of war surplus and sold for £90. Three years later, having seen the potential of hydraulics, Bamford introduced what was probably the world's first two-wheeled hydraulically-operated tipping trailer. In 1951 JCB launched the *Master-Loader* for use on farms and two years later developed the *Mark I* – the world's first backhoe loader. 1954 saw the introduction of the powerful *Hydra-Digger*, heralding the true commercial beginnings of JCB. It was the launch of the *3C* backhoe loader in 1963, however, that reinforced the company's position as market leader. Two years later, JCB entered a new market with the development of its first 360° crawler excavator. By 1969, 50% of its production was being exported and JCB continued to grow through its commitment to design and engineering excellence. In 1976 Anthony Bamford took over the chairmanship of the company from his father and in 1980 oversaw the successful launch of the

JCB *Mark I* excavator and major loader, 1953

JCB 3CX – a new backhoe loader to replace the earlier 19-year-old model, of which the company had sold an astonishing 72,000 units. The 3CX took ten years to develop and was an immediate commercial success. By 1984, JCB had a turnover in excess of £150 million and captured 17% of the total world backhoe market, with exports accounting for 65% of production. Throughout the 1990s JCB continued producing significant innovations, including the world's first 4 x 4 x 4 backhoe loader. In 1996 the company won a **Design Council** Award for Innovation for the JCB *Landpower*, while its JCB *Teletruk* (telescopic lifting truck) was named "Innovation of the Year" by the Institute of Transport Management in 1997. In total, JCB has won over 50 major awards for engineering excellence, design, exports, marketing, management and environmental concerns, including 14 Queen's Awards for Technology and Export achievement. As a consequence of the company's innovative research and design, extensive testing and stringent quality control, JCB machines have become renowned all over the world for their performance, value and reliability.

JCB 3CX excavator, late 1990s

JEEP (WILLYS-OVERLAND)

FOUNDED 1909
TOLEDO, OHIO, USA

Willys-Overland
Jeep, 1940 in action

Described by General George C. Marshall as "America's greatest contribution to modern warfare", the Jeep is an iconic design with a name that has become generic for small four-wheel drive vehicles. Following the First World War, the US military realised the need for an all-terrain reconnaissance vehicle that was both lightweight and fast, to replace motorcycles. To this end, in 1940 the US Army issued specifications for such a vehicle and invited 130 automobile manufacturers to present working prototypes within just 49 days. Only three manufacturers responded to this challenge, and it was Willys-Overland, which had been founded in 1909 by John North Willys (1873–1935), whose *MB* Jeep with its powerful "Go Devil" engine ultimately won the Army contract. Over 700,000 Jeeps were subsequently manufactured under licence by **Ford** and 368,000 were produced by Willys-Overland. This versatile four-wheel drive vehicle, which was capable of carrying a 660-pound payload, was used for ferrying troops and carrying the wounded. It was also used as an all-purpose pick-up truck and "limousine" for top-ranking officers. One of its greatest assets was undoubtedly its robust design, allowing it to be easily disassembled for transportation and quickly re-assembled for rapid deployment. There are several theories as to how the name Jeep came about, but whatever its origin, Willys-Overland managed to obtain the Jeep trademark in 1950. Civilian Jeeps (known as CJs) have also been manufactured in various models since 1945, including the famous *CJ-5*, which was introduced in 1955 and remained popular well into the 1980s. Other landmark Jeep sport utility vehicles include the J-series *Wagoneer* (1962), the J-series *Cherokee* (1974) and more recently the *Grand Cherokee* (1993). Willys-Overland was purchased by Kaiser Manufacturing in 1953 and became known as Willys Motors and then later as the Kaiser Jeep Corporation. This company was purchased by the American Motors Company (AMC) in 1970, which was in turn acquired by the **Chrysler** Corporation in 1987, which still owns the rights to the Jeep name.

Willys-Overland
Jeep, 1940

JAKOB JENSEN

BORN 1926
COPENHAGEN, DENMARK

Jakob Jensen studied industrial design at the Kunsthandvaerkerskolen, Copenhagen, graduating in 1952. Between 1952 and 1959 he was the chief designer for Denmark's first industrial design consultancy, founded in 1949 by **Sigvard Bernadotte** and **Acton Bjørn**. From 1959 to 1961 Jensen lived in the USA, where he established a design office with Richard Latham (b. 1920) and others. During this period he also taught at the University of Illinois, Chicago. On returning to Copenhagen in 1961, Jensen founded his own industrial design office and three years later began designing audio equipment for **Bang & Olufsen**. His sleek, high-performance designs set the aesthetic and technological standards for audio systems. His *Beogram 1200* hi-fi (1969) was awarded a Danish ID prize for its harmonious balance between "apparatus" and "furniture", while his *Beogram 4000* record player (1972) was notable for incorporating a tangential pick-up arm. Although best known for his pioneering designs for Bang & Olufsen, Jensen has also designed products for other Danish manufacturers, including the *E76* pushbutton telephone (1972) for Alcatel-Kirk, an office chair (1979) for Labofa, ultrasound scanning equipment (1982) for Bruel & Kjaer and a wristwatch (1983) for Max René. One of Scandinavia's most famous industrial designers, Jensen has received numerous awards for his innovative work.

*Beogram 4000
record player, 1972*

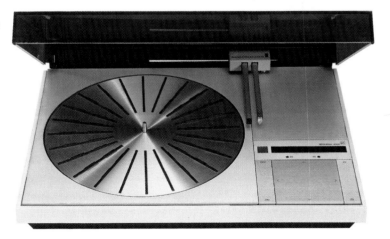

Junghans catalogue,
1898

One of the first
wristwatches manu-
factured by Junghans,
1928

Erhard Junghans (1823–1870) imported straw from
America to make hats, but by the beginning of the American Civil War in 1861 the material had become pro-
hibitively expensive. Junghans turned instead to building clocks in his native Black Forest, a region traditionally famed for its clock-making. The industry was being
hit, however, by the cheaper mass-produced clocks
flooding the market from America. Junghans' response
was to manufacture "American clocks according to the
American principle" (i. e. of industrialized mass production). Junghans' inexpensive clocks gained a reputation
for their accuracy and after his untimely death in 1870,
his son, Arthur travelled to America to study modern
production methods at first hand. On his return to Germany in 1872, Arthur implemented a rationalized system of production at the Junghans factory. He developed the first mass-produced alarm clock in 1875 and
by 1878 was producing 100,000 clocks per annum. By
1903 Junghans had become the world's largest clock
manufacturer and that same year launched its first
pocket watch. Arthur's son, Oskar, later developed the
first radium-illuminated clock-face (1912). Junghans
survived the difficult 1920s and 1930s with its "class for
the masses" motto that translated into good quality yet
inexpensive designs. Junghans published its first post-
war export catalogue in 1949. In 1956 the company was
acquired by the Diehl group; the following year, **Max Bill**
designed his famous wall clock and wristwatches for
Junghans, which were celebrated for their functional
clarity. Remaining committed to innovation, in 1969
Junghans launched the very first quartz watch and in
1986 introduced the first solar-powered watch and
the first radio-controlled analogue clock.

Max Bill, wall clocks,
Model 367/6046 &
Model 367/4047,
1957

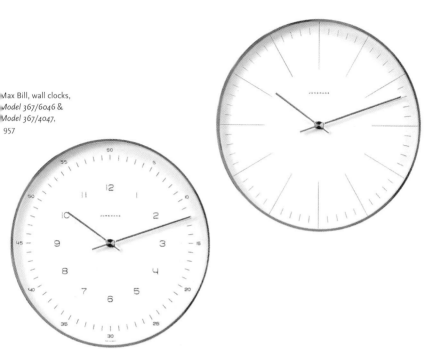

Advertising stand
for Junghans alarm
clocks, 1936

WILHELM KÅGE

BORN 1889 STOCKHOLM, SWEDEN
DIED 1960 STOCKHOLM, SWEDEN

Liljeblå (Blue Lily)
tureen, 1917

Wilhelm Kåge studied painting and graphic design before being recruited by **Gustavsberg** in 1917, following a campaign by the Svenska Slöjdföreningen (Svenska Form) to encourage manufacturers to employ the services of artists, and thereby improve the design of their products. Kåge's subsequent *Liljeblå* (Blue Lily) dinner service (1917), with its folk-inspired decoration, was so affordable that it became known as the "Workers' Service". For the 1930 Stockholm exhibition, which heralded the new "funktionalmen" movement in Sweden, Kåge designed the *Pyro* range that was the first oven-to-table ware produced by Gustavsberg. His later *Praktika* earthenware service (1933), however, was a far more revolutionary design. But while it was extremely practical with stackable, different sized elements, *Praktika* was not a commercial success – the result, no doubt, of its severe utilitarian aesthetic. Design critics nevertheless deemed *Praktika* "the" functionalist service of the era. In 1937 Kåge designed the less austere *Soft Shapes* service. With its subtle organic forms and simple grey-striped pattern, *Soft Shapes* sold extremely well for more than 20 years. While Kåge was also celebrated for his studio **ceramics**, his designs for industrial production were of far greater significance as they predicted the basic functional shapes that have come to dominate large-scale ceramics manufacture.

Praktika ware for
Gustavsberg, 1933

Gino Colombini,
KS 1171 plastic dish
rack with plastic
dishes, 1954

KARTELL

FOUNDED 1949
MILAN, ITALY

Kartell, a **plastics** consumer products manufacturing company, was founded by Giulio Castelli (b. 1920) in 1949. Castelli started from the view that: "The public is willing to accept new forms for machines performing new functions, but when century-old objects like a spoon, a chair, etc. are involved, it's not easy to make new aspects accepted. If men fear novelty, let's give them something even more novel." The first product launched by Kartell was an innovative ski-rack designed by the architect and industrial designer, Roberto Menghi (b. 1920). This was followed by a range of household articles, many of which were designed by **Gino Colombini**. From lemon squeezers and dustpans to washing-up bowls and baby's baths, Kartell transformed every-day objects into sleekly designed polyethylene products that were startlingly forward-looking. The company's design and materials innovations were widely celebrated and its products won numerous Compasso d'Oro awards at Milan Triennale exhibitions. Kartell also produced furniture, including a metal and plastic sectional cupboard system (1956) designed by Gino Colombini and Leonardo Fiori, but it was not until the 1960s that it became widely known in this field, in particu-

Anna Castelli Ferreri,
4870 stacking chair,
1987

lar through its manufacturing of **Marco Zanuso** and **Richard Sapper**'s *Model No. 4999/5* stacking child's chair (1961–1964), **Joe Colombo**'s revolutionary *Model No. 4860 Universale* (1965–1967) which was the first adult-sized fully injection-moulded plastic chair, and the injection-moulded ABS *4953–54–55–56* stacking storage cylinders (1970) designed by Anna Castelli Ferrieri (b. 1920). Kartell also produced lamps by Sergio Asti (b. 1926), Marco Zanuso and **Achille** and **Pier Giacomo Castiglioni**. The company maintained a high profile in the 1980s through its production of furniture designed by **Philippe Starck**, such as the characterful *Dr. Glob* chair (1988). In the 1990s Kartell's translucent *Mobil* drawer unit (1995), designed by Antonio Citterio (b. 1950) and Glen Oliver Löw and manufactured in injection-moulded PMMA, was awarded a Compasso d'Oro. In 1997 the company began producing **Ron Arad**'s highly successful *Book Worm* shelving in extruded and injection-moulded technopolymers. Kartell's in-house design studio, Centrokappa, which is directed by Giulio Castelli's wife, Anna Castelli Ferrieri, has also produced some notable designs, including the *5300, 5312, 5320* system of children's furniture. When he founded Kartell in 1949, Castelli stated that he "intended to arrive at the difficult synthesis between technology and design, between economy and the response to social demand". Having successfully struck this balance over its 50-year history, Kartell today can boast that one of its products is sold somewhere in the world every 30 seconds.

Antonio Citterio,
Mobil drawer
system, 1995

KAWASAKI

FOUNDED 1896
HYOGO, JAPAN

*X600-G2
motorcycle, 1999*

In 1878 Shozo Kawasaki (1837–1912) founded a shipyard, which was incorporated as the Kawasaki Dockyard in 1896. In 1906 the yard diversified and began manufacturing locomotives, freight and passenger carriages. The following year, marine steam turbines were also introduced. In 1918 the company expanded into aircraft, building the first metal Japanese aeroplane, and it was the Kawasaki Aircraft Co. that eventually produced the first Kawasaki motorcycle in 1961. By the time the company was restructured in 1969 to form Kawasaki Inc., it had gained a firm foothold in the American motorcycle market, and by the early 1970s was a major force within the industry. 1973 saw the launch of the legendary 900 cc Z1, which was voted "Machine of the Year" by *Motorcycle News* for four successive years. During the 1970s and 1980s, Kawasaki introduced several off-road models and enjoyed numerous racing triumphs. The company also developed other sports vehicles, most notably the *Jet Ski* launched in 1973, which although not the first personal watercraft, was the most successful. Kawasaki is also known for its ATVs (All Terrain Vehicles), off-road utility vehicles and loaders. The **GK Design Group** has been responsible for many of Kawasaki's product designs.

*Kawasaki Jet Ski,
late 1990s*

Volksempfänger
(People's Radio)
VE301 receiver, 1936

Kleinempfänger,
1938 – a smaller
version of the
Volksempfänger

Walter Maria Kersting studied at the Technische Hochschule in Hanover from 1912 to 1914. He worked as a graphic designer and subsequently as art director for a printing works in Hanover. From 1922 to 1932 he worked as an architect, first in Weimar and then in Cologne, and also designed lighting and furniture. Recognizing the exciting potential of **Bakelite** as a material highly suited to the mass production of radio casings, in 1928 Kersting designed the first version of his well-known plastic-housed radio. His choice of material was inspired, as was his design for the layout of the controls, which he believed should be easy to operate and understand. Inexpensive and reliable, Kersting's radio came to be known as the *Volksempfänger* (People's Radio) – and with sales estimated to have topped 12.5 million by 1939, there was indeed a Kersting radio in virtually every German home. Manufactured by Hagenuh, the radio was produced in several versions, including the smaller *Kleinempfänger* and the model *VE 301*, which commemorated the date (30 January 1933) that Adolf Hilter became Chancellor of Germany.

Kersting's concept of an affordable, standardized radio was subsequently co-opted by the National Socialist government as a means of delivering state propaganda to German households; the fact that the somewhat utilitarian radio was only powerful enough for home broadcasts and could not receive Allied transmissions was simply an added advantage. In the 1950s Kersting went on to design a wide range of products, from telephones and sewing machines to lighting and industrial machinery.

KING – MIRANDA

FOUNDED 1975
MILAN, ITALY

The Milan-based design consultancy King – Miranda
was founded in 1975 by the British designer Perry King
(b. 1938) and the Spanish designer Santiago Miranda (b. 1947). After studying at the Birmingham School of Art, King worked briefly in Britain before
moving to Italy in 1964. He worked for **Olivetti**, for whom he created office
equipment such as the *Valentine* typewriter (1969), which he co-designed
with **Ettore Sottsass**. From 1972 he oversaw the development of the company's **corporate identity**. During this period, King also designed a corporate
identity programme for C. Castelli, dictaphones for Süd-Atlas Werke and
electrical equipment for Praxis.

Santiago Miranda studied at the Escuela de Artes Aplicadas y Oficios Artisti-
cos in Seville, graduating in 1971. He moved to Italy in 1975 and set up in
partnership with King that same year. King – Miranda subsequently designed graphics, furniture, lighting, glassware and power tools for Marcatré,
Disform, **Flos**, Murano, **Black & Decker** and Arteluce among others, and

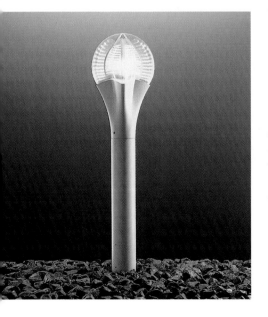

also developed computers and key-
boards for Olivetti. Their *Lucerno*
lighting system (1991) was designed
specifically for the public areas at
Expo '92 in Seville. While King –
Miranda's product designs are primarily informed by technology, they
also possess poetic qualities, as
their *Borealis* bollard light (1996)
for **Louis Poulsen** demonstrates.
The partners believe that: "Design
clearly must comprehend not only
technology and economics, but so-
ciology, too. This leads us to the
conclusion that designers must
have a political or philosophical
model on which to base their work."

KODAK

FOUNDED 1881
ROCHESTER, NEW YORK, USA

George Eastman

At the age of 24, George Eastman (1854–1932) decided to visit Santo Domingo and bought some photographic equipment with which to record his travels. In those days the paraphernalia needed for photography was extensive – an enormous camera and heavy tripod, a dark-room tent and a large assortment of chemicals. In the end Eastman never made the trip, but he became captivated by photography and determined to simplify the process. Having read that British photographers were making and using gelatine emulsions that, when dry, could be exposed at leisure, Eastman made his own using a formula outlined in a magazine. After three years' experimentation, in 1878 he perfected a dry-plate gelatine emulsion. The following year he developed an emulsion-coating machine, which allowed him to begin the

Brownie camera,
c. 1902

Springtime is Kodak time

Advertisement for Kodak cameras, c. 1910

commercial mass production of his dry plates. In 1881 he founded the Eastman Dry Plate Company together with Henry A. Strong (b. 1919). In 1884 the firm introduced Eastman Negative Paper and developed a roll holder for the papers, and in 1885 pioneered the first transparent photographic film. Three years later, the first Kodak camera was launched; costing just $25, it was marketed under the slogan: "You push the button – we do the rest." This revolutionary and easy-to-use camera heralded the advent of amateur photography around the world. In 1889 Eastman marketed the first commercial transparent and flexible film roll, paving the way for the subsequent development of **Thomas Alva Edison**'s motion picture camera (1891). The company was renamed the Eastman Kodak Company in 1892, and in 1895 introduced the *Pocket* Kodak camera, which used roll film. 1900 saw the launch of the famous *Brownie* camera, which was constructed of a pressed cardboard box with a wooden end. Inexpensive and cheap to use, the *Brownie* took perfectly acceptable photographs and made photography accessible to virtually everyone. In 1908 Kodak pioneered the first commercial safety film using a cellulose acetate rather than a flammable cellulose nitrate base. Four years later, it established one of the first industrial research laboratories in America, dedicated to the evolution of new cameras and films. 1923 saw the introduction of 16mm-reversal film, the 16mm *Cine-Kodak* motion picture camera and the *Kodascope* projector, all of which made amateur cinematography possible. 16mm Kodacolour film was launched five years later, together with the first microfilm system for the storage of bank records. Through the 1930s and 1940s, Kodak continued to develop new products, including the first commercially successful colour slide film (1935) and the highly successful *Bantam* camera range, which included the streamlined Kodak *Bantam Special* camera designed by **Walter**

Bantam Colorsnap
camera, 1955–1959

Kodak *Instamatic*
50 camera, 1972

Kenneth Grange,
Kodak *Instamatic*
130, 230 and *330*
cameras, 1975

Dorwin Teague. Its cameras were often developed by outside design consultants in conjunction with the Kodak **Styling** Division and the engineering
department, and grew increasingly compact and easier to use. Thus the Ban-
am *Colorsnap* camera of 1954, with its "dial-the-weather" exposure setting,
simplified colour photography to such as extent that it was nicknamed
"Auntie's Camera". Advances in colour film meant that lightweight snap-
shot cameras were being manufactured by the early 1960s. Kodak *Instamatic*
cameras, which first appeared in 1963, required only very simple exposure
controls – just sunny or cloudy flash settings – and featured easy-to-load
cartridge film. By 1970, more than 50 million *Instamatics* cameras had been
sold, revolutionizing popular photography. By using narrower cartridge film,
Kodak further reduced the size of the *Instamatic* so that it could fit inside a
pocket. Designed by **Kenneth Grange**, the Kodak *Pocket Instamatic* was extremely popular and over 50 million cameras were sold within seven years.
In 1987 Kodak introduced its first one-time-use disposable camera, the Kodak
Fun Saver, and by 1995 over 50 million of these, too, had been sold. 1995 also
saw the launch of the Kodak Digital Camera *40*, the first full-feature digital
camera to retail for less than $1000. Kodak's historic success has been the
result of its "everyman" approach to photography. With film-based photography increasingly becoming a thing of the past, it seems certain that Kodak
will continue to be at the forefront of affordable digital camera technology.

ERWIN KOMENDA

BORN 1904 JAUERN, AUSTRIA
DIED 1966 GERLINGEN, GERMANY

Porsche 356
cabriolet, 1952

One of the most influential automotive stylists ever, Erwin Komenda initially worked for Daimler-Benz as chief designer in the body development section. He later joined the automotive design office of Ferdinand **Porsche** I in Stuttgart, where his first brief was to design a body for Auto Union's world record-breaking *Type A* racing car (1934). Obviously influenced by earlier aerodynamic studies undertaken by **Paul Jaray** and Wunibald Kamm (1893–1966), Komenda later designed a streamlined body for the original **Volkswagen** *Beetle* (1934), which was first produced in 1938. Komenda also undertook his own **aerodynamics** research in the wind tunnels at the **Zeppelin** works in Friedrichshafen and at Kamm's in Stuttgart. His interest in this field was translated into streamlined forms that were designed for maximum efficiency rather than just styled to look beautiful. For the **styling** of Porsche's first car, the *356* (1949), Komenda was asked by Ferdinand "Ferry" Porsche II to ensure that the car's nose was kept as low to the ground as possible. The resulting compact and unified form was extremely influential upon later car design. Komenda designed a number of other models for Porsche, concluding with the great design classic, the *911* (1963–1964).

Porsche *911*, 1964

KOMPAN

FOUNDED 1970
RINGE, DENMARK

M572 Water Tapper,
1988

M581 Sandworks,
1988

Kompan is one of the world's leading manufacturers
of playground equipment and was founded by Tom
Lindhardt Wils and Hans Mogens Frederiksen in 1970.
One of its earliest products was a seat in the shape of
a stylized animal mounted on a rocking spring, called
the *M100/Classic* (1970). Another early product was the *M401 Sprout* wave-
like ladder (1972) – a relatively simple yet ingenious design that required
only a small area for its installation. An improved version of this design was
later introduced with two platforms and a small crawl space at its bottom.
Among Kompan's most successful designs is the *Sandworks M581* (1988).
This multi-activity play centre was designed for children between the ages of
three to eight years old and has numerous features, such as two crane-like
elements which can be manipulated to scoop, pour and lift sand. Its sculptural form and cheerful combination of primary colours encourages interaction and provides play that is both fun and educational. For older children

aged between six and twelve years old, Kompan launched the *Sirius GXY 8011* "jungle-gym"
in 1998. Constructed of galvanized steel and
polyurethane, it has a more sombre aesthetic
than the equipment for younger children, yet
still invites interactive play. Guided by a design philosophy that recognizes play as an important part of life, Kompan enters the world
of play from the child's perspective in order to
work with issues of great importance to the
child: "Play Value, Play Rights, Play Safety and
Play Together". In recognition of its innovative
products, Kompan was awarded a European
Community Design Prize in 1994 and an Industrial Designers Society of America award
in 1999.

Jörg Glasenapp, *3 Mix* hand-blender, 1960

3 Mix 3 7007 Pro hand-blender, 1996

The German domestic appliance manufacturer Krups was founded in 1846 by Robert Krups, initially to produce rotating scale and spring balance weighing machines. Over the decades, the company remained under family control and continued specializing in these types of products. It was not until 1956 that it began to diversify its product line with the introduction of an electric coffee grinder, which it launched under the Onko brand in collaboration with the well-known Bremen-based coffee company, Hag. Within a year, over one million of these coffee mills had been sold. In 1958 Krups began the production of electric motors, and a year later began manufacturing electric hand-mixers. 1960 saw the lauch of the *3 Mix* mixer designed by Jörg Glasenapp, which became an icon of 1960s German design. A year later, the company introduced an electric coffee-maker that was a forerunner of the hugely successful *T8* ma-

Tea Time tea-maker,
late 1990s

hine of 1970. During the 1960s, Krups began producing a wide range of lectrical appliances, including hair dryers, meat slicers, clocks, toasters, gg boilers, juicers, yoghurt makers and razors. In 1983 it launched its first lectric espresso makers. In 1991 Krups was acquired by the Moulinex Group, which sells 45 million products a year (or one product every second). n 1996 new versions of the famous *3 Mix* mixer and *Nespresso Programatic* offee machine were launched. Krups is renowned for household products vhich exemplify the attributes of German design: functionality, logic and uperb build quality. Its commitment "to make good products even better" as led Krups to become one of the leading brands not only in Germany, ut also in other European countries and in the USA.

LAMBORGHINI

FOUNDED 1960
RENAZZO DI CENTO, ITALY

Bertone, Lamborghini
Miura P400, 1966

After the Second World War, Ferruccio Lamborghini (b. 1916) began conver
ing surplus military vehicles for civilian use. In 1949 he founded Lamborgh
ini Tractors to produce high-powered agricultural vehicles, and in 1960 established Lamborghini Bruciatori to manufacture boilers and air-condition-
ers. By now a leading Italian industrialist, he next decided to construct his
own car. The engineer Giotto Bizzarrini subsequently designed a powerful
V-12 engine, while a body was styled by Franco Scaglione, resulting in the
350 GTV (1963). In 1965 Lamborghini began building a new supercar styled
by **Bertone**, the *Miura*, named after a legendary Spanish fighting bull. It had
aggressive modern lines and was the fastest series production car in the
world. In 1971 the Bertone-styled *Countach* was launched at the Geneva Mo
tor Show and caused a sensation. Futuristic, masculine and angular, the
Countach was a missile-like testosterone-driven dream car. Its successor,
the *Diablo* (1987), designed by Marcello Gandini, was more curvaceous yet
retained the strong and menacing character synonymous with the marque.
Lamborghini is now a subsidiary of **Chrysler**.

Bertone, Lamborghini
Countach LP400, 1973

Cover of *Scooter* magazine, August 1961

LAMBRETTA

FOUNDED 1926
ROME, ITALY

Lambretta's origins can be traced back to a steel pipe manufacturing business founded in Rome by Ferdinado Innocenti (1891–1966). In 1933 the company (then known as Innocenti Brothers) began manufacturing a revolutionary scaffolding system designed by its founder. The same year it moved to the Lambratte district of Milan and eventually began manufacturing automotive components. After the Second World War, the company decided to develop an inexpensive means of transport, and so Innocenti, together with the company's director Giuseppe Lauro and the engineer Pierluigi Torre, set about designing a scooter. Taking its name from the local district, the resulting Lambretta *Model A* was launched at the 1947 Paris Motor Show and boasted a maximum speed of around 70 kph. It was followed in the 1950s and 1960s by numerous Lambretta scooters in a wide range of colours, amongst them the *Grand Prix* model launched in 1969, which was styled by **Bertone**. Despite their having become synonymous with youth culture, production of Lambretta scooters ceased in 1971.

Lambretta *LD150* scooter, 1957

Learjet *Model 23*
on its first flight,
7 October 1963.
The jet was prod-
uced from 1964
to 1966.

In the late 1950s, William Lear (1902–1978) founded the
Swiss American Aviation Corporation (SAAC) in Switzer-
land to develop a small corporate jet. In 1963 the operation was moved to
Wichita, Kansas, and in 1964 was renamed the Lear Jet Corporation. The
first compact jet, the Learjet *Model 23*, made its maiden flight that same year
and in 1965 became the first business jet to be put into production. 1966
saw the launch of the Learjet *Model 24*, which circled the globe in an impressive flying time of 50 hours 20 minutes. The Learjet *Model 55* launched in
1983 broke six altitude and "time-to-climb" records in its weight class. In
1985 Learjet's aerospace division built components for the main booster
rocket on **NASA**'s Space Shuttle. The Learjet Corporation was purchased
by **Bombardier** in 1990, and four years later the Learjet 60 (1990) became
the best-selling mid-sized corporate jet in the world. In 1998 the company's
super-sleek eight to nine-passenger Learjet 45 (1992) became the first business aircraft to receive certification from the European Joint Aviation Authorities (JAA).

Learjet *Model 45*,
1992

Duplo figure,
introduced in 1983

LEGO

FOUNDED 1932
BILLUND, DENMARK

In 1932 Ole Kirk Christiansen (d. 1958) founded a workshop to manufacture stepladders, ironing boards and wooden toys. Two years later he came up with a name for both his toys and workshop – Lego, a contraction of the Danish words "leg godt" meaning "play well". During the 1930s and 1940s, the company produced well-built wooden toys which reflected the company motto that "Only the best is good enough". The Second World War accelerated research into new materials and production-methods, including the injection moulding of **plastics**. In 1947 Lego became the first Danish company to invest in this new technology for high-volume plastic manufacture. Two years later it brought out its *Automatic Binding Bricks*, which were forerunners of today's Lego bricks but which were only sold in Denmark. In 1953 these were renamed "Lego Bricks" and a year later the "Lego" name was officially registered. The "stud-and-tube" brick-coupling system was invented and patented in 1958. This new means of connecting the bricks offered greater stability for models and opened up unlimited building possibilities. Over the succeeding years, the range was expanded to include sloping roof bricks (1958), wheels (1961), the Lego family (1974) and mini figures (1978).

Lego bricks, first launched in 1947

For pre-school children, Lego developed the larger and easier to use brick systems *Duplo* (1967) and *Primo* (1995). Lego is a unique and hugely successful toy – between 1949 and 1998, approximately 203 billion Lego elements were moulded. While appealing to both genders and to a wide range of ages, Lego provides stimulating play that teaches children the fundamental principles of construction and, therefore, design.

Advertisement for Oskar Barnack's revolutionary *Leica I* camera, 1930

Leica's origins can be traced back to 1849, when a young mathematician, Carl Kellner, founded an "Optical Institute" to develop and market lenses and microscopes. In 1865 the mechanic Ernst Leitz became a partner in the firm. He later took over the company and expanded it under his own name. Around 1914 **Oskar Barnack** perfected the world's first 35mm camera, the *Ur-Leica*, but its production was hampered by the outbreak of the First World War. After his father's death in 1920, Ernst Leitz II decided to begin mass-producing Barnack's design in 1924 (although it was not marketed until the following year). Known as the *Leica I* – its title being derived from LEItz CAmera – it was the first commercially successful 35mm camera and proceeded to revolutionize the art of still photography. The Leitz works earned widespread regard for its precision-engineered cameras and the exceptional quality of its lenses. In 1930 the company was first to pioneer a system of interchangeable accessories and

Leica camera, 1914

Leica camera, 1955

screw-fit lenses that were compatible between camera models. The *Leica II*, launched in 1932, incorporated a built-in rangefinder that enabled faster and more accurate focusing. The *Leica M3* (1954) introduced a bayonet fitting that allowed for even quicker lens changes. In response to Japanese competition, Leica began producing single lens reflex (SLR) cameras from 1964. Throughout its history, Leica has consistently stood for design and manufacturing excellence. Today, the company continues to produce precision-engineered cameras and high optical performance lenses which rank among the very best.

Kelly Johnson
standing in front
of a *U-2*, 1955

LOCKHEED MARTIN

LOCKHEED	MARTIN
FOUNDED 1916	FOUNDED 1912
SANTA BARBARA,	LOS ANGELES,
CALIFORNIA, USA	CALIFORNIA, USA

Glenn Martin, an early aviation pioneer, flew his first aeroplane in 1909 and three years later performed the world's first major over-ocean flight in his own hydroplane. That same year he established his own aircraft manufacturing company in Los Angeles. Among the firm's earliest contributors were the renowned design engineers James **McDonnell**, Donald **Douglas** and Chance Vought. A similar venture was set up in Santa Barbara in 1916 by the brothers Allan and Malcolm Loughead, who had developed and flown their own seaplane in 1913. In 1918 their Loughead Aircraft Manufacturing Company (renamed the Lockheed Aircraft Corporation in 1926) produced the *F-1* Flying Boat, while Glenn Martin constructed the first twin-engine bomber, the *MB-2*, that same year. During the 1920s and 1930s, both companies enjoyed high profiles – Amelia Earhart became the first woman to fly the Atlantic solo in 1932, piloting a

Lockheed *F-104 Starfighter*, 1958

SR-71 Blackbird, 1964

F-117A Stealth
Fighter, c. 1988

Lockheed *Vega* designed by John Northrop (1895–1981), while Pan-Am used
the Martin *M-130* (nicknamed the "China Clipper") to make the first scheduled trans-Pacific flight in 1935. During the war, Martin's *B-26 Marauder*
(1940) became renowned for its survivability rate, while Lockheed's *XP-80
Shooting Star* (1944) became the first American jet fighter. The *XP-80* was
developed at Lockheed's experimental development facility, the famous
"Skunk Works" established in 1943 by the company's legendary chief engineer, Clarence "Kelly" Johnson (1910–1990). The Skunk Works (now officially known as the Lockheed Martin Advanced Development Programme)
was, according to Johnson, "a concentration of a few good people solving

F-22 Raptor
Advanced Tactical
Fighter, 1991

problems far in advance – and at a fraction of the cost – by applying the simplest, most straightforward methods possible to develop and produce new products". It was set up to develop a fighter that could compete with the **Messerschmitt** *Me 262* jet aircraft (1943). The resulting *XP-80* was the first American aircraft to exceed 500 mph. Located in Palmdale, California, Johnson's Skunk Works was responsible for some of the greatest advances in military aviation, including the *F-104 Starfighter* (1958), the first operational aircraft to sustain a speed of Mach 2; the *U-2* Spy Plane (1955), which is still in operation; the *SR-71 Blackbird* (1964), which is still the world's fastest and highest flying aeroplane with a maximum speed of Mach 3.5 and an operational ceiling of over 100,000 feet; the *F-117* Stealth Fighter (c. 1988), the first operational aircraft with a cross-section invisible to radar; and the *F-22 Raptor* (1991) designed to ensure American air-dominance in the 21st century with its thrust vectoring system. Apart developing aircraft, Lockheed also introduced the first FAA-approved "Black Box" flight data recorder in 1958. The Martin Company also specialized in military development programmes – notably the *Pershing* Missile project begun in 1958 – and worked extensively on **NASA** space programmes. In 1995 a "merger of equals" took place between Lockheed and Martin, creating one of the world's largest aerospace and defence companies.

→ Design for *L300*
refrigerator for
Electrolux, 1939

RAYMOND LOEWY

BORN 1893 PARIS, FRANCE
DIED 1986 MONACO

Raymond Loewy was the greatest pioneer of industrial design consulting
and is still remembered for the humorous adage that summed up his approach to design – "Never leave well enough alone". At the age of 15, Loewy
designed, built and flew a toy model aeroplane that won the then-famous
James Gordon Bennett Cup. Around this time, he also designed and patented a model plane powered by rubber bands. Having subsequently sold the
rights to this toy, called the *Monoplan Ayrel*, to a company that marketed it
across France, Loewy learnt that "design could be fun and profitable". With
the money he earned from this venture, Loewy was able to study at the Uni-
versité de Paris and later at the École de Lanneau, from where he received
an engineering degree in 1918. During the First World War he served in the
French army as a second lieutenant. After his demobilization, he travelled

First redesign of the
Gestetner duplicator
machine, 1929

Coldspot refrigerator
for Sears, 1934

to America in 1919. Arriving in New York, he was utterly amazed by what he later described as "the chasm between the excellent quality of American production and its gross appearance, clumsiness, bulk and noise". He initially found work as a window dresser for Macy's, Saks Fifth Avenue and Bonvit Teller. From around 1923 until 1928, he enjoyed a "pleasant but superficial career" as a fashion illustrator for *Vogue*, *Harper's Bazaar* and *Vanity Fair* among others. In 1923 he also designed the trademark for the Neiman Marcus department store. Loewy eventually left the fashion world to set up his own industrial design office in New York in 1929. Always a self-promoter, Loewy had a card printed that read: "Between two products equal in price, function and quality, the better looking will outsell the other", and sent it to everyone he knew. Soon afterwards he received his first brief, namely to redesign the casing of Sigmund Gestetner's duplicator machine. For this project, he used modelling clay to create a sleek form – a technique he later employed to great effect for his automotive designs. The resulting machine not only looked better but, thanks to its simplified form, was also easier to manufacture and maintain than earlier models. In 1932 Loewy designed his first car, the *Hupmobile*, which was less boxy than existing automobiles. His improved, tapering model of 1934 featured innovative integrated headlamps and predicted the streamlined forms for which he later became so famous. That same year, Loewy also designed the *Coldspot* refrigerator for **Sears Roebuck**, which was the first domestic appliance to be marketed on the

←Design for floor-polisher for Electrolux, 1939

Design for vacuum cleaner for Electrolux, 1939 – this model was never put into production

strength of its aesthetic appeal – contemporary advertisements invited consumers to "Study its Beauty". The *Coldspot* was also alluring to customers because its improved design had reduced its cost of manufacture, and this was reflected in its competitive retail price.

Also in 1934, the Metropolitan Museum of Art in New York displayed a mock up of Loewy's glamorous design office, thus illuminating his rise in status from product designer to celebrity industrial design consultant. From 1935, Loewy was commissioned to re-plan several large department stores, including Saks Fifth Avenue. During this period he was also designing aerodynamic locomotives, such as the *K4S* (1934), the *GG-1* (1934) and the *T-1* (1937), and in 1937 published a book entitled *The New Vision Locomotive*. In 1946 he famously remodelled coaches for Greyhound and in 1947 designed his innovative *Champion* car for Studebaker – a precursor of his later European-styled *Avanti* car, also for Studebaker. Not only could **streamlining** make transportation look more appealing, it also often improved performance and efficiency because of the effect of **aerodynamics**. Loewy regarded the design of a casing or sheath as an opportunity to allow a machine or a product to express itself. Unlike so many Modernists who allowed form to be completely dictated by function, Loewy balanced engineering criteria with aesthetic concerns in order to achieve what he believed to be the optimal solution. Loewy also became celebrated for his **corporate identity** work, and in particular for his repackaging of Lucky Strike cigarettes (1942). His presti-

Greyhound
Scenicruiser bus,
1954

Interior of a
Greyhound bus,
c. 1946

Logo for Exxon, 1966

→Logo for Shell, 1967

Redesign of Lucky Strike packaging for the American Tobacco Company, 1942

→Redesign of Coca-Cola dispenser *Dole Deluxe*, 1947

gious clientele included Coca-Cola, Pepsodent, the National Biscuit Company, British Petroleum, Exxon and Shell. By 1939 his design office had branches in Chicago, São Paulo, South Bend and London. In 1944 he founded Raymond Loewy Associates with four other designers. In 1949 Loewy became the first designer to be featured on the front cover of *Time* magazine; his picture accompanied by the memorable caption: "He streamlines the sales curve." That same year, Loewy expanded his operations and founded the Raymond Loewy Corporation to undertake architectural projects. During the 1960s and 1970s, Loewy worked as a design consultant to the United States Government, most famously redesigning Air Force One for John F. Kennedy and designing interiors for **NASA**'s *Skylab* (1969–1972). Always putting the consumer first, Loewy's MAYA (Most Advanced, Yet Acceptable) design credo was crucial to the success of his products. Undoubtedly the greatest pioneer of streamlining in the 20th century, Raymond Loewy clearly demonstrated that the success of a product depends as much on aesthetics as it does on function. Few design consultants have been as influential or prolific as Loewy, nor as misunderstood; for although he was a **styling** genius, he also skilfully improved the design of many products and pioneered numerous design innovations. Signficantly, Raymond Loewy glamorized the practice of design and in so doing raised the status of industrial design as a whole.

→Stowable hammock for Habitablity Study for Saturn Five Space Station, c. 1972

Exploded view of
Skylab for NASA,
c. 1970

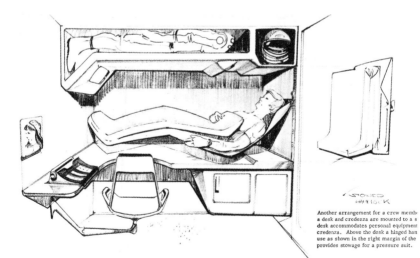

STOWED
HAMMOCK

Another arrangement for a crew membe
a desk and credenza are mounted to a s
desk accommodates personal equipment
credenza. Above the desk a hinged han
use as shown in the right margin of the
provides stowage for a pressure suit.

SATURN FIVE SPACE STATION
HABITABILITY STUDY
RAYMOND LOEWY/WILLIAM SNAITH

HEINRICH LÖFFELHARDT

BORN 1901 HEILBRONN, GERMANY
DIED 1979 STUTTGART, GERMANY

Heinrich Löffelhardt trained as a silversmith at P. Bruckmann & Söhne in Heilbronn before moving to Berlin, where he studied sculpture under Georg Kolbe from 1924 to 1928. In 1934 he designed a service of canteen china for the "Schönheit der Arbeit" (Beauty of Work) council. Having moved to Stuttgart, from 1937 to 1941 he worked under **Wilhelm Wagenfeld** for the Vereinigte Lausitzer Glaswerke, Weißwasser. After the war, Wagenfeld arranged for Löffelhardt to be appointed director of the design departement of the Landesgewerbeamt (District Trade & Craft Offices) in Stuttgart. In 1950 Löffelhardt designed glassware and ceramics for Gral-Glaswerkstätten and others. From 1952 to 1971 he was artistic director of the **Arzberg** and Schönwald porcelain factories. His elegant *Form 2000* service (1954) was inspired by **Hermann Gretsch**'s earlier design for Arzberg, while his *Form 2050* service (1959) was awarded a grand prize at the 1960 Milan Triennale. Löffelhardt also designed utilitarian pressed glassware for the Jenaer Glaswerke Schott & Gen. in Mainz and the Vereinigte Farbenglaswerke in Zwiesel, which by 1978 had sold 240 million of his *Neckar* wineglasses.

Form 2000 service
for Arzberg, 1954

LONDON TRANSPORT
STANDARD "UNDERGROUND" BULLSEYE DESIGN

LONDON
UNDERGROUND
TRANSPORT

All reproductions of this symbol must be strictly proportionate in every detail to the design shown hereon.

D^a N^o A.N. 7000

LONDON TRANSPORT

FOUNDED 1933
LONDON, ENGLAND

Edward Johnston, Standard Underground Bullseye design, post-1918

Enid Marx, *Shield* upholstery textile designed for London Transport, 1948

Renowned worldwide for its distinctive **corporate identity**, London Transport's integrated design programme was pioneered by Frank Pick (1878–1941), who joined the Underground Group in 1906 and became its commercial manager in 1912. Pick believed that public transport should be both functional and aesthetically pleasing, and so began instituting a unified and standardized corporate identity that was based on a "fitness for purpose" ethos. In 1908 he commissioned the typographer Edward Johnston (1872–1944) to design a new typeface for the Underground Group, which resulted in an easy-to-read sans serif font. Johnston also redesigned the Underground's eye-catching "roundel" symbol, which had first appeared in 1908. Pick also commissioned posters from well-known artists and engaged the architect Charles Holden (1875–1960) to modernize existing stations and design new ones. In total, Holden designed over 50 London Underground stations, which were influenced by Modern Movement architecture in Scandinavia and the Netherlands and represented a high point in English rational design. He also designed shelters, signs, lamp-standards and platforms for LT. In 1932 Pick became president of the Design and Industries Association and in 1933 Vice Chairman and Chief Executive Officer of the newly formed London Transport company, which controlled London's bus, tram and underground train networks. Pick continued to pursue his "house-style" design mission for London's transport systems, and the same year invited the engineering draftsman, Henry Beck (1903–1974), to redesign the London Underground map. Ingeniously, Beck used a diagrammatic approach for his new, highly legible map, which showed the spatial relationship rather than the geographical distance between the stations. In 1937 Enid Marx (1902–1993) designed a range of hard-wearing upholstery textiles with Modern patterns for London Transport, which further enhanced the visual coherence of its trains, trams and buses. 1938 saw the design of new Underground rolling stock which

Design for approved
colour scheme for
STL-type bus, 1940s

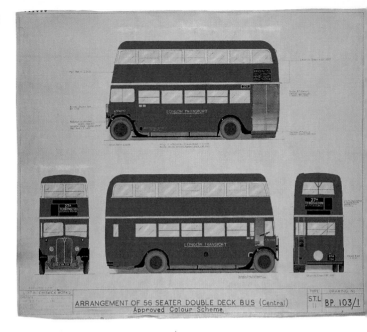

also fitted Beck's "form and function" criteria. The best-loved and most famous LT design, however, must be the red *Routemaster* bus designed by
Douglas Scott in 1954. While red double-decker buses had been used in
London from the 1930s, it was the *Routemaster* that came to symbolize the
capital. This was LT's last in-house bus design and while today it no longer
designs its own vehicles, LT remains committed to integrated design.

STL-type bus, 1940s

Lotus *Elise*, 1999

LOTUS

FOUNDED 1952
HORNSEY, ENGLAND

While studying engineering at University College, London, Colin Chapman (1928–1982) established a second-hand car business and began modifying and improving cars prior to selling them. After this business was disbanded, Chapman transfigured a 1937 *Austin 7* into the first Lotus, the *Mark 1*, with the addition of a new body, a modified engine and improved suspension. He subsequently designed a number of highly innovative racing cars, including the Lotus *Mark 6* (1952), which had a light yet robust multi-tubular body frame. Chapman's success on the track led him to set up the Lotus Engineering Company in 1952, in order to commercialize some of the technology he was pioneering. Between 1960 and 1981 the Lotus Team dominated Formula 1 racing. The Lotus *25* (1962) was the first race car to have a stressed monocoque shell, and the later Lotus *33* (1965) became one of the most successful racing cars ever. The revolutionary Lotus *43* (1966–1967) had a truncated monocoque chassis with the engine and transaxle carrying the rear suspension – a design that every F1 race car has since followed. Lotus has also produced several classic road cars, including the Lotus *Elite* (1957), *Elan* (1962), *Esprit* (1976 – styled by **Giorgetto Giugiaro**) and *Elise* (1999). It is true to say that every single car on the road and racetrack today owes some part of its design and engineering to the genius of Colin Chapman.

Lotus *33* at the
French Grand Prix,
1965

Poster showing a
PH lamp designed
by Poul Henningsen,
late 1920s

Poul Henningsen,
PH4–3 table lamp,
1966

The well-known Danish lighting company Louis Poulsen & Co evolved from a wholesale cork-trading and general ironmongery business established by Ludwig R. Poulsen, which from 1892 had an "electrical department" that supplied carbon arc lamps. Louis Poulsen, the founder's nephew, headed the business from 1906 and two years later took over premises in Nyhavn. The turning-point for the company occurred in 1911, when Sosphus Kaastrup-Olsen acquired a 50% stake in Poulsen's electrical division. Olsen was a true believer in **electric power** as well as a political radical. In 1917 Louis Poulsen retired and Olsen became the sole proprietor of the company. Almost immediately, he began benefiting from the increase in demand brough about by the end of the First World War and the return of the long over-looked Southern Jutland region to Denmark. By 1918, Louis Poulsen & Co. had a considerable turnover of 5 million kronen, of which 600,000 kronen was clear profit. Although the 1920s were a financially turbulent period for the company, it gaine valuable market share after Olsen came into contact with the lighting designer Poul Henningsen (1894–1967). Henningsen produced lighting designs for the company from 1924 and in 1925 some of his designs won a gold medal at the Paris "Exposition Internationale des Arts Décoratifs et Industriels Modernes" A year later, he designed the famous *PH* series of lamps, which remain in production today. The *PH* lamps were startlingly modern for their day and appealed in particular to the avant-garde – Alvar Aalto (1898–1976), for example, incorporated

them into his design for the Auditorium at Turku in Finland. These "shadow-free" lamps featured three highly effective light-diffusing shades, and were used in large public buildings such as Cologne railway station (1929) as well as in smaller domestic and commercial premises. By 1931, 30,000 PH lamps had been sold around the world. Louis Poulsen & Co. managed to survive the Second World War and continued to expand during the post-war years, establishing international subsidiaries and commissioning other well-known designers, such as **Arne Jacobsen**, **King – Miranda** and Alfred Homann (b. 1942), to design domestic and contract lighting products for the company. By 1997 Louis Poulsen & Co. had a workforce of 1,090 and wholly owned subsidiaries in Germany, Sweden, Norway, Finland, Britain, Holland, France, Switzerland, Australia, Japan and the USA.

Louis Poulsen,
Enamel pendant
lamp, 1930

Alfred Homann,
Nyhavn pendant
lamp (maxi), 1933

ROSS LOVEGROVE

BORN 1958
CARDIFF, WALES

→ *Pod* lights for
Luceplan, 1996–1997

Ross Lovegrove studied industrial design at Manchester Polytechnic and later trained at the Royal College of Art, London, receiving an MA in 1983. Between 1983 and 1984 he worked for the industrial design consultancy **Frogdesign** in Altensteig, where he was assigned to projects that included the design of Walkmans for **Sony** and computers for **Apple**. From 1984 to 1987 he worked as an in-house designer for Knoll International in Paris, for whom he created the successful *Alessandri* office system. Having returned to Britain in 1986, Lovegrove established a design partnership with Julian Brown (b. 1955) and designed the highly successful *Basic* thermos flask for Alfi Zitzmann (1990). After the partnership was dissolved in 1990, Lovegrove established his own London-based design office, Studio X. In addition to furniture for such clients as **Kartell**, Ceccotti, Cappellini, Fasem, Moroso, Driade and Frighetto, Lovegrove's designs include an innovative range of lighting for Luceplan, featuring the solar-powered *Solar Bud* (1996–1997). He has also designed luxury leather goods for Connolly Leather (1994), a kettle for Hackman (1999), *Handy* kitchen tools and knives for Fratelli Guzzini (1999), the *Fluidium* lava lamp for Mathmos (1999) and accessories for **TAG Heuer** (2000). Lovegrove has also worked on a proposal for **product architecture** involving a solar-powered "nomadic" structure known as a *Solar Seed*. By "seeking perfection in everyday tools for living", Lovegrove combines sculptural and ergonomic forms with state-of-the-art materials and industrial techniques, to create designs that are at once emotionally seductive and technologically persuasive. Inspired by the natural world and driven by ecological concerns, Lovegrove strives to achieve organic unity of design wherever possible while working within an essentialist formal vocabulary. Very much more a leader and risk-taker than a follower in industrial design, Lovegrove pushes materials, technology, function and even aesthetics to the limits – often breaking new ground – in his pursuit of new and better performing forms that touch the souls of the people who use them. Lovegrove's holistically derived and highly considered products reveal the direction of design for the 21st century.

Ammonite communicator (study project) for Apple Computer, 1995–1996

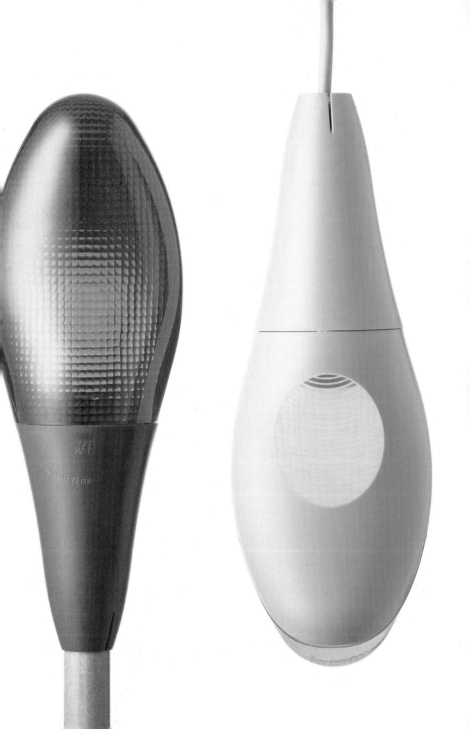

DIETRICH LUBS

BORN 1938
BERLIN, GERMANY

Dietrich Lubs studied shipbuilding and later trained as a designer at **Braun**. In 1962 he began working for the company's design department, headed by **Dieter Rams**. From 1971 he has been responsible for product graphics and has also been active as a product designer. He has designed numerous pocket calculators, alarm clocks and watches for Braun, all of which are distinguished by geometric forms and functional layouts. The majority of his clock designs have matt black plastic bodies, which allow the white and yellow numerals and hands to stand out so that they can be easily read. Since 1995 Lubs has been deputy director of the Braun Design Department. His many product designs are typical of the rational language of design pioneered by Rams at Braun, while also being characteristic of the attributes generally associated with German industrial design – functionalism, logic and high quality engineering.

Pocket calculator,
variant of *ET 44*
co-designed with
Dieter Rams, 1978

Quartz clock,
c. 1990

ABR 314 df time
control digital radio,
1997

MAG INSTRUMENTS

FOUNDED 1974
LOS ANGELES, CALIFORNIA, USA

Anthony Maglica

Born in New York during the Depression, Anthony Maglica was brought up in Europe, where he received machinist training before returning to the United States in 1950. Five years later, he invested $125 of his savings in setting up a small "job shop" in Los Angeles. With his well-honed skills as an experimental machinist, coupled with his unyielding perseverance, Maglica soon gained an excellent reputation for the industrial, military and aerospace components that he produced. During the 1970s he turned his attention to flashlights, which at that stage were generally considered poor-quality items, frequently thrown away because of their unreliability. Recognizing a market opportunity for a more durable and better-performing flashlight, in 1974 Maglica founded Mag Instruments and went on to produce a machined **aluminium** flashlight called the *Mag-Lite* (1978). Designed to be an essential tool for police officers and fire fighters, professions such as these were soon insisting upon the distinctive *Mag-Lite* flashlight as standard issue. Consumers also began buying into the design's superior reliability. In 1984 a smaller flashlight was introduced, the *Mini Maglite*, and in 1987 an even smaller version was launched, followed a year later by a micro-flashlight that was ideal for use on key chains. Like the **Airstream** trailer

2-cell AA Mini Maglite flashlight, 1984

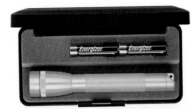

and the Zippo lighter, the water and shock-resistant *Mag-Lite*, with its cam-driven beam that can be transformed from spot to flood, is considered a classic example of American design.

3-cell D Black Mag-Lite flashlight, 1978

Salad bowls and
salad servers for
Stelton, 1986

ERIK MAGNUSSEN

BORN 1940
COPENHAGEN, DENMARK

Erik Magnussen trained as a ceramicist at the Kunst-handvaerkerskolen in Copenhagen until 1960. From 1962 he worked as a designer for the porcelain manufacturer Bing & Grøndahl, designing simple geometric tableware such as his *Form 25* service of 1965. The eleven pieces that made up this service could be stacked and employed a self-insulating double-wall construction. In 1967 Magnussen was awarded the prestigious Lunning Prize for his ceramic designs, which combined artistic sensibility with an understanding of industrial production techniques. The following year, he designed the **tubular metal** and canvas folding *Z-chair* for Torben Ørskov & Co., and subsequently created a range of school furniture in collaboration with Per Kragh-Müller. In 1969 he designed an innovative teapot, featuring a handle that was incorporated into the main body of the vessel. From 1976 Magnussen worked for **Stelton**, for whom he produced a number of well-known designs, including the classic *Thermal Carafe* (1976). From 1978 he also designed hollow-ware for Georg Jensen. Magnussen's designs from the 1980s and 1990s include the *301* kerosene lamp (1988) for Harlang & Dreyer, a clock and barometer for Georg Christensen, and a table and chair (1989) for Paustian, the frame of which comprised one continuous piece of bent tubular metal.

Thermal Carafe set,
1976

GERHARD MARCKS

BORN 1889 BERLIN, GERMANY
DIED 1981 BURGBROHL, GERMANY

Gerhard Marcks started teaching himself to sculpt in 1906, and from 1908 to 1912 trained in the Berlin workshop of sculptor Richard Scheibe (1879–1964). In 1914 his sculpture was featured in the entrance of **Walter Gropius'** installation at the **Deutscher Werkbund** exhibition in Cologne. Marcks subsequently joined the Novembergruppe, a radical art association founded in Berlin in 1918. After his military service, Marcks taught at the Berlin School of Applied Arts from 1918 to 1919. In 1919 he became a member of the Arbeitsrat für Kunst (Working Council for Art) in Berlin and also began teaching at the Weimar **Bauhaus**. In 1920 Gropius appointed him "Formmeister" (artistic director) of the Bauhaus' newly established **ceramics** workshop at Dornburg, some 30 kilometres from Weimar. His own style nevertheless remained relatively ornamental throughout his tenure. It was only in 1928,

Sintrax coffee-maker and stand for Jenaer Glaswerke, Schott & Gen., c. 1928

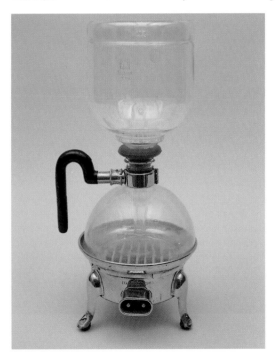

when he became director of the **Burg Giebichenstein** School of Applied Arts in Halle, that he shifted towards an increasingly rational style, one better suited to industrial manufacture. That year, he designed a prototype of his *Sintrax* coffee-maker, which was subsequently manufactured by the Schott & Gen. glassworks in Jena. This quintessentially Modern utilitarian design, which was constructed of heat-resistant glass, plastic, wood and metal and intended for large-scale mass production, was utterly in keeping with the "Form ohne Ornament" (Form without Ornament) ethos that had been promoted earlier by the Deutscher Werkbund. In 1930 **Wilhelm Wagenfeld** designed a new straight handle for Marck's *Sintrax* coffee-maker, giving it an

even more rational appearance. Marcks went on to design the *Tiergarten* tea
set (1932) for the Staatliche Porzellanmanufaktur (formerly the Königliche
Porzellanmanufaktur and still known as KPM) in Berlin. While rationally conceived with a number of interesting ergonomic features, the *Tiergarten* tea
set was never put into production. In 1933 Marcks was dismissed from his
post in Halle by the National Socialists and subsequently moved to Nieha-
gen, near Wustrow in the Mecklenburg region. From 1936 he attempted to
work as a freelance artist in Berlin but was banned from exhibiting. Marcks
suffered further persecution when his studio in the Nikolassee suburb of
Berlin was destroyed in 1943. After the Second World War, he taught sculpture at the Hamburg Landeskunstschule from 1946 to 1950 and then moved
to Cologne in 1950, where he worked as an independent artist. The Gerhard
Marcks Foundation was established in Bremen in 1971 and in 1987 a retro-
spective of his work was held in Bremen, Cologne and Berlin.

Guglielmo Marconi was educated in Bedford, England and Florence and later spent two years at a technical school, the Leghorn Lyceum, in Livorno. While there, he studied physics and undoubtedly investigated the work of scientific pioneers such as James Clerk Maxwell (1831–1879), Heinrich Hertz (1857–1894) and Sir Oliver Lodge (1851–1940) – all of whom had conducted research into either electro-magnetism or radio receivers. He later attended lectures at the University of Bologna by special arrangement and from 1894 began conducting his own experiments. Using relatively primitive apparatus Marconi was able to demonstrate that, by using an aerial, the range of a wireless signalling system could be extended by up to 2.4 km. He also discovered that if reflectors were attached to the aerial, the radiated electrical energy could be concentrated into a beam. Convinced that his discoveries would lead to a new system of wireless communications, but unable to generate any commercial interest in them in Italy, he travelled to London where he filed his first patent in 1896. Between 1896 and 1897, he conducted a series of successful demonstrations, some of which involved balloons and kites to increase the height of his aerials and thereby their signalling range. To exploit his invention, he established his own company (renamed Marconi's Wireless Telegraph Company in 1900) and on Christmas Day 1898 transmitted ship-to-shore telegraph signals over a 12-mile stretch between the moored East Goodwin Lightship and the South Foreland Lighthouse. In 1899 the lightship was involved in a collision with a steamer and the use of Marconi's wireless telegraph helped avert the

Marconi's first transmitter, 1897

Marconi ribbon
microphone, used
by the BBC from
1934 to 1959

loss of life. This incident did much to publicize the life-saving potential of his invention. The same year, the first wireless telegraph messages were sent across the English Channel from South Foreland to Wimereux near Boulogne. In 1900 Marconi went on to file his famous patent for "Improvements in Apparatus for Wireless Telegraphy". Although several eminent mathematicians believed that the curvature of the earth would limit the signalling range of wireless telegraphs to between 161 and 322 km, in 1901 Marconi succeeded in transmitting the first trans-Atlantic signal from St. John's in Newfoundland to Poldhu in Cornwall, thus heralding the beginning of global communications. Over the following years, Marconi continued increasing both the range of his wireless transmissions and the quality of their reception and received numerous accolades for his work, including the Nobel Prize for Physics in 1909. In 1920, the first advertised public broadcast – a song recital by Dame Nellie Melba – was transmitted from the Marconi works in Chelmsford. Two years later, Marconi and five other companies together established the British Broadcasting Company (BBC) and began broadcasting from London. In 1937 the company received an order from the British Government for "Chain Home" stations – the first air defence radar network established in Britain. Marconi subsequently became a world leader in defence radar systems, amalgamating with the English General Electric Company in 1968 and later merging with British Aerospace to form BAE Systems, the second largest defence contractor in the world. Marconi was the most important pioneer of radio broadcasting and his revolutionary communications designs quite literally changed the world.

ENZO MARI

BORN 1932
NOVARA, ITALY

Between 1952 and 1956, Enzo Mari trained at the Accademia di Belle Arte di Brera, Milan. While there, he published his first article on art and after his studies, became involved in the design of books and games. In 1957, he began working for **Danese** and one of his first projects was a simple yet ingenious toy – a child's puzzle made up of sixteen different stackable wooden animals (1957). In 1959, he began experimenting with **plastics** and these researches resulted in a number of products that were manufactured by Danese, including a PVC cylindrical umbrella stand (1962) and the *Pago-Pago* vase (1969) made of ABS. He joined the radical Nuove Tendenze movement in 1963 and taught at the Scuola Umanitaria, Milan. In 1970, his radical theories were published in *Funzione della ricerca estetica* and in 1974, he exhibited a number of "tecnica povera" (poor technology) furniture pieces. Mari has also designed several notable chairs including the wire framed *Sof Sof* (1971) for Driade, the *Box* (1975–1976), a knock-down design that was sold in a box, for Anonima Castelli and the elegant *Tonietta* (1980–1985) for Zanotta, which was awarded a Compasso d'Oro in 1987. As a lead-

EM02 colander for
Alessi, 1997

ing design theorist, Mari was president of the ADI (Associazione per il Disegno) from 1976 to 1979 and has taught in Milan, Rome, Parma and Carrara. Mari's most recent designs for **Alessi** such as the *EM02* colander (1997) and the *Titanic* lemon squeezer (2000) reveal his continuing search for new forms. Alberto Alessi has stated, "I am truly glad that Enzo Mari continues to exist and to work, a courageous and irreducible paranoic in a world … liable to become too meekly commercial."

Koichi Takahashi, MS4/M40 video camera for Panasonic, 1995

Konosuke Matsushita (1894–1989) founded the Matsushita Electric Housewares Manufacturing Works to produce an adapter socket that he had designed. He later developed an improved two-way socket that was better than any other on the market, and was soon flooded with orders. In 1922 he built a factory in Osaka and a year later developed a revolutionary bullet-shaped, battery-powered bicycle lamp that was marketed under the "National" name. In 1931 his factory also started manufacturing radio and dry cell batteries. Konosuke Matsushita was an industrialist with a social mission who believed that mass production would contribute to "the growth of human civilization". In 1935 the company began researching television technology and was incorporated as the Matsushita Electric Industrial Co. During the war, MEI constructed ships and aeroplanes for the Japanese military and was subsequently branded a *zaibatsu* company by the Allied powers. By 1950, however, such restrictions had been lifted and in 1951 Konosuke Matsushita travelled to America for the first time. MEI produced its first washing machine that same year, followed by its first black-and-white television (1952), refrigerator (1953), tape recorder (1958), colour television (1960) and microwave oven (1963). Today, Matsushita is the world's largest manufacturer of consumer electrical products, which it markets under the brand names of Panasonic, Quasar, Technics, Victor and JVC. The Matsushita concept of product design "incorporates both function and beauty, maximizing the interface between user and machine to enhance the comfort and convenience of people's lives". The company's basic principles of product design are: fun and ease of use, innovation, universality, environmental responsibility and facilitative of new lifestyles.

Model R-72S wristradios, 1969

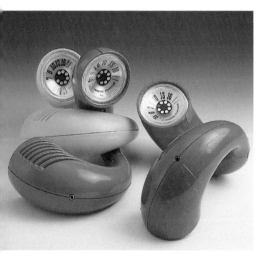

CYRUS HALL McCORMICK

BORN 1809 ROCKBRIDGE COUNTY, VIRGINIA, USA
DIED 1884 CHICAGO, ILLINOIS, USA

Cyrus Hall McCormick was the eldest son of Robert McCormick, a farmer, blacksmith and inventor. While the elder McCormick had developed several agricultural implements, he failed in his attempts to construct a viable reaping machine. At the age of 22, however, Cyrus succeeded in designing and building the first practical mechanical reaper, which he patented in 1834. This pivotal invention could cut a swath of wheat over 70 yards long in just over a minute. By providing an inexpensive means of gathering crops, the reaper helped turn the West into a "bountiful wheatland" and transformed the North's economy during the American Civil War, which contributed to the Union's victory over the Confederacy. It was not until the mid-1840s, however, that the reaper was perfected and fully commercialized. By the 1850s, McCormick's reaper was being mass-produced and sold throughout the United States. In 1851 it was exhibited at the **Great Exhibition** in London, where it was described by the *Times* newspaper as a "cross between an Astley chariot, a wheelbarrow and a flying machine". By 1856, McCormick was selling over 4,000 reapers per annum and in the process, revolutionized agricultural production.

Engraving showing a horse-drawn harvester, originally developed by Cyrus Hall McCormick in c. 1831

Douglas DC-3, 1935

McDONNELL DOUGLAS

FOUNDED 1967
ST. LOUIS, MISSOURI, USA

Douglas DC-3
operated by KLM
airlines, late 1930s

The McDonnell Douglas Corporation was formed in 1967 as a result of a merger between the McDonnell Aircraft Corporation, founded by James S. McDonnell (1899–1980) in St. Louis in 1939, and the Douglas Aircraft Company, founded by Donald W. Douglas (1892–1981) in Santa Monica in 1920. As a civil engineering assistant at the Massachusetts Institute of Technology's department of **aerodynamics**, Douglas helped develop one of the first wind tunnels for the testing of aircraft (1914–1915). In 1920 he founded his company and the same year designed the *Cloudster*, the first aerodynamically streamlined aeroplane and the first aeroplane able to carry a load that exceeded its own weight. Recognizing the passenger and cargo-carrying potential of aeroplanes over long distances, in 1932 Douglas began developing his famous *DC* (Douglas-Commercial) series of aircraft. This included the immortal *DC-3* (1935), which was the world's first successful commercial airliner. Powered by twin Pratt & Whitney engines, it could carry 21 passengers at a speed of 195 mph.

Assembly line
at the Douglas
Aircraft Company
in California,
early 1940s

FH-1 Phantom, 1948

During the Second World War, the *DC-3* was converted for military use and re-designated the *C-47*. In total, the Douglas Aircraft Company supplied 29,000 warplanes to the US military (including over 10,000 *C-47s*) – one-sixth of the airborne fleet. After the war, the company continued developing commercial passenger aircraft, including the advanced piston-engined *DC-7*, whose range made non-stop coast-to-coast services possible for the first time. With the advent of passenger jet aircraft, Douglas developed the *DC-8* (1958) and *DC-9* (1965), but nevertheless found itself lagging behind **Boeing**. As a result of this, the merger with McDonnell was sought.

Having been founded on the eve of the Second World War, the McDonnell Aircraft Corporation grew quickly and became a major US defence supplier. During the war it began research into jet propulsion, which led to the development of *FH-1 Phantom* (1946), the world's first operational carrier-based jet fighter. As the first US Navy aircraft able to attain a speed of 500 mph, the *FH-1* was the progenitor of a long and distinguished line of McDonnell fighter aircraft, including the *A-4 Skyhawk* (1954), the super-versatile Mach 2 *F-4 Phantom* (1958), the *F-15 Eagle* (1972) – still the US Air Force's premier fighter – and the *F/A-18 Hornet* (1978). Both McDonnell and Douglas were at the forefront of Space Age technology and made important contributions to the Mercury, Gemini and Saturn/Apollo space programmes. Today, McDonnell Douglas' commercial transports, such as the impressive *MD-11* wide-body tri-jet (1990), combat aircraft and space vehicles add new lustre to both companies' long history of aerospace accomplishments. In 1997 the McDonnell Douglas Corporation merged with Boeing.

F-4A Phantom, 1961

F-15C Eagle, 1978

F-15E Eagle, 1989

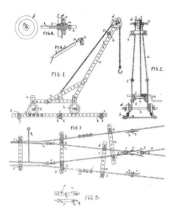

Frank Hornby, patent
drawing for Meccano
constructional toy,
1901

In 1899 Frank Hornby (1863–1936), the eventual creator
of Hornby train sets and Dinky toys, decided to make a
toy crane for his two small children. Unable to source
the necessary components to build it, he decided to fab-
ricate the parts himself. He later recalled: "I saw that
only interchangeable pieces could remedy this state of
affairs. This would require a new principle of **standardization** allowing for the assemblage of pieces in the system in multiple combinations. Then, happily, I realized that pieces perforated with tiny holes of like diameter at regular intervals would do the trick
admirably." He subsequently patented his "Mechanics Made Easy" constructional toy system in 1901, and together with a Mr. Elliot, who provided
the necessary capital, began production. The early sets sold for 7s 6d and
comprised 15 tinplate strips, nuts and bolts in a tin box. The sets may well
have been inspired by "active" teaching theories advocated most notably by
Marie Montessori and John Dewey, who urged for educational toys that captured the spirit of play. Over the following years, different sized sets were

Cover of an instruction
manual for Meccano,
France, 1960s

developed and new pieces, such as brass gears and wheels, were introduced. The company generously sponsored model competitions that in turn helped popularize the toy. In 1907 a new factory was built in Duke Street, Liverpool, and a year later Hornby bought out his partner and renamed the venture Meccano. By now the company was exporting its nickel-plate and brass components to Canada, Australia, New Zealand, India and other Commonwealth countries. In America, Meccano was manufactured for a short while in New Jersey, but faced strong competition from Erector, a similar constructional toy produced by the Gilbert Toy Company. Unbelievably, even the First World War failed to interrupt the production of Meccano, which in 1914 moved to a larger factory that would remain its home for the next 66 years. 1916 saw the first publication of the *Meccano Magazine*, followed after the war by the founding of the Meccano Club – all of which inspired a great deal of customer loyalty. Importantly, Meccano allowed the construction of models that were capable of movement; the Meccano *Inventor's Outfit* (1920), for example, had all manner of cranks, wheels, pinions and hinges that enabled children to share in the heyday of the engineering industry. 1926 saw the launch of *Meccano in Colours,* with pieces available in red, green and brass, followed in 1934 by the introduction of gold and blue Meccano strips. Due to material shortages, production ceased altogether during the Second World War and experienced a slow-down during the Korean War. During the 1960s and 1970s, Meccano faced increasing competition from toy manufacturers such as **Lego**, and in 1979 the company went into receivership. Meccano continued (and continues) to be manufactured, however, by the company's original subsidiary in France. Beautifully designed and made, Meccano still provides endless hours of entertainment with its components that embody the central principles of mass production: rationalism, **standardization**, prefabrication and adaptability.

Advertisement for Meccano, Great Britain, 1960s

ALBERTO MEDA

BORN 1945
LENNO TREMEZZINA/COMO, ITALY

← *Meda* chair for
Vitra, 1996

Alberto Meda studied mechanical engineering at the Politecnico di Milano, graduating with a MA in 1969. In 1973 he became technical manager for **Kartell**, where he was made responsible for furniture projects and also instigated a research programme into laboratory equipment made of **plastics**. In 1979 he began working as a freelance industrial designer, acting as a consultant to **Alfa Romeo**, Alias, Cinelli, Colombo Design, **Gaggia**, Ideal Standard, Mandarina Duck, Luceplan, **Philips**, Vitra and others. Between 1983 and 1987, Meda lectured on industrial technology at the Domus Academy and from 1995 taught industrial design at the Politecnico di Milano. He has received two Compasso d'Oro awards for his innovative lighting, co-designed with Paolo Rizzato (b. 1941) – for the *Lola* lamp in 1989 and the *Metropoli* lighting system in 1994, both manufactured by Luceplan. Meda has also produced several notable furniture designs, including his remarkably light yet strong *Light Light* chair (1987), which is constructed of a **carbon-fibre** matrix with a Nomex-honeycomb core. During the 1990s, Meda designed a range of elegant cast-**aluminium** seating for Alias, including the *Armframe* chair and *Longframe* lounger (1996). In 1997 his *Meda* office chair for Vitra was assigned "Best of Category" in *I. D.* magazine's design review. The structural rightness and visual coherence of Meda's work testifies to his engineering background. This and his talent for exquisite detailing and proportion make him one of the most interesting industrial designers working today.

Madi handle for
Colombo Design,
1990s

AVOIR SA BENZ!!

Benz poster, 1910

MERCEDES-BENZ

FOUNDED 1926
STUTTGART, GERMANY

Mercedes-Benz is the luxury car brand of the Daimler-Benz company, which was formed in 1926 through the merger of two pioneering German automobile companies, Benz & Co founded by Karl Benz (1844–1929) in 1883, and the Daimler-Motoren-Gesellschaft founded by Gottlieb Daimler (1834–1900) in 1890. Karl Benz initially established his company to manufacture stationary internal-combustion engines. In 1886, however, he launched the three-wheeled *Motorwagen*, which was the world's first practical car powered by an internal-combustion engine, and patented it the following year. Meanwhile, Gottlieb Daimler and Wilhelm Maybach (1846–1929) had also been developing internal-combustion engines from 1883 onwards, and in 1885 they patented a four-stroke version. That same year, they mounted one of their engines onto a wooden-framed bicycle and thus created the world's first motorcycle. In 1886 they motorized a previously horse-drawn carriage; this was a revolutionary design in that it was propelled by an "invisible" power source. In 1890 Daimler finally founded the Daimler-Motoren-Gesellschaft. In 1894

Mercedes car, 1904

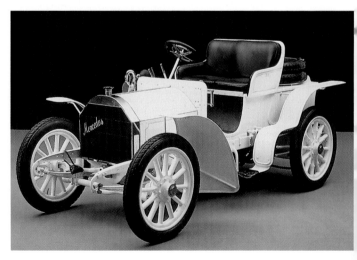

Mercedes-Benz 260
Stuttgart roadster,
1928

Mercedes-Benz
150 H, 1934

Mercedes-Benz
190 SL, 1955

the company launched a belt-driven car, featuring a four-speed belt transmission connecting the engine to the back wheels, which allowed smoother gear changing. Benz, meanwhile, had its own new car to launch that year, the *Benz Velo*, which was the world's first series production car. By 1899 it had developed a racing car, one of the finest and most reliable of its time. Daimler followed this in 1906 with its first Mercedes racing car. In 1926 the two companies merged to create the Daimler-Benz company, which continued to produce cars under the Mercedes-Benz name. These included the famous *SSK* sports car (1928) and the *260 Stuttgart* (1928), which was the company's mainstay during the late 1920s. During the 1930s, as well as producing elegantly styled saloons and sport cars such as the *159 H* roadster (1935), the company also designed and built aerodynamic racing cars such as the *W 25* (1934) and the *W 125* (1937), the latter scoring 27 victories. During the Second World War, Daimler-Benz's factories were turned over to military production. Car production resumed in 1946, but it was not until 1950 that the company launched a new model, the *170 S* cabriolet, which became a powerful symbol of Germany's economic recovery. 1955 saw the introduction of the *180* – the first Mercedes with a modern three-box body – and the elegant *190 SL*, which was described as "the ideal vehicle for sporting ladies". Over the succeeding decades, Daimler-Benz continued to produce cars, vans, trucks and buses and by 1965 it was the largest manufacturer of commercial vehicles in Europe. In 1988 Mercedes-Benz produced its ten millionth car. Having merged with the **Chrysler** Corporation in 1997 to become DaimlerChrysler, the company continues its long tradition of producing high-quality innovative vehicles – from the top-of-the-range *S-Class* to the compact *A-Class* – that exemplify German design at its best.

Messerschmitt
Me 109F, 1938

Wilhelm "Willy" Messerschmitt (1898–1978) studied engineering at the Munich Institute of Technology, graduating in 1923. Three years later, he became chief designer and engineer at the Bayerische Flugzeugwerke in Augsburg. While there, he designed the low-wing, single-engine, single-seat *Bf 109* monoplane, which was re-designated the *Me 109* when the company became the Messerschmitt-Aktien-Gesellschaft in 1938. The famous *Me 109* fighter was powered by a 1150-hp Daimler-Benz engine and had a top speed of 350 mph. During the Second World War, around 35,000 *Me 109*s were manufactured for the German Luftwaffe. Other wartime designs included the *Me 110* bomber/night fighter, the *Me 163* (1943), the first operational rocket-propelled aeroplane, and the *Me 262* (1943), which was the first operational jet-propelled plane. After the war, Messerschmitt diversified its production to include a sewing machine and pre-fabricated housing, and in the mid-1950s

Fritz Fend, Messer-
schmitt *KR 200*
bubble car, 1956

became a pioneer of bubble cars, such as the three-wheeled *KR 200* (1956) designed by Fritz Fend and the four-wheeled rocket-like *TG500* (1957).

Agip petrol station,
1996

MINALE TATTERSFIELD

FOUNDED 1964
LONDON, ENGLAND

The concept of multi-disciplinary design consultancy originated in the United States during the 1930s, with designers such as **Raymond Loewy**, **Norman Bel Geddes** and **Walter Dorwin Teague** producing designs across a wide range of fields – from heavy industrial to graphic design. Today, however, multi-disciplinary design consultancies are more common in Europe, while American design practices tend to specialize in specific areas such as communications design or **corporate identity**. One of the most interesting and successful multi-disciplinary design consultancies is Minale Tattersfield & Partners, which was founded by Marcello Minale and Brian Tattersfield in 1964. From the beginning, the office placed an emphasis on innovation and attracted high-profile clients such as Johnson & Johnson, Harrods and Thames Television. Since then, it has grown into a huge international consultancy. As well as working in the realms of corporate identity, **packaging** and multi-media, the firm also designs furniture, products and interiors. As a "design strategy group", the success of Minale Tattersfield & Partners lies in its ability to solve problems in a stylish and often witty way, while "expressing the maximum possible with the simplest means".

Packaging design for
Shark energy drink,
1996

MINTON

FOUNDED 1793
STOKE-ON-TRENT, ENGLAND

Christopher Dresser, artwork border for transfer-printed ceramics for Minton & Co., 1871

Having previously worked for Josiah Spode, Thomas Minton (1765–1836) founded his own factory in Stoke-on-Trent in 1793. Three years later, he began mass-producing transfer-printed earthenware and was responsible for the popularization of the well-known blue and white "Willow Pattern" decoration. The factory also became renowned for its fine bone china, introduced in 1821, and for its elaborate wares of the 1850s, which were influenced by a variety of historical styles. Herbert Minton took over from his father in 1836 and began instituting innovative methods of production. He also widened Minton's product range to include Parian, porcelain and majolica wares. The company also produced encaustic tiles that were exhibited at the **Great Exhibition** of 1851 and used in the interiors designed by A. W. N. Pugin (1812–1852) for the new Houses of Parliament. After Herbert Minton's death in 1858, the company was divided into two new ventures: Minton, Hollins & Co. (which produced encaustic floor tiles and mosaics) and Minton & Co. (which manufactured decorative tiles and chinaware). The latter was a pioneer of "art-manufactures" and commissioned leading designers such as **Christopher Dresser** to create wares suitable for serial production. Around 1900, Minton manufactured designs influenced by the Art Nouveau style. It was eventually acquired by the Doulton company in 1968.

C. Bevan (attributed), tile for Minton & Co., c. 1860

REGINALD MITCHELL

BORN 1895 TALKE, NEAR STOKE-ON-TRENT, ENGLAND
DIED 1937 SOUTHAMPTON, ENGLAND

Spitfire LFVc, 1936

Reginald Mitchell trained as an engineer prior to joining the Supermarine Aviation Works in Southampton in 1916. Three years later, he was appointed chief designer and engineer for the works, a position he held until his death. During his time there, Mitchell designed high-speed seaplanes for the Scheidner Trophy races (1922–1931) that influenced his later aircraft designs. He subsequently designed the legendary *Spitfire* (1936), a single-engine, single-seat, low-wing monoplane, according to Air Ministry Specifications. The *Spitfire* was both a day and night fighter and the model that first entered service in 1938 had a top speed of 360 mph. As one of the fastest fighters deployed in the Second World War, it was highly effective as a defence interceptor during the Battle of Britain. The *Spitfire* was continually modified throughout the war and served a variety of roles, including photo-reconnaissance. The *Spitfire XIV*, one of the last models of the war, had a remarkable top speed of 440 mph and an operational ceiling of 40,000 feet (12,200 m). Renowned for its aerodynamic sleekness and high manoeuvrability, the *Spitfire* remained in active service with the Royal Air Force until 1954.

Supermarine *Spitfire*
MK16e, c. 1940

JASPER MORRISON

BORN 1959
LONDON, ENGLAND

Jasper Morrison studied at the Kingston Polytechnic Design School from 1979 to 1982 and at the Royal College of Art in London from 1982 to 1985. He also studied at the Academy of Arts in Berlin on a scholarship in 1984. Morrison became well known in the early 1980s for his experimental furniture and in 1986 established his own London-based industrial design practice, Office for Design. His celebrated indoor/outdoor *Thinking Man's Chair* (1987) for Cappellini was followed by his *Ply-Chair* (1989) for Vitra, which predicted the de-materialistic approach to design that is characteristic of his more recent work. In 1992, in collaboration with James Irvine (b. 1958), Morrison organized the *Progetto Oggetto* for Cappellini – a collection of household items created by a group of young designers. The same year, Morrison published a book entitled *A World Without Words* and received a Bundespreis für Produktdesign award for his anodized **aluminium** door-handle range designed for **FSB** (1990). One of his most important commissions to date has been the design of the new Hanover tram *TW 2000* (1995–2000) for Expo 2000, which was awarded the IF Transportation Design Prize and the Ecology Award. Other recent projects include his stainless steel *Tin Family* kitchenware for **Alessi** (1998), the *Glo-ball* suspension light for **Flos** (1998), a flat-screen television and hi-fi system for **Sony** (both 1998) co-designed with John Tree, and furniture for the Tate Modern at Bankside (1999). Morrison's aesthetically pure and highly functional work exemplifies what has become known as the new **essentialism** in design.

Hanover tram
TW 2000 for
the Expo' 2000,
1995–2000

MOTIVATION

FOUNDED 1991
LONDON, ENGLAND

Wheelie training
in Bangladesh
with Motivation
wheelchair

Motivation is a charitable trust that initiates "self-sus-
taining projects to enhance the quality of life for wheelchair users, predominantly in third world countries".
The impetus for its foundation arose out of a tragic accident in 1982, when
David Constantine – a young British agriculture student – had a diving accident that left him paralysed from the chest down. He became fascinated by
design for disability, and after working as a computer programmer for two
years, enrolled on the industrial design course at the Royal College of Art,
London. While there, he co-designed with Simon Gue a competition-win-
ning wheelchair specifically for developing countries. Their design, with a
tubular metal frame, had a removable canvas seat that allowed it to be used
in a shower. In 1990 they visited Bangladesh, where they were asked to make
wheelchairs for the Centre of the Rehabilitation of the Paralysed in Dhaka.
The following year, they built their first wheelchair in Dhaka, incorporating
locally available rickshaw wheels. By 1999, Bangladeshi workers were assembling 150 of these wheelchairs a month at a production cost of £85 each.
Considerably less expensive than Western-made wheelchairs, the design
is also infinitely better suited to the region because replacement parts are
readily available. Motivation later developed the *Mekong* wheelchair with a

Meekong wheelchair,
1993

wooden frame specifically for the
land-mine victims of rural Cambo-
dia. In addition to its activities in
other developing countries such
as Afghanistan and Nicaragua,
Motivation has also developed the
Moti wheelchair, which addresses
the special needs of children with
cerebral palsy. Manufacturing and
distributing over 3,000 wheelchairs
around the world each year, Motivation has clearly shown how good
design can make a real difference
to the quality of people's lives.

ALEX MOULTON

BORN 1920
BRADFORD-ON-AVON, ENGLAND

Raleigh *Moulton MK3* bicycle, 1970

Richard Grigsby of the Bath Cycling Club riding a Moulton bike, 1970s

Alex Moulton studied mechanical engineering at Cambridge University and subsequently worked as technical director of his family's **rubber** manufacturing company until 1956. The same year, he founded Moulton Developments, which worked in close collaboration with Dunlop developing rubber automotive components, including the innovative Hydrolastic fluid suspension system used in **Alec Issigonis'** *Mini* and Morris *1100*. Between 1956 and 1960, Moulton endeavoured to design and manufacture "a better bicycle system". His first series of Moulton bicycles was launched in 1962, and that same year he founded Moulton Bicycles Ltd. to produce his small-wheeled bicycle design, which featured high-pressure tyres and a full suspension.

The Moulton bicycle was claimed to be easier to control because of its low centre of gravity and increased stability. Over 100,000 of these high-performance mini-bikes were built and exported to widespread acclaim. In 1965 Raleigh began producing an imitation of the Moulton bike and two years later acquired the licence to produce it (an agreement that was terminated in 1975). Undeterred, Moulton launched a range of deluxe folding bicycles in 1983 and in 1998 introduced the *New Series* Moulton as "more pleasurable to have and effective to use than any other". With a price tag of over £2,700, however, it is intended only for the seriously committed bicycle *aficionado*.

GERD ALFRED MÜLLER

BORN 1932 FRANKFURT/MAIN, GERMANY
DIED 1991 ESCHBORN, GERMANY

SM 3 electric shaver, 1960 (co-designed with Hans Gugelot)

→ *Blender MX 32* liquidizer for Braun, 1962

Gerd Alfred Müller apprenticed as a joiner prior to studying interior design at the Werkkunstschule in Wiesbaden. Between 1955 and 1960 he worked in the prestigious design department at **Braun** headed by **Dieter Rams**. While there, he designed many of the company's best-known household appliances and electric shavers. Typical of his work for Braun are his *Blender MX 3/MX 32* liquidizer (1958/1962) and *Multimix KM 3/KM 32* food processor (1957/1964), whose designs are distinguished by a remarkable clarity and underlying logic that made them easy to use and at the same time aesthetically pleasing. The neutral and rational styling of these products exemplified Braun's essentialist approach to design, which was founded on Rams' belief that it is the responsibility of the designer to create order in a world of visual noise and that products should therefore be as unobtrusive as possible. In 1960 Müller established his own office in Eschborn and began working as a freelance industrial and graphic designer. The solidity and purity of Müller's products were highly influential upon the design of succeeding generations of household appliances.

Multimix KM 3 food processor for Braun, 1957

PETER MÜLLER-MUNK

BORN 1904 BERLIN, GERMANY
DIED 1967 PITTSBURGH, PENNSYLVANIA, USA

Plastic table radio
for Sylvania Electric
Products Inc.,
c. 1950

Peter Müller-Munk trained as a silversmith under Waldemar Rämisch at the Kunstgewerbeschule, Berlin. In 1926 he emigrated to the United States and worked briefly as a silversmith for Tiffany & Co. His silver designs were shown at the 1929 "Modern American Design in Metal" exhibition held at the Newark Museum. That same year, he published an article entitled *Machine-Hand* in which he asked: "Why does nobody ever try to define the process of the machine in action and deduct the forms and decorations most closely related to it?" In the same article, he also urged for "a new art resulting from the harmony of technique and object". Around this time, he established his own design practice in New York, which specialized in the design of products for industrial manufacture. Most notably, he designed metalware for the Revere Copper & Brass Company, including the well-known *Normandie* pitcher (1935), named after the luxurious French ocean-liner. The simple streamlined form of this design was dictated both by functional requirements and by its suitability to industrial production. The clean lines and strong geometric form that are typical of Müller-Munk's work were strongly influenced by the European Modern Movement. Although best known for his metalware designs, such as his copper-bottomed saucepans for **Ekco** (c. 1950), he also designed various appliances, including a plastic table radio for Sylvania Electric Products Inc. (c. 1950), a retractable ball-point pen and a retractable lip-brush for the B-B Pen Company (c. 1950), a portable sander for the Porter-Cable Machine Company (c. 1950) and the well-known Waring blender. He established another office in Chicago in the 1930s and his work was included in an exhibition at Metropolitan Museum of Art, New York, in 1939. Müller-Munk also designed interiors, exhibitions and vehicles, but is best remembered as an early pioneer of American product design who made the transition from skilled craftsman to professional industrial design consultant.

Normandie pitcher
for Revere Copper &
Brass Inc., 1935

Apollo 1 spacecraft, 1961

NASA
(NATIONAL AERONAUTICS AND SPACE ADMINISTRATION)

FOUNDED 1958
WASHINGTON, DC, USA

The National Aeronautics and Space Administration (NASA) was founded in 1958 in direct response to the Soviet Union's launch of the *Sputnik 1* satellite the previous year. Having evolved out of the National Advisory Committee for Aeronautics, which had been established in 1915, NASA was (and still is) an independent US government agency committed to the research and development of space flight, aeronautics, space science and space applications. The US space programme was boosted in 1961 by President John F. Kennedy's pledge to put a man on the moon before the end of the decade. The resulting *Apollo* programme, after several setbacks and fatalities, eventually managed to achieve this goal with the *Apollo 11* mission on July 20, 1969. For this sensational historic event to be made possible, NASA drew technical expertise from the cream of American industry. In 1975 NASA launched the unmanned *Viking I* and *Viking II* spacecraft to investigate Mars, while its *Voyager* probes launched in 1977 provided further insights into Jupiter, Saturn, Uranus and Neptune. More recently, the *Galileo* unmanned mission launched in 1989 has brought even greater understanding of Venus, Jupiter and the outer solar system. Apart from its exploratory

Apollo 11 – Edwin (Buzz) Aldrin descending the Lunar Excursion Module's ladder to become the second man on the moon, 20 July 1969

missions, NASA has also undertaken cutting-edge aeronautics research into **aerodynamics** and wind shear, as well as research into space habitability and earth-orbital space stations, such as the *Sky-Lab* and the *Saturn Five* Space Station (both of which **Raymond Loewy** helped design in the 1970s). 1981 saw the launch of the first Space Shuttle, heralding the advent of partly re-usable space vehicles and opening up the future possibility of commercial space travel. NASA has also developed and launched numerous communication and weather satellite systems, such as *Echo*, *Telstar* and *Syncom* in the 1960s and *Landsat* in the 1970s. With nearly 40 years "blue sky" research behind it, NASA's programmes have not only fundamentally altered man's perspective of the universe but also changed his material environment on earth. NASA continues to push the frontiers of space and technology in its quest for knowledge. Significantly for the realm of industrial design, the advanced technologies pioneered by NASA continue to result in "spin-offs" across wide-ranging scientific, technical and commercial fields of endeavour.

Arthur Jones built his first exercise machine in 1948, but it took over two decades of experimentation before he perfected the design. His *Pullover Torso Machine* (1970) was the first design he actually sold. The following year, Jones established Nautilus Sports/Medical Industries, the name of which was inspired by the shell-like form of the original spiral pulleys. Nautilus' exercise equipment completely revolutionized sports training and American National Football League teams were among its first users. During the 1970s the company developed a number of "Time Machines" that provided athletes with a source of "total exercise", and by the mid-1980s Jones had developed more sophisticated multi-gyms in accordance with medical findings gleaned from Nautilus' own research into body-building. The ancestry of today's high-tech exercise equipment can be directly traced to the highly innovative Nautilus machines of the early 1970s.

Pullover exercise machine with *Pulldown* attachment, early 1970s

Weight assisted chin/dip exercise machine, 1992

GEORGE NELSON

BORN 1907 HARTFORD, CONNECTICUT, USA
DIED 1986 NEW YORK, USA

George Nelson was one of the foremost design practitioners, theorists and communicators of the 20th century. After studying architecture, he became an associate editor at *Architectural Forum* and wrote profiles of leading Modernist architects for *Pencil Points*. In 1941 Nelson joined the architecture faculty at Yale University, where he developed numerous innovative concepts, including the pedestrianized shopping mall in his "Grass on Main Street" proposal (1942). He also pioneered the concept of built-in storage with his *Storagewall* (1944). In 1946 Nelson became design director of **Herman Miller** – a position he held until 1972. During his time there, Nelson designed Modern furnishings for the company, including the *Comprehensive Storage System* (1957) and the revolutionary *Action Office I* system (1964–1965). As a prolific industrial designer, Nelson also created the *Prolon* melamine line of dinnerware for the Pro-Phy-Lac-Tic Brush Co. (1952–1955), numerous

Prolon line of melamine dinnerware for the Pro-Phy-Lac-Tic Brush Co., 1952–1955

wall and table clocks for the Howard Miller Clock Company (late 1940s and early 1950s), the *Bubble* lamps made of self-webbing plastic (1947–1952) and the *Omni* extruded **aluminium** pole system for Dunlap. Nelson was also interested in **product architecture** and designed a modular plastic-domed *Experiment House* (1957). Nelson's concepts were extremely influential and forward-looking – in 1978 he correctly predicted that the advances in computer technology would result in greater "**miniaturization**, ephemeralization, dematerialization". In 1979 he stated: "Human needs are variable and frequently unpredictable; they certainly are not quantifiable; they are complex, subtle and mysterious."

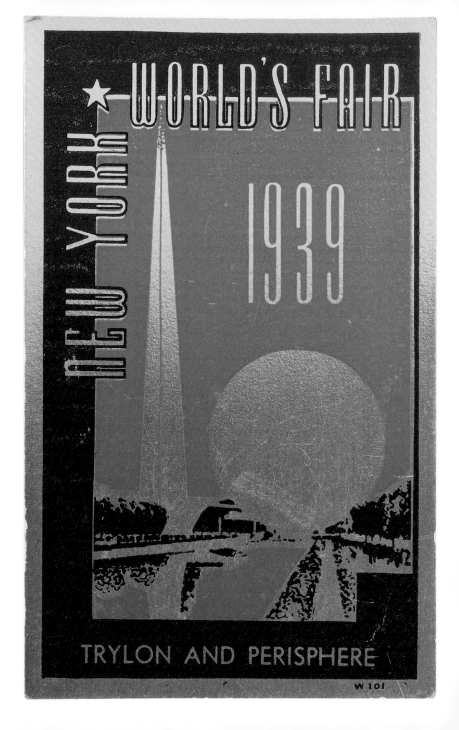

NEW YORK WORLD'S FAIR

1939
NEW YORK CITY

Elektro, the mechanical man with his pal *Sparko*, the world's first robotic dog, which could perform 27 different tricks, at the New York World's Fair

← Postcard showing the *Trylon & Perisphere* at the New York World's Fair, 1939

With the effects of the Great Depression still fresh in people's minds, the New York World's Fair of 1939 offered an opportunity to look towards a better future.

At this time there was also an overwhelming desire to assert America's cultural identity before a global audience. Many of the exhibitors were among the country's leading corporate giants; they included AT&T, **Chrysler**, Dupont, Eastman **Kodak**, Firestone, **Ford**, **General Electric**, **General Motors**, **RCA** and **Westinghouse** – all of whom had major exhibits that showed the "wonders of modern research" and used the machine as a potent symbol of the future. General Motors' towering Automobile Salon had glamorous Hollywood connotations, while the "No.1 hit of the Fair" was the Futurama, which provided "a magic flight through time and space into the world of 1960". **Norman Bel Geddes'** *City of the Future* and **Henry Dreyfuss'** *Democracy* offered utopian visions of futuristic lifestyles that included robots, space travel, televisions and freeway systems. Among the most memorable symbols of the fair, however, were the enormous *Perisphere* and soaring *Trylon* structures, which reflected America's optimism and faith in the future.

Giant *National Cash Register* (designed by Walter Dorwin Teague) at the New York World's Fair, showing daily attendance figures

Tailwind trainer, 1979 – Nike's first air-cushioned shoe

In 1957 Bill Bowerman, a coach, and Phil Knight, an athlete, met at the University of Oregon, Eugene. Knight subsequently attended Stanford Business School, while runners at the University of Oregon broke records wearing shoes that had been handcrafted by Bowerman. In 1962, under the name of BRS (Blue Ribbon Sports), Knight started importing athletic shoes made to Bowerman's specifications by the Onitsuka Tiger Company in Japan. Bowerman calculated that, over a distance of a mile, every ounce that could be reduced in a runner's shoe would reduce the total weight lifted by 200 pounds. To this end, in 1971 he began developing prototypes of a new and highly innovative running shoe using **rubber** moulded in his wife's waffle iron. The following year, after a dispute with Tiger, BRS launched the new Nike brand (named after the Greek goddess of victory) and by 1974 the now fully developed *Waffle Trainer* became the best-selling training shoe in America. In 1978 the company signed up the star tennis player, John McEnroe, in the first of many high-profile sports sponsorship deals. A year later, the landmark *Tailwind* shoe was launched and took over as the best-selling running shoe in America. The *Tailwind* was the first trainer to incorporate Nike-Air cushioning – a system of sponge rubber wedges (or mid-soles) sandwiched between the upper and outer sole that act as shock absorbers. The stress put on the body by sport can be immense. A basketball jump shot, for example, will exert a force ten times that of the player's weight, while a runner's foot will endure over 25,000 heel strikes when running a 26.2 mile marathon. 1987 saw the introduction of Nike-Air shoes,

Tuned Air Max trainer, 1999

which had a revolutionary cushioning system that comprised flexible urethane bags filled with a pressurized gas that reduced the force of impact. After absorbing the force, the air-filled bags quickly returned to their original form and volume in readiness for the next impact. Also in 1987, Nike sold its five millionth pair of *Air Pegasus* shoes (introduced in 1983) and launched

Air Warp Flex in-line
skates, 1999

Triax watch, c. 1998

the high-profile "Just Do It" advertising campaign. By 1989 Nike's annual
net income had risen to $167 million, largely as a result of its product en-
dorsements by top sports personalities. In 1995 it launched a new high-
performance cushioning system, called *Zoom Air*, featuring an 8mm gas
cushion which provided the necessary impact protection yet which kept
the foot close to the ground for increased feel and response. Today, Nike
is the world's number one sports brand and its product range includes
footwear, equipment and apparel.

Giorgetto Giugiaro,
Nikon *F3* SLR
camera, 1980

NIKON

FOUNDED 1917
TOKYO, JAPAN

In 1917 Japan's three leading optical manufacturers merged to form the Nippon Kogaku company and subsequently began researching and producing optical glass, resulting in the issue of its miniature *Mikron* binoculars (1921) and its *Joico* microscope (1925). In 1932 the company adopted the brand name Nikkor for its camera lenses and around 1945 began manufacturing cameras. From 1946 it marketed its 35mm cameras under the Nikon name and two years later introduced the *Nikon I* camera. Gaining a reputation for its superior cameras and lenses, the company founded the Nikkor Club to promote photography and in 1953 established Nikon Inc. in America. 1957 saw the launch of the *Nikon SP* rangefinder model, which was an important precursor of the landmark *Nikon F* which appeared in 1959. The *Nikon F* quite literally revolutionized photojournalism and marked a shift from rangefinders (RF) to single lens reflex (SLR) cameras as the professionals' preferred choice. With its reflecting mirror, the SLR system offered much more precise framing than the rangefinder system, which only gave an approximate idea of what the image would look like. Although not the first SLR camera on the market – the Contax *S* had been available since 1948 – the *Nikon F* was far lighter than earlier Contax models and had the added advantage of being much easier to load

Giorgetto Giugiaro,
Nikon *F5* SLR
camera, 1996

with film. The *Nikon F* – or *The Nikon*, as it became known – not only ushered in widespread consumer acceptance of "Made in Japan" cameras, but also heralded the adoption of 35mm film among professionals and a move towards complete camera systems. *The Nikon* subsequently became the ultimate professional camera and in 1971 the improved *F2* was launched. This was followed in 1980 by another landmark model, the *F3*, designed by **Giorgetto Giugiaro**. Featuring a number of automatic

rather than manual functions, the *F3* led automation to become accepted among professional photographers. The design, however, retained Nikon's concept of modularity and could therefore be "stripped" and customized with a wide array of accessories to suit photographers' needs. Three years later, Nikon introduced its first autofocus SLR camera based on the *F3*. Described as "a non-offensive tool to a photographer on duty because it serves more than commands", the *F3* became a modern classic celebrated for its ease of control and handling, extraordinary reliability and system compatibility. Giugiaro also designed the *F4* (1988), the *Nikonos RS* (1992), the first underwater SLR with autofocus, and the now top-of-the-range super-fast *F5* (1996), which has an eight frames per second capability and an automatic focus tracking system. Giugiaro declared of the *F5*: "this is a top-class camera designed with thorough emphasis on ease of use. I was convinced that originality could be produced from within a condensed form." Over 2,300 products currently carry the Nikon name, including a range of digital still cameras known as *Coolpix* (1998) and the state-of-the-art professional *D1* digital SLR camera (1999) which, like its illustrious predecessors, offers high-performance to "pro-sumers".

↖Nikon *F60* SLR camera, 1990s

↑Nikon *D1* digital SLR camera, 1999

↖Nikon *Nuvis S* ultracompact camera, 1996

↑Nikon *Pronea S* autofocus SLR camera, 1996

E-Connect 126K3
built-in down-lights
for Wila Leuchten,
1995

Peter Krouwel (b. 1952) trained as an industrial designer at the Technical University in Delft, graduating in 1978. A year later, together with fellow student Wolfram Peters (b. 1952), he founded the industrial design practice Ingenieorsbureau Peters en Krouwel. In 1985 the office was joined by a third partner, Bruno Ninaber van Eyben, who had previously studied at the Maastricht Art Academy. Renamed Ninaber, Peters & Krouwel Industrial Design, the practice designed a wide range of products for industrial manufacture, including desk accessories for Randstad (1991), the *Yachtboy* world radio receiver for Grundig (1992), light fittings for Wila Leuchten (1995), the *Quin* stacking chair for Kembo (1995) and the *Discovery* child's bicycle seat for Hamax (1995). Ninaber, Peters &

Plano hospital bed
and bedside cabinet,
1991

Krouwel also produced award-winning "public design" such as road **signage** for ANWB (the Royal Dutch Touring Club), a bus station in Leiden and a postbox for the Dutch post office (PTT Post). In 1997 the practice was renamed NPK. Today it has some 50 employees and is widely regarded as the leading Dutch industrial design consultancy. The firm's areas of activity include all stages of the design process, from conception through development to pre-production. NPK has received over 100 design awards, including nine Gute Industrieform prizes, and its work is featured in the collections of the Museum of Modern Art, New York and the Stedelijk Musem, Amsterdam.

Sno Family sledge
for Hamax AS, 1996

↓Redesign of ANWB
(the Royal Dutch
Touring Club) sign-
posts, 1997

Nissan assembly
line, 1990s

NISSAN

FOUNDED 1934
YOKOHAMA, JAPAN

After the launch of the original *Dat* car by the Kwaishinsha Company in 1914, in 1931 the Tobata Casting Co. began producing a new generation of cars which it christened Datsun ("son of Dat"). Two years later, another company, founded by Yoshisuke Aikawa (1880–1967), took over their manufacture. In 1934 this new venture was renamed Nissan Motor Company and its founder, who had a grand scheme of mass-producing 10,000–15,000 cars per annum, began putting his plan into action. The first small-sized Datsun rolled off the assembly line at Nissan's plant in Yokohama just one year later, symbolizing Japan's rapid pre-war industrialization. During the Second World War, production switched to military trucks and engines for aeroplanes and torpedo boats. Nissan resumed production of non-military vehicles in 1945 and two years later recommenced manufacturing Datsun cars. In 1958 Nissan began exporting Datsun cars to the United States and sold its first Datsun compact pick-up truck there a year later. During the 1960s and early 1970s, Nissan continued to make inroads into the American automotive market, with annual sales rising to 255,000 cars in 1971. The oil crisis massively boosted American sales as consumers began opting for smaller, less expensive and more fuel-efficient automobiles such as those exported by Nissan, **Honda** and **Toyota**. By 1973, one million Datsun vehicles had sold in the United States, and two years later the company became the top vehicle exporter to North America. During the 1980s, Nissan established two strategic manufacturing bases in America and Britain

Nissan *Micra*, 1992

and in 1983 began marketing cars under the Nissan name, such as the hugely successful Nissan *Sunny* launched in 1985. By 1987 its cumulative exports had exceeded 20 million units. In the early 1990s, Nissan became one of the first automotive manufacturers to adopt **soft design** with its introduction of the *Micra* (1992), which was named Car of the Year in Europe in 1993.

Elettrosumma 14
adding machine
for Olivetti, 1946

Marcello Nizzoli studied painting, architecture and graphic design at the Scuola di Belle Arti, Parma. In 1914 he exhibited two small paintings and four embroideries with the Futurists in the first "Nuove Tendenze" exhibition in Milan. After the war, he produced a series of tapestries – an early foray into the applied arts. In 1918 he founded his own design office in Milan and during the 1920s became associated with the Rationalists. His designs for applied art brought him widespread recognition when they were exhibited at the first "Esposizione delle Arti Decorative e Industriali Moderne" held in Modena in 1923. Subsequently, Nizzoli received a diversity of commissions ranging from the decoration of silk shawls to exhibition architecture. During this period, Nizzoli also designed advertising posters, most notably for Campari in 1926. From 1934 to 1936 he worked in partnership with the architect Edoardo Persico (1900–1936), with whom he designed two stores for **Parker** Pen in Milan (1934) and several exhibitions, including the 1934 "Aeronauti-

Mirella sewing
machine for Necchi,
1957

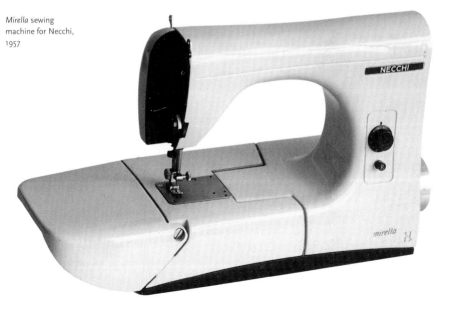

Lettera 22 typewriter
for Olivetti, 1950

cal Exhibition" held in Milan. Between 1931 and 1936 Nizzoli also worked
in collaboration with Giuseppe Terragni (1904–1943). From 1931 he worked
freelance for **Olivetti** as a graphic designer, and subsequently joined the
company's advertising office in 1938. While there, Nizzoli became Olivetti's
chief product designer and produced landmark typewriter designs such as
the *Lexikon 80* (1948) and the *Lettera 22* (1950), which won a Compasso d'Oro
award in 1954. He also designed a number of adding machines, including
the *Elettrosumma 14* (1946) and the *Divisumma 14* (1948), as well as workers'
housing for the company. Beyond his work for Olivetti, Nizzoli also produced sculptural designs for Necchi, such as the *BU* sewing machine (c. 1953),
which was awarded a Compasso d'Oro in 1954, and the *Mirella* sewing machine (1957), which won the Premio Compasso d'Oro the same year. In 1959
Nizzoli designed pocket lighters for Ronson and, a year later, petrol pumps
for Agip. As one of the most important post-war Italian designers, Nizzoli's
rational yet sculptural work was immensely influential upon later Italian
product design.

NORTH AMERICAN

FOUNDED 1928
DELAWARE, USA

F-100 Super Sabre,
1953

Having decided in 1934 to concentrate on manufacturing military aircraft, the first North American aeroplane, the *NA-16* trainer, was introduced in 1935. A year later, the company's manufacturing operation was moved to Los Angeles and in 1937 the twin-engine *NA-21 Dragon* bomber was launched. During the Second World War, North American built a staggering 41,000 aircraft and between 1935 and 1967 manufactured more military aeroplanes than any other American company. Among its most famous aircraft were the *P-51 Mustang* fighter (1940), which was described by the US Government as "the most aerodynamically perfect pursuit plane in existence", and the *B-25 Mitchell* medium bomber, which was the first bomber to be deployed in all combat theatres of operation during the Second World War. After the war, North American developed advanced jet fighters such as the *F-86 Sabre* jet (1949), the first swept-wing design in the US fighter inventory, and the *F-100 Super Sabre* (1953), which established the world's first supersonic speed records. North American also produced the legendary *X-15* rocket-propelled research aircraft (1959), which flew beyond the earth's atmosphere and could reach speeds in excess of Mach 6.5.

P-51 Mustang, 1940

ELIOT FETTE NOYES

BORN 1910 BOSTON, MASSACHUSETTS, USA
DIED 1977 NEW CANAAN, CONNECTICUT, USA

Eliot Fette Noyes studied architecture at Harvard University and at the Harvard Graduate School of Design. In 1939 he joined **Walter Gropius** and **Marcel Breuer**'s architectural practice. Upon Gropius' recommendation, he was appointed the first director of industrial design at the Museum of Modern Art, New York – a position he held from 1940 to 1942 and 1945 to 1946. Between 1946 and 1947 Noyes was design director of **Norman Bel Geddes'** industrial design practice, which consulted to **IBM**, and in 1947 established his own office in New Canaan, Connecticut. From 1956 to 1977 he was corporate design director of IBM, during which time he designed several revolutionary products, most notably the *Selectric I* golf-ball typewriter (1961). Noyes helped establish a strong **corporate identity** for the company not only through the integrity of his own product designs, but by commissioning graphics from Paul Rand (1914–1996) and buildings from Breuer. Noyes worked as a design consultant to many other companies, including **Westinghouse**, Mobil, **Xerox** and Pan Am. As one of the most influential advocates of Good Design, Noyes reshaped entire corporations and set new standards for product design in America.

Executary secretarial transcribing machine, *Model 212*, for IBM, 1961

Selectric 1 golf-ball typewriter for IBM, 1961

Detail showing golf-ball element of the *Selectric 1*, 1961

ANTTI NURMESNIEMI

BORN 1927
HÄMEENLINNA, FINLAND

Antti Nurmesniemi initially worked in a metal workshop and an aircraft factory. Having been inspired by the designs shown at the Museum of Modern Art's 1945 "America Builds" exhibition held in Helsinki, in 1947 he enrolled at Central School of Applied Arts, Helsinki, where the "Nordic Line" approach to design concentrated on social and technical issues. From 1951 to 1956 he worked for the architect Viljo Revell, designing interiors and furniture, including his well-known horseshoe-shaped *Sauna* stool (1951–1952) for the Palace Hotel. Between 1954 and 1955 he worked in Giovanni Romano's design practice in Milan, where he was inspired by functional yet stylish products designed by **Marco Zanuso** and **Roberto Sambonet** among others. In 1956 he established a Helsinki-based design office and a year later designed his well-known enamelled coffeepots for Wärtsilä. He went on to design furniture for Artek, Tecta and Cassina, kitchenware for Arabia and Kymi, trains for the Helsinki metro and the Valmet Oy company. Among his most successful and characteristic designs were the *Antti* and *Yleispuhelin* telephones (both 1984) for Fujitsu, which reflected his belief that, "An object should not be designed simply to fulfil a purpose (not to mention manufacturing), but for the user ... A simply designed object is wrong if its use is complicated ... A designed object does not need to reflect the times. It is much better if it offers a hint of the future."

Yleispuhelin telephone for Fujitsu, 1984

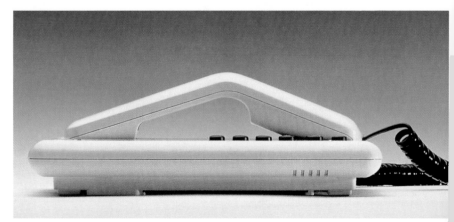

A Frame snow
goggle, 1999

OAKLEY

FOUNDED 1975
FOOTHILL RANCH, CALIFORNIA, USA

Founded by Jim Jannard in 1975, Oakley has a worldwide reputation for the superlative design of its eyewear products. Since its inception, Oakley has defied convention while focusing on new designs, new materials and new production techniques. In 1984 Oakley came up with the concept of "sculptural physics" and subsequently introduced designs, such as the *XYZ Optics*, that exemplified style and performance. Designs such as these completely redefined the eyewear market, which had been stultified by a handful of major manufacturers who believed they "knew it all". According to Oakley, "an idea is born in the depths of our design bunker, given form with **CAD-CAM** modelling, brought to life with SLA liquid laser prototyping, tested with spectrophotometers, environmental simulation chambers and ANSI Z87.1 impact rigs" before undergoing rigorous field testing by athletes for whom the designs are originally intended. The form of Oakley's eyewear is dictated by performance: the *Racing Jacket* (described as "chiselled adrenaline"), for example, has a vented and hingeless frame that maximizes peripheral vision and provides de-fogging ventilation, while its "nosebomb" and "earsock" features ensure that the lightweight frame remains securely in place. Oakley has managed to position itself as a major global brand and as well as its innovative eyewear – from sunglasses to snow goggles – it also manufactures sports apparel, footwear and wristwatches.

Racing Jacket
eyewear, 1999

OLIVETTI

FOUNDED 1908
IVREA, ITALY

Camillo Olivetti

←Poster advertising
Olivetti's *M1*, 1912

Born in Ivrea, Samuel David Camillo Olivetti (1868–1943) studied engineering at the Politecnico di Torino under Galileo Ferraris, who was the discoverer of rotating magnetic fields. After graduating, Olivetti went to London to continue his studies. In 1893 he accompanied Ferraris to the Electricity Congress held in Chicago, and later took a physics course at Stanford University, where he also worked as an assistant electrical engineer. After returning to Italy in 1894, he established a small factory by the name of C. G. S. (Centimetre, Gram, Second) to manufacture electric measuring instruments, which he moved to Milan in 1903. Four years later, he decided to move back to Ivrea and subsequently established Italy's first typewriter factory there in 1908. Aware of American mass production methods, Olivetti began industrially manufacturing typewriters, and in 1909 Ing. C. Olivetti & C. SpA introduced its first typewriter, the *M1*, which was described by a contemporary critic as "robust and elegant" and was praised for its faster carriage speed and smoother key movements. The company grew rapidly, and in the 1920s Camillo's son Adriano Olivetti (1901–1960) was sent to the United States to observe first-hand the latest mass-production techniques, with a view to adopting them back in Ivrea. In 1931 Olivetti launched the *M40* typewriter and the following year unveiled its first portable model, the

Camillo Olivetti, *M1*
typewriter, 1910–1911

MP1. As a moral crusader and social reformer, Camillo Olivetti established a foundation in 1932 that offered his workers social security benefits on a level unprecedented in Italy. Other far-sighted employee incentives were later provided, including seaside summer camps for the workers' children, a cafeteria, a kindergarten, a library and housing. In 1933 Adriano Olivetti was appointed managing director and began diversifying the company's product line, launching its first teleprinter in 1937 and its first adding machine in 1941. He also raised the profile of the company by developing a strong **corporate identity** through product design, architecture, exhibition design, advertising and graphics. Indeed, during the immediate post-war period, Olivetti was one of only a handful of companies worldwide with a truly Modern image. As a consequence, it was also exceptionally influential upon the design of other firms' corporate identities. As an "intelligence co-ordinator", Adriano Olivetti commissioned leading designers to produce cutting-edge product designs, such as the sculptural *Lettera 22* typewriter (1950) by **Marcello Nizzoli**. He also commissioned leading graphic design-

Marcello Nizzoli,
Lexikon 80 typewriter,
1948

Ettore Sottsass &
Hans von Klier,
Editor 4 typewriter,
1964–1969

Mario Bellini,
Quaderno lap-top
computer, 1989

Michele De Lucchi,
M4–82 Modulo
computer, 1993

Michele De Lucchi,
OFX 500 phone/
facsimile machine,
1998

ers, such as Giovanni Pintori (b. 1912), to produce eye-catching posters and advertisements. Although Olivetti introduced Italy's first electronic computer in 1959 – the *Elea 9003* designed by **Ettore Sottsass** – the company was obliged for financial reasons to sell its Electronics Division after the death of Adriano Olivetti in 1960. The company continued its research into electronic processing systems, however, and in 1965 it launched the *P101* programmable desktop computer, an innovative forerunner of the personal computer. During the late 1960s and early 1970s the company launched other notable products, including the *Editor 4* typewriter (1964–1969) designed by Ettore Sottsass and Hans von Klier (1934–2000) and the *Divisumma 18* calculator designed by **Mario Bellini** (1973). In 1969 Olivetti also introduced the bright red *Valentine* portable typewriter designed by Ettore Sottsass and Perry A. **King**. A year later, the Olivetti trademark, which is still in use today, was designed by the famous graphic designer Walter Ballmer (b. 1923). While Olivetti experienced financial difficulties in the late 1970s, it nonetheless developed a number of key products, including the company's first electronic typewriter in 1978 and its first personal computer in 1982. During the 1980s Olivetti expanded its operations in the IT field and in the 1990s concentrated on its telecommunications activities. Today, the Olivetti Group is made up of twelve companies operating in either the information technology or telecommunications fields. For Olivetti, "industrial design is a range of activities that not only creates a visual image, but above all contributes to the development of the project as a whole. The designer, in other words, is no longer regarded simply as an expert in aesthetics or style, but as a specialist in the relationship between man and machine."

OLYMPUS

FOUNDED 1919
TOKYO, JAPAN

Olympus *μ (mu)*
Zoom camera, 1993

Having been founded in 1919, the Takachiho Seisa-kusho company produced the first Japanese microscope in 1920 and a year later adopted the Olympus brand name. In 1930 it launched its first metallurgical microscope and five years later opened its Optical Research Centre. It went on to develop the first *Zuiko* photographic lens for its first camera, the bellows-style *Semi-Olympus I* (1936). With increasing demand, the firm introduced a new camera design, the *Olympus 6* (1940). The *Olympus 35* introduced in 1948 was the first Japanese 35mm camera with a lens shutter system. In 1949 the company was re-named the Olympus Optical Co. and subsequently unveiled its first medical camera, the *Gastrocamera* (1950). During the late 1950s, the *GTF* gastrocamera with its fibrescope started being used for new medical procedures. Alongside its innovative scientific products, Olympus also released the *Olympus Wide* in the mid-1950s, which inspired a boom for wide-angle lens photography in Japan. In 1969 the company unveiled the world's first microcassette tape recorder, the *Zuiko Pearlcorder*. During the 1960s and early 1970s Olympus pioneered the development of compact SRL cameras, such as the *Olympus Trip 35* (1968), which was produced for two decades and sold over ten million units, making it one of the most successful 35mm cameras ever. Olympus' compact cameras cleverly incorporated the light-meter cells around the lens, which made their bodies more unified and easier to use. The *Olympus 35-ECR* (c. 1970), for example, had automatic exposure control and focusing that enabled even amateurs to take good quality photographs. As well as continuing to produce state-of-the-art cameras and medical optics, Olympus also manufactures blood-typing systems (80% of all the blood held in North America's blood banks is typed using Olympus machines), along with other diagnostic systems that are used to perform around a quarter of all chemical tests in the US.

Olympus *35-ECR*
camera, c. 1970

Opel *Doktorwagen* automobile, 1909

In 1898 the five sons of Adam Opel (1837–1895) converted his sewing-machine and bicycle factory in Rüsselsheim into a car manufacturing plant. Having introduced its first vehicle in 1902 under its own name, Opel AG, the company was beginning to emerge as a major force in the dawning automotive industry when, in 1911, the plant was completely destroyed by fire. Opel took advantage of this calamity by rebuilding and retooling from scratch with the most modern equipment available. In 1913 it introduced the *Rennwagen* racing car and began producing trucks. During this period, Opel also manufactured a distinctive roadster, which was nicknamed the *Laubfrosch* (tree frog) because of its unusual green colour. After the First World War, however, the company's production was restricted by the Allies. Opel nevertheless became one of the first European automotive manufacturers to institute an efficient assembly-line system in 1924, which enabled it to become the largest producer of cars in Europe prior to the Second World War. During the 1920s Opel brought out a number of interesting models, such as the

Opel *Olympia Bi*, 1938

Opel *Corsa*, 1993 (first version launched in 1982)

small, two-seater *4/16PS* roadster (1928). As a consequence of the hyperinflationary economic conditions of the era, however, the Opel family lost control of the company, and it eventually became a wholly-owned subsidiary of **General Motors**. It later manufactured some notable streamlined cars, such as the *Olympia Bi* (1938) and the *Kapitän* (1953). Although some of Opel's plants were lost to the Soviet Union after the Second World War, the company managed to recover and remains one of the largest European automotive manufacturers, accounting for around 25 % of Germany's vehicle production.

PARKER PENS

FOUNDED 1892
JANESVILLE, WISCONSIN, USA

George S. Parker

George Parker (1863–1937) was a teacher at the Valentine School of Telegraphy. To supplement his income, he became a part-time representative of the John Holland Fountain Pen Company and sold pens to his students. The firm's pens were of poor quality and Parker's customers frequently complained to him about them. He realised that the problems arose from the design of the feed-shaft, and so with a few simple tools, such as a scroll saw and a file, he began making shafts for the pens he was selling, which allowed the air to travel up at a steadier pace and so improve writing performance. Impressed with the results, Parker decided to manufacture his own pens and with his meagre savings began purchasing components from jobbing manufacturers. In 1886 the first Parker pens were made in the bedroom of the hotel in which he was living. He subsequently patented his fountain pen in 1889 and found a firm to produce them by the gross. In 1892 he established the Parker Pen Company and designed the first "jointless" pen, which was patented in 1899. Advertised as "the greatest improvement ever made in fountain pen construction", this design had "no screw to break, no old-fashioned nozzle, no joints to leak", and according to the first British advertisement for Parker pens, was retailed by over 6,000 dealers worldwide. 1929 saw the introduction of the *Duofold*, which could be converted from a pocket pen to a desk model with the addition

Advertisement for the *Jointless Lucky Curve* fountain pen, 1889

f a barrel extension. In 1932 the famous arrow clip was introduced, having
een designed by the New York artist, Joseph Platt. Around 1935, the com-
any launched the *Vacumatic* pen that incorporated an innovative rubber
iaphragm and a mechanism that allowed the ink to be sucked up. The re-
owned *51* fountain pen, with a sleek profile and Lucite barrel designed to
esist the corrosive effects of the alkali in the ink, was introduced in 1939 –
1 years after the first Parker pen was introduced. Rated as one of the
vorld's best-designed consumer products, the *51* became one of the com-
any's all-time best sellers and was highly influential upon later pen designs.
oday, Parker retails twelve different product lines, including fountain pens,
ollerballs, ball-point pens and pencils, which range in price from £5.95 to
5,600.

STEPHEN PEART

BORN 1958
DURHAM, ENGLAND

Stephen Peart studied industrial design at Sheffield City Polytechnic and later at the Royal College of Art, London, where he graduated with a Master's Degree in industrial design. After working briefly in London, he joined the international design consultancy **Frogdesign** in Germany. When the office began working with **Apple Computer** in 1982, Peart was one of four members of staff who went to America to establish Frogdesign's office in Campbell, California. He subsequently worked on a number of Apple products, including the *ImageWriter II*, which gave Peart "an early and thorough understanding of the technical future". Having decided to remain in Campbell, he founded his own design consultancy, Vent, in 1987 and has since worked for Apple Computer, **Nike**, GE Plastics, **Herman Miller**, the Knoll Group, Sun Microsystems, Visioneer, COM21, Jetstream, Plantronics and O'Neill. Peart's *Animal* wetsuit (1990) for O'Neill demonstrates his mastery of high-tech materials and processes and his broad understanding of **ergonomics**. The suit is made of high-pressure injection-moulded foamed neoprene, and its highly innovative "accordion pleats", which are designed to facilitate greater freedom of movement, were the result of studies into the kinetics of the human body. Peart has also worked with **Ross Lovegrove** on

Animal wetsuit for
O'Neill, 1990

several projects, most notably the *Surf Collection* for Knoll (1994), which includes a wave-shaped ergonomic keyboard wrist rest and a chair lumbar support. Peart and Lovegrove also collaborated on the *EMMA* system (1995–1999), a state-of-the-art work environment concept for Herman Miller which features an interconnected floor system for electronic services management, and workstations with injection-moulded honeycomb-celled polycarbonate work surfaces Peart is also involved in the design of high-tech advanced communica-

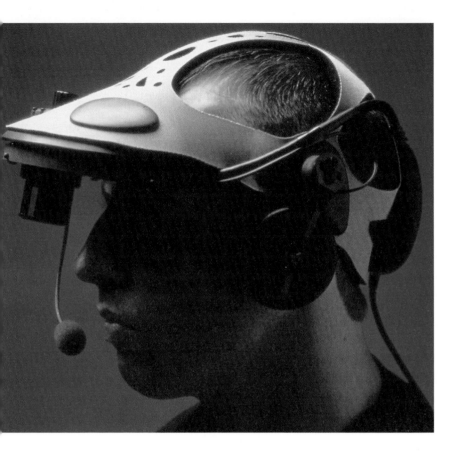

ions products. His *Phone in Your Ear* for Plantronics (1993), *Dri Projector* for
Digital Reflection Inc (1996) and *Persona* for Sun Microsystems Inc. (1998)
all encase technology in seductive three-dimensional visions of the future.
Central to all of Peart's design work is the idea of "human interaction", and
while he is fascinated by what he describes as the "touchy-feely part of technology", he never forgets that "humans should always be the goal".

Computer Cap for
Virtual Vision, 1996

PENGUIN BOOKS

FOUNDED 1936
LONDON, ENGLAND

Jan Tschichold,
Re-design of the
Penguin logo, 1947

Allen Lane (1902–1970) began his publishing career at Bodley Head, the progressive publishing house owned by his uncle, John Lane. He became a director of the company in 1925 and was appointed its chairman in 1930. In 1935 he started independently publishing paperback reprints and a year later founded Penguin Books – to "put into the hands of people like myself the books they would have read if they had gone to university". Alongside reprints of classics, he also published new titles that were sold at the modest price of 6d (12 US cents) each. Many of these directly promoted the cause of 1930s British Modernism, in particular Anthony Bertram's *Design* (1938) and works by the renowned design historian, Nikolaus Pevsner (1902–1983). Penguin books were extremely well designed and used easy-to-read *Gill Sans* typography. The standardized layouts designed by Jan Tschichold (1902–1974) also gave the imprint a strong visual identity. As the chief pioneer of paperback publishing in Britain, Lane is also credited with having improved printing and binding techniques. Described as having the "noblest list in the history of publishing", Penguin Books attracted a completely new readership who avidly consumed the company's works of quality fiction and non-fiction.

Jan Tschichold,
Penguin book covers,
1946–1949

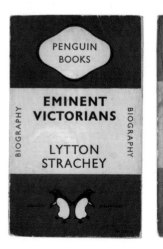

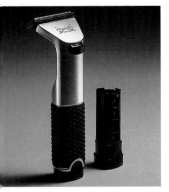

Kenneth Grange,
Compakt razor for
Wilkinson Sword,
1998

In 1962 Alan Fletcher (b. 1931), Colin Forbes (b. 1928) and Robert Gill (b. 1931) formed a progressive graphic design partnership based in London. Not wanting to become a fully blown advertising agency, the partners decided to expand into three-dimensional design and in 1965, when Gill left, were joined by the architect Theo Crosby. The office subsequently received two large commissions from Cunard and British Petroleum, the latter of which included the design of petrol stations. For these, they approached the industrial designer **Kenneth Grange** to assist in the design of pump equipment. Eventually, Grange and the graphic designer Mervyn Kurlansky (b. 1936) became officially affiliated with the consultancy and the resulting five-man team was renamed Pentagram in 1972. During the 1970s and early 1980s, John McConnell, David Hillman, Peter Harrison and David Pelham also became partners. Pentagram became noted for its bold and direct **corporate identity** designs for British Petroleum, Reuters, Faber & Faber, Watneys, the Victoria & Albert Museum, the Tate Gallery, *The Guardian* and Prestel, among others. It also developed typography for many large clients, including **Xerox**, **IBM** and **Nissan**. The Pentagram "house-style" was further perpetuated through its product design, which

Kenneth Grange,
*TT350, TT750 &
TT950* toasters for
Kenwood, 1993

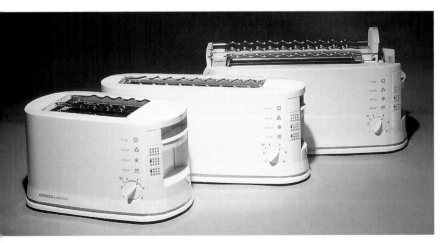

Kenneth Grange,
Power Car Exterior
for the 125 High
Speed Train, British
Railways Board, 1976

was solely undertaken by Kenneth Grange. Under the direction of Crosby, Pentagram has also undertaken exhibition design, including the British Industry Pavilion at Expo '67 in Montreal and the "Environment Game" (1973) and "British Genius" (1977) exhibitions in London. Central to the group's activities is the belief that design should permeate all areas of life, and this idea has been promoted through the publication of several books, including *Pentagram: The Work of Five Designers* (1972), *Living by Design* (1977), *Ideas on Design* (1986) and *The Compendium* (1993). The consultancy has also issued a series of pamphlets, entitled the *Pentagram Papers*, which highlight "curious, entertaining, stimulating, provocative and occasionally controversial points of view". Its consumer product designs are distinguished by a utilitarian directness which is tempered by subtle yet sophisticated **styling**, and are based on the belief that "product design is not simply a matter of juggling variables, but also a process of discovery where the connections can produce their own surprises." With offices in London, New York and San Francisco, Pentagram has maintained its position as a leading international design consultancy whose partners share "a passion for design, an entrepreneurial spirit, an interest in business, and an interest in the kind of collaborative efforts that we find in a community of designers". Pentagram's longevity and success can be attributed to its collective structure and multi-disciplinary approach to design, which enhances creative synergy and innovative problem solving.

Stainless steel
saucepans for
Kooperativa
Förbundet, 1978

Sigurd Persson trained under his father, Fritiof Persson, as a silversmith in Helsingborg from 1928 to 1937. He gained a journeyman's diploma in 1937 and subsequently studied at the Akademie für Angewandte Kunst, Munich, from 1937 to 1939 and at the Konstfackskolan, Stockholm, from 1939 to 1942. He established his own workshop in Stockholm in 1942 and began designing cutlery and jewellery. In 1949 Persson founded an industrial design practice in Stockholm and subsequently designed sculptural stainless-steel vegetable dishes for Silver & Stål (1953) and the *Jet Line* range of cutlery (1959) for the Kooperativa Förbundet (Co-operative Society of Sweden). In 1959 he won a competition to design in-flight cutlery for the Scandinavian Airlines System (SAS) and later executed matching grey plastic tableware for the airline. Between 1968 and 1982 Persson worked as a freelance consultant to the Kosta glassworks, producing brightly coloured, minimally decorated designs. In 1976 he designed uniform badges for the Swedish Army and, with nine other silversmiths, opened the Argentum gallery in Stockholm. In 1978 he designed a range of strong and boldly formed saucepans for the Kooperativa Förbundet which exemplified his approach to the design of practical, everyday objects. Successfully bridging the worlds of handcraftsmanship and industrial design, Persson believed that design had to be truthful in order to be successful. "In every well-designed object", he felt, "it is not just the function that is described in the form. Something else is there, too – we might call it an expression of the times."

Grey plastic in-flight
service for the airline
SAS, 1959

Workers putting the finishing touches to the newly launched *Model 444* broadcast receiver (soon afterwards renamed the *People's Set*) at the Philco radio factory in Perivale, 1936

Philco, *Model 444* broadcast receiver, 1936

Philco was established as the Helios Electric Co. and initially produced batteries. In 1909 it was re-named the Philadelphia Battery Storage Company (Philco) and in 1927 began manufacturing radios. As early as 1929, it was mass-producing radios using assembly-line techniques and rapidly became one of the "big three" radio manufacturers. In 1932 Philco made inroads into the British market by establishing a custom-built manufacturing facility in Perivale. In 1936 Lord Selsden's Ullswater Committee investigated the high price of radios in Britain and challenged the whole industry to produce better and cheaper models. That same year, Philco responded by launching its one-piece **Bakelite** *Model 444* radio, which sold for just six guineas. Like **Walter Maria Kersting**'s earlier *Volksempfänger*, the radio had a standardized design that was specifically intended for mass production. Initially it did not sell well, but when it was renamed the *People's Set* it became extremely popular and over 500,000 units were sold. During the 1930s, Philco also financed

the researches of the American Philo Taylor Farnsworth (1906–1971), an early pioneer of television. From the late 1940s to the mid-1950s, the company produced a wide range of technically sophisticated televisions housed in historicist cabinets. In the late 1950s, however, Philco began using futuristic **styling** for its television sets. Known as the Philco *Predictas*, these Space-Age televisions were the most distinctive sets ever designed in America. Despite the use of highly innovative forms and features such as swivelling picture screens, Philco was dogged by the poor picture quality of its televisions and eventually went out of business in 1962.

PHILIPS

FOUNDED 1891
EINDHOVEN, THE NETHERLANDS

In 1891 Gerard Philips founded a company to "manufacture incandescent lamps and other electric products". The firm concentrated on the production of carbon-filament lamps and by 1900 had become one of the largest manufacturers of its kind in Europe. As a means of stimulating product innovation, in 1914 Philips established a research department that explored both physical and chemical phenomena. In 1918 the company manufactured its first X-ray tube and from 1925 began conducting experiments into television. Two years later, Philips launched its first radio and by 1932 it had sold over one million. The following year, the company produced its 100 millionth radio valve and began manufacturing X-ray equipment in the United States. Around 1950, having previously pursued an only loose approach to design, Philips' appointed the architect Rein Veersema to oversee the design of the company's shavers, radios, televisions and record players. During his 14-year tenure at Philips, Veersema instituted a systematic design strategy that addressed every aspect of the design process, from **ergonomics** research to product costings. He also promoted the idea of developing product families so as to project a coherent **corporate identity**. Many of the designers employed by Philips had worked previously in the **Braun** design department and were thus particularly adept at designing products within the parameters of a house style. During this period, Philips became a leading

pioneer of sound recording technology and in 1963 introduced its landmark Compact Audio Cassette, which was soon after adopted as an international standard format. In 1972 the company had also developed a laser videodisc, but by the time this was finally introduced in 1978 the market was already saturated with video tape systems. In the course of developing the videodisc, however, it was discovered that it offered significantly better

sound reproduction than conventional vinyl records. This led directly to
the introduction of Philips' revolutionary compact disc (CD) in 1983, which
in a stroke rendered the gramophone record virtually obsolete and seriously
threatened the future existence of audio cassettes. From 1980 Philips' design team was headed by the American industrial designer and architect
Robert Blaich, who espoused the concept of "global design" and championed "product semantics" in an attempt to compete with Japanese companies. Since then, Philips has continued to strengthen the visual identity of
its products with an increasingly "high-tech" aesthetic. Through its commitment to design, Philips has developed an exceptionally strong brand
that is guided by its corporate motto: "Let's Make Things Better".

PIAGGIO

FOUNDED 1884
SESTRI PONENTE, ITALY

Vespa motor
scooters launched
by Piaggio in 1946

Corradino d'Ascanio,
Vespa 125 motor
scooter, 1951 –
original version
designed in c. 1945

Rinaldo Piaggio (1864–1938) founded the Società Rinaldo Piaggio in 1884. Subsequently renamed Piaggio & C., the company initially specialized in luxury ship fitting but eventually began manufacturing railway carriages, engines, coaches, vans, trams and truck bodies. During the First World War, production included aeroplanes and seaplanes, and the company later manufactured several notable aircraft, including the futuristic-looking *P7* seaplane (1929) and the all-metal *P16* bomber (1935). Piaggio's plants were virtually all destroyed in the Second World War, but the company decided to rebuild and address the pressing need for an inexpensive means of transportation. This led to the development of a scooter, which was designed by the aircraft engineer Corradino D'Ascanio (1891–1981) and whose now famous name was coined by Enrico Piaggio, who when he first saw the new design exclaimed: "It looks just like a wasp" (wasp = *vespa*). After being patented, the *Vespa* went into production and became an immediate success. By 1956, one million *Vespas* had been produced and the scooter became synonymous with the Italian Reconstruction and Latin high-life. Today, the *Vespa* is enjoying an unprecedented revival of popularity as a stylish, practical and affordable form of transport.

Lancia *Aprilia Coupé*,
1936–1937

PININFARINA

FOUNDED 1930
TURIN, ITALY

Battista "Pinin" Farina (1893–1966) worked in his brother's bodyshop, the Stabilimenti Farina, in Turin from the age of eleven. In 1920 he visited Detroit in order to meet Henry **Ford** and learn about the new industrial car production techniques which he was pioneering. In 1930 he established his own bodyshop, the Carrozzeria Pinin Farina. His goal was to develop coachbuilding into an independent industry, and with this in mind he instituted an assembly line that was able to produce seven to eight car bodies a day. Throughout the 1930s and 1940s, Farina produced a number of innovative car designs, such as the **Alfa Romeo** *6C* (1931) and the aerodynamic Lancia *Aprilia Coupé* (1936–1937). He also designed the landmark *Cisitalia* (1947), which became the first car to go on permanent display at the Museum of Modern Art in New York as "one of the eight outstanding cars of our time". It was described as the best expression of beauty and simplicity in the automotive field and set new standards for post-war car design. From 1952 Farina began styling exquisite models for **Ferrari**, such as the *250 SWB* (1961) and the *Dino Berlinetta* (1965). In 1958 the Carrozzeria Pinin Farina built a much larger plant to increase production and three years later became known officially as Pininfarina. 1966 saw the opening of the Pininfarina Study and Research Center, which subsequently developed many influential concept and production cars, including the Ferrari *Modulo* (1970) and the **Jaguar** *XJS* (1978). In 1972 Pininfarina constructed the first full-scale

Cisitalia, 1947

wind tunnel in Italy for its researches into **aerodynamics**. During the 1980s, Pininfarina lived up to its reputation as Italy's premier carrozzeria with designs such as the Ferrari *Testarossa* (1984). From 1981 the company was also assembling major components and building complete vehicles – today it produces around 40,000 car bodies per annum. Reviewing Battista Pinin Farina's life and work, the committee of the Gran Premio Nazionale Compasso d'Oro stated in 1957: "After a long period and rigorous elaboration of form, he has created the plasticity of line so influential in industrial production and in the determination of aesthetic standards for the automobile." Today, as throughout its history, Pininfarina's research into new areas of design, aerodynamics, new technologies and new materials, as well as safety and environmental protection, has allowed the company to offer innovative ideas to the automotive industry and to make practical contributions to the evolution of automotive engineering and design.

Metal table fan for
Zerowatt, 1954

EZIO PIRALI

BORN 1921 ITALY

An electromechanical engineer, Ezio Pirali was the
managing director of the Italian electrical appliance
manufacturer, Zerowatt. He designed and developed
numerous products for the company and forged a distinctive house style that relied on sleek, essentialist
forms and precision engineering. In 1954 he received
one of the first Compasso d'Oro awards and won a
silver medal at the Milan Triennale for his VE505 table
fan (1954). The motor of this unusual fan was cased in highly polished **aluminium**, and its **rubber** blades rotated within a chrome-metal wire cage
that also served as a structural support. The VE505, which was fabricated
by Elettriche Riunite of Milan for Zerowatt, was characteristic of the smooth
yet rationalist industrial aesthetic of Pirali's work and was produced from
1954 to 1960. Ezio Pirali's product designs were later shown at the 1979
"Design & Design" exhibition in Milan and at the "Design Since 1945" exhibition held in Philadelphia (1983–1984). Pirali's work reflected the sheer in-
ventiveness of post-war Italian design, an attribute born of necessity that
contributed significantly to the country's ensuing economic reconstruction.

Zerowatt VE505 table
fan for Fabbriche
Elettriche Riunite,
1953

POLAROID

FOUNDED 1937
BOSTON, MASSACHUSETTS, USA

Polaroid *Land
Camera* being used
by the US military,
c. 1952

Polaroid *Land
Camera Model 95,*
1948

In 1932, while at Harvard University, Edwin Land (1910–
1991) developed the first polarizing filters made from
synthetic material. That same year, Land and his physics
professor, George Wheelwright, established a company
to develop applications for this new polarizing technology, including sunglasses, cameras and car headlamps.
The Polaroid trademark was officially registered in 1935 and the Polaroid
Corporation founded two years later. The company licensed its polarizing
technology to American Optical and Bausch & Lomb, and in 1939 Polaroid
sold over one million pairs of its anti-glare sunglasses. Land's revolutionary
concept of instant photography was born during a Christmas holiday in
Santa Fe in 1943, when his daughter asked: "Daddy, why can't I see the picture now?" Land developed his idea over the next three years and in Febru-
ary 1947 presented the first Instant Photograph to the Optical Society of
America. The first Polaroid *Land Camera*, the *Model 95*, was subsequently

launched in 1948, and by 1956 production of this groundbreaking design had reached one million units.
With its concertina lens, the camera
was nevertheless relatively bulky
and could only take black-and-white
images. 1957 saw the introduction
of the first Polaroid slide-system,
which enabled images to be projected within a two-minute waiting
period. In 1963 Polaroid introduced
the first instant colour film, representing the culmination of 15 years'
research and development. By 1964,
over five million *Land Cameras* had
been manufactured and in 1971 the
MiniPortrait camera was introduced,
designed for taking multiple pass-
port photographs. By incorporating

an automatic ultrasound rangefinder, Land was able to reduce the dimensions of his cameras, which resulted in the compact *SX-70* (1978) designed by **Henry Dreyfuss** Associates. This design served as a blueprint for many subsequent models, such as the *Spectra System* (also known as *Image System*) launched in 1986. This computerized camera used an advanced film that required two different chemicals for its processing. In 1996 Polaroid introduced its first digital camera, the *PDC2000*, and a year later launched the *SpiceCam* instant camera (named after the Spice Girls pop band). These new products have neverthess failed to restore the embattled company's diminishing fortunes, which have suffered considerably from the emergence of digital photographic technology.

GIO PONTI

BORN 1891 MILAN, ITALY
DIED 1979 MILAN, ITALY

→Office chair for
Kardex Italiano, 1938

Giovanni (Gio) Ponti studied architecture at the Politecnico di Milano and
subsequently worked for the architects Emilio Lancia and Mino Fiocchi.
From 1923 to 1930 he was art director of the Richard Ginori **ceramics** factory,
where he designed porcelain wares decorated with neo-classical motifs in
the Novecento style, some of which were awarded a Grand Prix at the 1925
Paris "Exposition Internationale des Arts Décoratifs et Industriels Modernes". Ponti also designed low-cost furniture for the La Rinascente department store in the mid-1920s, and from 1925 to 1979 was the director of the
Monza Biennale exhibitions. In 1925 Ponti designed his first building – his

Coffee machine for
La Pavoni, 1949

own neo-classical style house on the Via Randaccio in Milan – and in 1926
he established an architectural practice in partnership with Emilio Lancia in

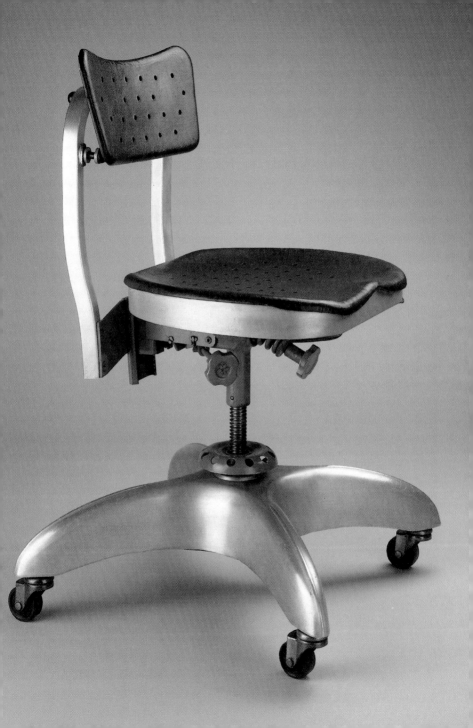

Series P sanitary
fixtures for Ideal
Standard, 1953

Milan, which continued until 1933. In 1928 Ponti launched *Domus*, the prestigious design journal established to promote the Novecento Movement and counter both "the fake antique" and "the ugly Modern" in architecture and design. Between 1933 and 1945, Ponti worked in partnership with the engineers Antonio Fornaroli and Eugenio Soncini in Milan and during this period undertook many architectural commissions for both private and public buildings. From 1930, he designed lighting and furniture for the Fontana company and in 1933, with Pietro Chiesa (1892–1948), was made artistic co-director of its subsidiary, Fontana Arte. During the 1940s, Ponti contributed extensively to *Stile* magazine, produced sets and costumes for La Scala opera house (1947), created multi-coloured glass bottles, glasses and a chandelier for Venini (1946–1950) and designed the famous La Pavoni coffee machine (1949). During the late 1940s and throughout the 1950s, he collaborated with Piero Fornasetti (1913–1988) on several furniture designs and interiors, including those for the Casino at San Remo (1950). Alongside several prestigious building commissions, such as the second Montecatini Building (1951) and Pirelli Tower (1956), both in Milan, Ponti also designed flatware for Krupp Italiana (1951) and Christofle (1955), sanitary ware for Ideal Standard (1953) and the widely celebrated *Superleggera* chair (1957) for Cassina, which possessed the timeless classicism that was so characteristic of his work. Ponti skilfully balanced the rational with the poetic to create designs that exemplified the Italian Line. Ponti's forward-looking products were extremely influential and reflected his belief that: "Happiness, man's last hope, is a dream. But nothing ever came to pass without being dreamed of first."

PORSCHE

FOUNDED 1931
STUTTGART, GERMANY

Professor Ferdinand
Porsche

The design practice that would later evolve into Porsche AG was founded in Stuttgart in 1931 by Ferdinand Porsche (1875–1951), who went on to design the hugely successful **Volkswagen** *Beetle* (1934–1938). During the Second World War, Porsche designed several military vehicles, including the *Tiger* tank. After the war, his son "Ferry" used Volkswagen components and a flat four-cylinder engine to create the 356 roadster (1948), which became the first vehicle to carry the Porsche name. The streamlined body of this sports car was designed by **Erwin Komenda** not only to look beautiful, but to be as functionally efficient as possible. By 1958, over 10,000 356s had been manufactured. In 1961 the company began work on a new model with a rear-mounted, air-cooled six-cylinder engine and a body designed by Komeda and Ferry Porsche's son **Ferdinand Alexander Porsche**. This resulted in the immortal 911, which was first produced in 1964 and went on to become one of the most famous sports cars of all time. Still in production today, numerous evolutions of the 911 were manufactured, including the 911 *Carrera RS* (1972), which was the first production car with a front and rear spoiler, and the 911 *Turbo* (1974), which was the world's first production car with an exhaust gas turbocharger. Porsche's first front engine transaxle configured car, the 924, was introduced in 1976 and was followed by the similarly configured 928 in 1977, which was conceived – controversially to purists – as the successor to the 911. By the mid-1990s, however,

Evolution of the
Porsche 911 *Carrera*,
1972–1998

Porsche 959, 1985

Porsche 959,
1985 – rear view

both these models had been discontinued, despite healthy sales. As well as producing highly acclaimed road cars, such as the limited edition 959 supercar (1985), Porsche has also produced numerous racing cars, including the 917 (1970) and the 956 (1982) – two of the most successful racing cars ever. In 1993 Porsche launched a new concept car at the Detroit Auto Show – the six-cylinder mid-engine *Boxster*. Production of this roadster began in 1996 and it became an immediate success. Through its commitment to design, engineering and technological excellence, Porsche has established one of the strongest brands in Germany and its name remains synonymous worldwide with the sports car. In the company's own words: "For Porsche, the ultimate test is to exhaust all possibilities for excellence even at the cutting edge of technology. We regard every idea as an opportunity. It is this perception that has created our company and what guides us still. It is the very essence of Porsche".

Porsche *Boxster*,
1997–1998

FERDINAND ALEXANDER PORSCHE

BORN 1935
STUTTGART, GERMANY

Ferdinand Alexander "Butzi" Porsche is the grandson of Professor Ferdinand Porsche (1875–1951) who founded **Porsche** AG. He apprenticed as
an engineer at **Bosch** in Stuttgart and from 1957 studied at the Hochschule
für Gestaltung in Ulm. The following year, he began working under **Erwin
Komenda** in the Porsche design department, which he later headed from
1961 to 1972. In this position, he designed the bodies of several cars, including the famous 911 (1964). In 1972 he established the independent Porsche
Design Studio, now based in Zell am See, Austria. The experience he gained
in automotive design proved useful in the creation of "lifestyle" products.
His sleek and highly engineered products, which balance function, technology and **styling**, epitomise German product design. The Porsche Design
Studio has produced work for a prestigious international clientele, ranging
from the *Cobra* motorcycle for Steyr-Puch (1976) to ergonomic high-tech
sunglasses for Bausch & Lomb (1997). During the late 1990s, the studio
also produced designs for transit systems for the cities of Singapore, Bang-
kok, and Vienna and developed a proposal for a computer-aided taxi system
(CATS). While most of the products designed by the studio are produced
and marketed under the brand name of Porsche Design, the creative origin
of other products designed for some of the studio's customers is not always
made evident.

Singapore *Airport
Express* train for
Siemens SGP, 1998

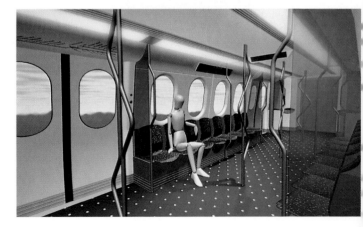

Diver's watch for
IWC, 1983

→Motorcycle helmet
for Römer, 1976

Telephone for NEC,
1981

TC 91Lz11d1.5100
coffee machine, *TW
91100* cordless kettle
& *Tt 91100* toaster
for Siemens, 1997

Trend porcelain tea service for Thomas, 1981

Lord (David) Queensberry (b. 1929) studied **ceramics** at the Central School of Art & Design, London, and design and technology at North Staffordshire College of Technology. He later studied industrial ceramic design at the Royal College of Art in London before working as a designer in the ceramics industry in Stoke-on-Trent. From 1955, he worked as a freelance for the German company, **Rosenthal**. During the early 1960s, Queensberry designed glassware for Webb Corbett that was distinguished by bold geometric forms and repetitive patterning. In 1960 he was appointed Professor of Ceramics at the Royal College of Art and four years later founded the Queensberry-Hunt Design Group with Martin Hunt (a fellow student at the RCA). Believing in "good design for the mass-market", the practice designed industrially produced ceramics and glassware for Australian Fine China, **Corning** Glass Europe, Curver, Dartington Crystal, Ideal Standard, Le Creuset and Poole Pottery. One of Queensberry-Hunt's most successful designs is the *Trend* dinner service (1981) which is still manufactured by Thomas – a subsidiary of Rosenthal. After many years' experimentation, in 1982 Rosenthal finally perfected the manufacture of "Queensberry Marble", a material that requires the simultaneous processing of two different colours of porcelain or ceramic slip. Queensberry-Hunt's restrained designs, which balance practicality and versatility with attractive yet simple forms, are ideally suited to mass production. The practice was later joined by Robin Levien (b. 1952), who was nominated as a Royal Designer for Industry, and renamed Queensberry Hunt Levien.

Kyomi washbasin for Ideal Standard, 1997

DIETER RAMS

BORN 1932
WIESBADEN, GERMANY

Dieter Rams is Germany's most important post-war designer. During his
40-year career at **Braun**, he transformed the way consumer products are
designed. Guided by deeply held moral convictions, Rams pioneered an approach to design that was based fundamentally on logic. From an early age,
Rams was exposed to construction techniques in his grandfather's cabinet-
making workshop in Wiesbaden, which produced well-made undecorated
furniture without the use of machinery. He went on to study architecture
and interior design at the Werkkunstschule in Wiesbaden from 1947 to 1948.
His father, an electrical engineer, insisted that he gain practical experience,
however, and so he served a carpentry apprenticeship in Kelkheim from
1948 to 1951. On returning to Wiesbaden, Rams resumed his training under
Professor Söder, who taught both architecture and interior design. After
graduating in 1953, Rams worked briefly for an architect who designed public sector buildings, which Rams detested for their shoddy quality and lack
of honesty. He eventually managed to get a job in the Frankfurt architectural

Phonosuper SK 4
radio-phonograph
for Braun, 1956
(co-designed with
Hans Gugelot)

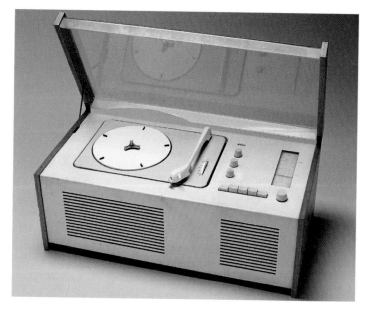

Audio 2 TC 4S hi-fi for Braun, 1964

practice of leading German Modernist Otto Apel (1906–1966). Apel's office had an affiliation with Skidmore, Owings & Merrill, and Rams became inspired by the "can-do" attitude of the American practice and learnt from it "the simple essence of industrial construction". During this period, Rams applied for a position at **Braun**. He was subsequently summoned to the company for an interview, during which he met Erwin Braun and was shown designs by **Hans Gugelot**, which he found astoundingly Modern. In 1955 Rams began working as an architect and interior designer for Braun and the following year began designing products for the company, most notably the *Phonosuper SK 4* hi-fi system (1956) which he co-designed with Hans Gugelot. Rams also designed other audio equipment for Braun, including the portable *Transistor 1* radio (1956) and the combination phonograph and pocket radio (1959), all of which embodied the practical and ordered approach to oversee the design of that was promoted by the **Bauhaus** and the Hochschule für Gestaltung, Ulm (with which Braun maintained close links). In 1961 Rams was appointed head of the company's design department and went on to oversee the design of the *KM 2* kitchen appliances, the *Sixtant* electric razor, calculators, radios, hifi systems and a cylindrical table lighter. These products were characterized by a pared-down rationalist aesthetic and exemplified the attributes of good design. Also in the 1960s, Rams designed the *606* shelving system and *620* and *601* modular seating ranges

for the furniture manufacturer Vitsoe, all of which shared the rational clarity
of his product designs for Braun. In 1968 Rams became director of design
at Braun and the same year was elected an Honorary Designer for Industry
by the Royal Society of Arts, London. By the late 1960s, the idea of "product
aesthetics" and in particular the Modern vocabulary of functionalism was
coming under increasing attack, as it was seen by many as a promotional
sales tool. Rams, however, remained unswayed by such criticism and continued to promote **essentialism** in design through efficient, well-designed
and executed products that incorporated state-of-the-art technology. Rams
believes that the central responsibility of designers is to instil order in contemporary life. His approach to design has always been based on his "Ten
Commandments on Design" – "Good Design ... is innovative, makes a
product useful, is aesthetic, helps to understand a product, is unobtrusive,
is honest, is durable, is consequential to the last detail, is concerned with
environment ... Good design is as little design as possible." His rallying cry
of "Back to purity, back to simplicity" and his declaration that the "aesthetic
requirement of an industrial product is that it should be simple, carefully
made, balanced and unobtrusive" have ensured that his work has maintained an unparalleled consistency and quality throughout his career. Rams
officially retired from Braun in 1995, but his legacy continues to powerfully
influence the ethos of the company.

RCA
(RADIO CORPORATION OF AMERICA)

FOUNDED 1919
NEW YORK, USA

ridge R. Johnson,
lking Machine,
00 – an improved
rsion of Emil
erliner's earlier
amophone

Victrola phono-
aph, 1906

On entering the First World War in 1917, the US government decreed that all American wireless stations were to be taken over by the US Navy, including **Marconi**'s operations. After the war, the Assistant Navy Secretary, Franklin D. Roosevelt, felt that all patents pertaining to broadcasting equipment should remain in America and blocked **General Electric**'s planned sale of equipment to the British-owned subsidiary of the Marconi Wireless Telegraph Company. Instead, GE took over Marconi's American assets and established the Radio Corporation of America (RCA) in 1919. Headed by David Sarnoff (1891–1971), the company initially marketed radios manufactured by GE and **Westinghouse**. In 1920 RCA obtained the first commercial broadcasting licence and the following year made the first-ever sports broadcast. The company subsequently transmitted the first radio photograph across the Atlantic Ocean in 1924 – an astonishingly early precursor of the facsimile machine. In 1926 Sarnoff established the National Broadcasting Company (NBC) to take over responsibility for RCA's radio broadcasting, such as the first coast-to-coast broadcast of the Rose Bowl game in 1927. Two years later, RCA acquired the Victor Talking Machine Company and began manufacturing its own radios and phonographs. Victor had been founded by Elridge Johnson around 1900 and manufactured **Emil Berliner**'s gramophone and the highly popular *Victrola* phonograph (1906). With the acquisition of Victor, RCA also obtained its famous "His Master's Voice" (HMV) trademark.

During the 1930s, RCA became a household name not only through its ubiquitous cathedral-style table-top radios, but also through its associations

athedral-style table-
p radios rolling
own the massive
ssembly lines at the
CA Victor Company's
anufacturing head-
uarters in Camden,
ew Jersey, 1934

with Hollywood and the construction of Radio City in New York, its corporate headquarters and the home of NBC. The company also developed its first experimental television set, which it unveiled at the 1939 **New York World's Fair**. The development of commercial television was halted, however, as America entered the Second World War and

The first factory-fitted automobile radios in America, 1936

RCA Victor experimental television receiver featuring a 21" tri-colour tube, 1951

RCA began manufacturing equipment for military use. After the war, RCA immediately resumed radio production and, having benefited from its wartime research into picture tubes for the US Navy, began manufacturing the black-and-white *Model 630TS* 10" television set in 1946. Although not the first colour television on the market, the company's all-electronic black-and-white-compatible colour television, which was developed in 1953, was recognized as the US Standard. Despite the fact that the 12" set cost a massive $1000 when it was first marketed in 1954, within six years there were 500,000 colour televisions in use across America and 60% of NBC's evening programmes were broadcast in colour. It would be RCA cameras that later transmitted the historic images of **NASA**'s space flights into millions of homes. The company's investment in space technology considerably assisted its development of solid-state colour televisions, which were introduced in 1970. In 1977 the company launched the first-ever four-hour VCR and the following year began marketing the first programmable VCR and colour home recording cameras. In 1979 RCA produced its 100 millionth television picture tube. In 1986 it was acquired by its founder and rival GE, which subsequently sold its consumer electronics interests to Thomson Consumer Electronics the following year. In 1989 RCA manufactured its 50 millionth colour television and in 1994 Thomson entered the digital home entertainment market with the launch of the RCA *Satellite System*, the first high-power direct broadcast satellite system. Featuring DIRECTV programming, the system ushered in the era of HDTV (High-Definition Television) – a technology with vast implications for the future of multimedia. Today, Thomson is the fourth largest producer of consumer electronics in the world and is continuing to expand with the launch of state-of-the-art products marketed under the famous RCA brand.

John Vassos (attributed), portable phonograph *Model M, RCA Victor Special*, c. 1937

C. L. Sholes & C.
Glidden, Remington
No. 1 typewriter, 1873

REMINGTON

FOUNDED 1816
ILION GORGE, NEW YORK, USA

In 1816 Eliphalet Remington (1793–1861) fashioned a flintlock rifle that became renowned for its accuracy, and that same year turned his family's forge in Ilion Gorge, New York, into a gun manufacturing business. The venture grew, and in 1828 Remington established a large factory alongside the Erie Canal. Remington and his son Philo subsequently pioneered numerous innovations in arms manufacturing, including a lathe for cutting gun stocks and the reflection method used to straighten the barrels of guns. They also developed the first American practical cast-steel drilled rifle barrel and in 1847 supplied the US Navy with its first breech-loading rifle. As a pioneer of large-scale industrial production, it is not surprising that the company eventually diversified into other areas of manufacture. In 1873 E. Remington & Sons developed the first-ever commercial typewriter, the Remington *No. 1*. This landmark product was designed by Christopher Latham Sholes (1819–1890) and was based on an earlier, cruder model which Scholes had patented in 1868. It was Sholes who first coined the word "typewriter", and his Remington *No. 1* featured the "QWERTY" keyboard layout that is still the standard today. Indeed, many features of this revolutionary machine remained standard for over a century: the cylinder, with its line-spacing and carriage-return mechanism; the escapement mechanism, which moves the carriage along between each letter; the actuation of the typebars by means of key levers and wires; and printing through an inked ribbon. 1878 saw the introduction of the improved Remington *No. 2*, which was the first typewriter to have a shift-key mechanism and was a much better seller than the Remington *No. 1*, which could only print capital letters. Around this time, the company further diversified its product range to include sewing machines. In 1886 E. Remington & Sons sold its typewriter business and became the Remington Arms Company, which supplied a significant number of small arms to the US government during the First and Second World Wars and remains in operation today. Meanwhile, the Remington

Advertisement for
Remington sewing
machines, c. 1880

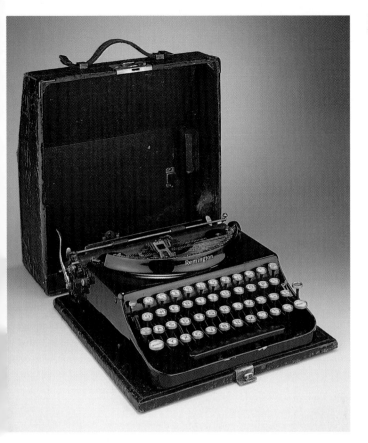

typewriter manufacturing company merged with the Rand Kardex Company
in 1927 and subsequently became a leading supplier of office equipment. In
1936 it established an electric shaver division and the following year launched its innovative *Close Shaver*. At the 1939 **New York World's Fair**, Remington
introduced the idea of dry shaving to literally millions of Americans, and a
year later launched its first dual-headed shaver, the Remington *Dual*. In 1960
Remington brought out the first cordless shaver, the *Lektronic*, employing
rechargeable nickel cadmium energy cells, and in 1975 adopted the *Soft Touch*
flexible foil cutting system. Four years later Victor Kiam formed Remington
Products Inc., stating: "I liked the shaver so much I bought the company".
In 1981 the first Remington retail store was opened and over the succeeding
decades the company has become a leading developer of shaving, grooming, personal care and travelling appliances.

Renault *R4*, 1961

RENAULT

FOUNDED 1898
BOULOGNE-BILLANCOURT, FRANCE

In 1898 Louis Renault (1877–1944) built his first automobile, the *Voiturette*, in a small workshop on his fam-
ily's estate in Billancourt. This early design had a direct transmission – an
invention that Louis Renault patented. The same year, he founded the So-
ciéte Renault Frères with his two brothers, Fernand and Marcel. Together
they designed pioneering racing cars which won numerous competitions,
but when Marcel was tragically killed during a Paris-Madrid race in 1903, the
company decided to concentrate its efforts on production vehicles. During
the First World War, Renault manufactured a tank that was used as a troop
escort vehicle. After the war, Renault's production capacity was increased
and during the 1930s the company manufactured a vast array of vehicles, including taxis. Although Louis Renault knew how to make successful cars, he
was not so politically astute and continued to produce military equipment
during the German occupation of
France. Afterwards, he was impris-
oned for collaboration and his company was nationalized in 1945. In
1961 the company issued the highly
successful Renault *R4* (1961), which
in turn became the basis of the
landmark Renault 5, introduced in
1972. Designed by Michel Boué
(1936–1971), the stylish Renault 5
hatchback was specifically intended
for young people and women and
created a completely new category
of car– the "supermini". By 1975
Renault was producing 1.5 million
cars per annum, of which 55% were
exported. Much of this success can
be attributed directly to the Renault
5, which became the best-selling
French car of all time. A new version
of the Renault 5 was launched in

Renault *5 TL*, 1972

Renault *Espace 2000 TSE*, 1986

Renault *Twingo*, 1992

1982 and in 1986 the company introduced another landmark design, the *Espace*, which similarly created a new market sector – the "people carrier". During the 1990s, the *Twingo* (1992) was one of the first automotive expressions of **soft design**, while the compact *Scenic* (1996) – voted Car of the Year in 1997 – was the first monospace design to appear in the lower-medium (M1) segment of the market. Today, Renault remains the leading French car manufacturer and continues to be one of the world's most innovative automotive companies.

Tapio Wirkkala,
Kurve cutlery, 1963

ROSENTHAL

FOUNDED 1879
SELB, GERMANY

During the 1870s, Philipp Rosenthal (1855–1937) worked in a porcelain factory in America. He eventually returned to Bavaria in the late 1870s and in 1879 established Rosenthal Porzellan AG, a workshop in Erkers-
reuth, near Selb, specializing in the decoration of porcelain. Experiencing
difficulties in acquiring blanks, he finally established his own factory in Selb
in 1891. In the early years of the 20th century, the company produced **ceramics** in a simplified Jugendstil style, which brought it widespread acclaim.
These wares, which included Philipp Rosenthal's *Darmstadt* service (1905)
and *Donatello* service (1907), were influenced by the design reform advocated by the Darmstädter Künstlerkolonie (Darmstadt Artists' Colony) and
the **Deutscher Werkbund**. In his *Isolde* service of 1910, Rosenthal subsequently attempted to break away from conventional forms altogether by integrating the handle into the overall shape. Rosenthal initially manufactured
such ceramics with undecorated finishes in order to emphasize their strong
graphic forms, but they were later made available with simple underglazed
hand-painted decoration, such as small boughs of cherries. The factory also
produced wares based on 18th-century patterns, as well as Art Nouveau-

Raymond Loewy &
Richard Latham,
Form 2000 coffee
service, 1954

Jasper Morrison,
Moon porcelain
service, 1997

tyle figurines, including several of bathing beauties designed by R. Marcuse around 1914. **Wilhelm Wagenfeld** designed a range of Modern ceramics for the company in the 1930s, and in the 1950s **Raymond Loewy** and Richard Latham designed the well-known *Form 2000* service (1954). By commissioning leading designers such as these, the company gained much recognition and as a result became the most prominent German ceramics manufacturer. In 1960 Rosenthal opened its first "Studio-Haus" to retail not only its own products but also those of other progressive factories and studio potteries. All the designs were approved by a panel of advisers that included both artists and design critics, and further Studio Houses were subsequently established in most major European cities. In 1969 Rosenthal introduced **Walter Gropius'** *TAC1* service, which epitomized the company's aesthetic – modern, sculptural and functional. Apart from its porcelain, Rosenthal has also produced glassware as well as silver and stainless steel products, such as Tapio Wirkkala's *Kurve* range of cutlery (1963). Among the many other designers and design practices that have designed products for the company's Studio-Line range are Timo Sarpaneva (b. 1926), Lino Sabbatini (b. 1925), Michael Boehm (b. 1944), **Queensberry-Hunt**, Michael Young (b. 1966) and **Jasper Morrison**. Rosenthal also has its own design studio based in Selb and a subsidiary, Thomas, that produces similar but more utilitarian wares.

RUBBERMAID

FOUNDED 1920
WOOSTER, OHIO, USA

All-plastic dustpan,
1956, and 2001
spatula, 1957, with
RM.33 dustpan and
brush, 1934

→*Roughneck* refuse
container, 1990s

Rubbermaid
advertisement,
1950s

The Wooster Rubber Company was founded in 1920 initially to produce Sunshine-brand rubber balloons. Around the same time, James Caldwell (1896–1977), a **rubber** company executive, noticed that his hanging metal dustpan was chipping the paint off his walls. He solved the problem by inventing the rubber dustpan, and was so impressed by the quality of the product that he set about designing rubber draining boards, soap dishes and sink strainers. Caldwell subsequently christened his new enterprise Rubbermaid. In 1934 the company joined forces with the Wooster Rubber Company to begin the large-scale manufacture of Rubbermaid products, and Caldwell was appointed the firm's president. During the Second World War, rubber was restricted to military use and so the company produced components for self-sealing aircraft tanks as well as life jackets and tourniquets. After the war, Rubbermaid resumed manufacturing housewares, although as colourants were still restricted its products were available initially only in black. In 1956 the company introduced its first all-plastic product, a dishpan (washing-up bowl), and around the same time began running advertisements that appealed to the new generation of homemakers. Over the following decades Rubbermaid expanded rapidly and today operates in numerous sectors, owning major brands such as Curver, Goody, Little Tikes, Rotring and Rolodex. Remarkably, the company manufactures over 5,000 products and introduces on average one new product a day. No other household products company can equal Rubbermaid's 97% brand recognition in the United States nor can claim to have its products in 9 out of 10 American homes. Rubbermaid's success can be directly attributed to its approach to design. The company's high-quality yet affordable utilitarian products, such as its *Roughneck* refuse container, are extremely well conceived in terms of function, durability and suitability to high-volume mass production.

Sixten Sason,
Saab 92, 1947

SAAB

FOUNDED 1937
TROLLHÄTTAN, SWEDEN

The Svenska Aeroplan Aktiebolaget (Swedish Aircraft Company) was founded by the financiers Marcus Wallenberg and Axel Wennergren in response to the Swedish government's re-arming initiative prior to the outbreak of World War Two. Having recruited some 40 engineers from **Boeing**, the company initially manufactured versions of Northrop and **North American** aircraft, before developing its own designs, such as the Saab 17 bomber, the Saab 21, featuring the world's first ejector seat, and the Saab 29, Europe's first swept wing fighter. In 1945 the Swedish Air Force reduced its purchasing of aircraft so Saab decided to produce automobiles instead. Its first car, the Saab 92 (1947), was a revolutionary streamlined design styled by **Sixten Sason** and engineered by Gunnar Ljungström. Inspired by aircraft design, it had sweeping aerodynamic lines and was the first small car to be produced in Sweden. The Saab 92 reflected Sason's design philosophy that form should follow function and that products should not be too expensive nor look too utilitarian. The Saab 92 evolved over the following decades into other Saab models such as the well-known Saab 95 estate car (1959). Sason and Ljungström went on to develop a new car, the Saab 99, which was compact yet spacious inside, aerodynamically efficient and economical to run. Launched in 1969, the Saab 99 became the blueprint for later Saab cars, including the Saab 90 (1985) and the Saab 9–5 SE (1997).

Sixten Sason,
Saab 95, 1959

Saab 90, 1985

Saab 9–5 *SE*, 1997

Drop light fixture for
Arteluce, 1993

MARC SADLER

BORN 1945
INNSBRUCK, AUSTRIA

Marc Sadler studied industrial design at E. N. S. A. D. in
Paris and in 1968 became a member of Design Center I,
a French organization for designers. The following year
he founded Deco Design, an innovative interior design
and furnishings company based in the south of France.
Sadler subsequently joined the France Design group in
1970 and around this time began studying plastic products and the technologies used to make them. From 1971 he began working
with the Italian company, Caber, designing plastic ski boots with patented
asymmetrical shells. Spending increasing amounts of time in Italy, Sadler
opened a studio equipped with **CAD** and three-dimensional modelling technology in Asolo, and gathered together a team of model-makers and technicians to assist him in the development of state-of-the-art products for his
growing clientele. After opening branches of his studio in New York and
Venice, he finally settled in Milan, where he continues to work today. Sadler
is most noted for his sports equipment design, such as his motorcyclist's
Bap back protector (1992) designed for Dainese, which was exhibited in the

Bap back protector
for Dainese, 1992

Museum of Modern Art's 1995 "Mutant Materials" exhibition. This innovative design features an anatomically shaped, expanded polyethylene compo-

nent, which conforms to the rider's
body and acts as a shock absorber,
and an outer shell of flexible polypropylene. His *Drop 1* and *Drop 2*
light fixtures (1993) for Arteluce
similarly incorporate state-of-the-
art materials and production technology, with their soft injection-
moulded silicon-elastomer diffusers
and rigid injection-moulded Lexan
polycarbonate supports. In 1994
Sadler was awarded a Compasso
d'Oro for these designs and was
named Createur de l'Année for 1997
by the Salon du Meuble in Paris.

ROBERTO SAMBONET

BORN 1924 VERCELLI, ITALY
DIED 1995 MILAN, ITALY

Stainless steel
cutlery for
Sambonet, 1959

Roberto Sambonet studied architecture at the Politecnico di Milano from 1942 to 1945. Between 1948 and 1953 he worked at the Museo de Arte Moderna in São Paolo, collaborating with Pietro Maria Bardi. In 1953 he worked in Alvar Aalto's office, before returning to Italy, where he designed products, graphics and exhibitions for La Rinascente. In 1954 his family firm in Vercelli began manufacturing his stainless steel designs, such as his elegant fish-serving dish *Pesciera* (1956) with a bivalve shell form. With his later flatware (1959) and eight-piece nesting *Center Line* cookware (1964), Sambonet pioneered new forms while expressing an essentialist aesthetic. He won Compasso d'Oro awards in 1956 for a series of oval trays and in 1970 for his series of stainless steel containers (including the fish server and *Center Line* cookware). He was also art director of *Zodiac* magazine from 1957 to 1966 and designed glasses for Baccarat (1971) and porcelain dishes for Richard Ginori (1979). As an advocate of "objective" design, Sambonet distrusted consumer decadence and during the 1980s became interested in the social aspects of town planning.

Center Line cookware
for Sambonet, 1964

14T-183 black-and-white television, 1955

SANYO

FOUNDED 1947
OSAKA, JAPAN

The name Sanyo means "three oceans", which reflects the company's global vision and its management's commitment to "become an indispensable element in the lives of people all over the world". The Sanyo Electric Works was originally founded in Osaka in 1947 by Toshio Iue, the brother-in-law of Konosuke **Matsushita**. Within three years the firm had become Japan's leading manufacturer of generator lamps for bicycles. In 1952 Sanyo began manufacturing and marketing plastic-housed radios and a year later launched its "whirlpool"-action washing machines – an event considered a milestone in Japan's post-war consumer electronics boom. In 1955 the firm began mass-producing televisions and by 1959 it had become Japan's largest exporter of transistor radios. In 1963 the company introduced the Sanyo *Cadnica* battery, which was the first of many batteries developed by Sanyo and which heralded the era of cordless electrical appliances. In response to the oil crisis of the early 1970s, in 1979 Sanyo developed amorphous silicon solar batteries, which it began mass-producing from 1982. In 1990 the company developed a new refrigeration system for air-conditioning units which was based on hydrogen rather than ozone-damaging CFCs. As well as pioneering and commercializing "clean technologies", Sanyo is also one of the world's largest consumer electronics companies. It designs and manufactures a wide array of products – from digital cameras and televisions to videos, audio equipment, mobile phones, refrigerators, vacuum cleaners and microwaves, all of which voice Sanyo's corporate motto: "Precision craftsmanship to be proud of the world over".

MCD X77 portable CD, late 1990s

RICHARD SAPPER

BORN 1932
MUNICH, GERMANY

Richard Sapper studied mechanical engineering and economics from 1952 to 1954 at the University of Munich. From 1956 to 1957 he worked in the car **styling** department of **Mercedes-Benz** in Stuttgart. He later moved to Italy and worked in the Milan studio of Alberto Rosselli (1921–1976) and **Gio Ponti** from 1957 to 1959. In 1959 he designed his *Static* table clock for Lorenz (which was awarded a Compasso d'Oro in 1960) and from 1959 to 1961 worked for the in-house design department of La Rinascente. Around this time, he also began working in the studio of **Marco Zanuso**. In 1970 Sapper established his own design office in Stuttgart and acted as a design consultant to **Fiat** and Pirelli, among others. He continued collaborating with Zanuso until 1975 and together they produced an impressive series of landmark designs, including the *Lambda* chair for Gavina (1963); the injection-moulded polyethylene *No. 4999/5* stacking child's chair for **Kartell**

Details of *Tizio* lamp for Artemide, 1972

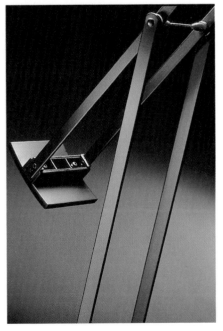

(1961–1964); the *Doney 14* television (1962); the *TS 502* radio (1964) and the *Algol* and *Black 201* portable televisions for **Brionvega** (1965 and 1969); and the *Grillo* folding telephone for **Siemens** (1966). In 1972 Sapper independently designed the highly successful *Tizio* task lamp for **Artemide**, which with its blatantly High-Tech design rhetoric became a cult object in the 1980s. In 1972, together with Gae Aulenti (b. 1927), he established the Urban Transport Systems Study Group, whose research culminated in an exhibition held at the XVIII Milan Triennale in 1979. Sapper has been a product design consultant to **IBM** since 1980, and since the late 1970s has also created a number of notable designs for **Alessi**, in which his earlier High-Tech style is synthesized with Post-Modernism. Such designs include the 9090 espresso coffee-maker (1979), which was the first Alessi product designed for the kitchen since the 1930s; the 9091 *Bollitore* whistling kettle (1983), which was Alessi's first "designer" kettle; the *Uri Uri* watch (1988); the elegant *Bandung* automatic teapot (1995); and the *Cobán* range of domestic espresso machines (1997–1998). According to Alberto Alessi, Sapper is a designer who has the "ability to transgress the rules of force in industry", allowing him to open up creative possibilities within the rigid constraints of mass production. His products – which Sapper describes as "transitional objects" – transform the cultural identities of the companies that he works for. He has also designed furniture for Knoll, Unifor, Molteni and Castelli. Sapper's technically sophisticated yet aesthetic design solutions spring from his extraordinary originality of thought. As he himself recognizes: "You cannot learn anywhere how to have an idea."

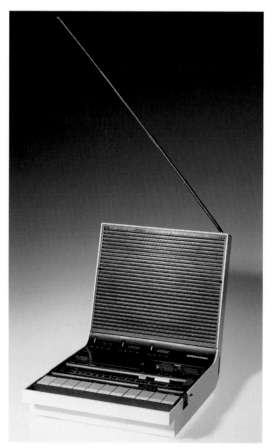

Soundbook cassette player for Brionvega, 1974

Richard Sapper &
Marco Zanuso,
Grillo telephone for
Siemens, 1966

Richard Sapper &
Marco Zanuso,
TS 502 radio for
Brionvega, 1964

SIXTEN SASON

BORN 1912 SKOVDE, SWEDEN
DIED 1967 SWEDEN

Design for an
industrial washing
machine for
Electrolux, 1950

→Design for vacuum
cleaner attachments
for Electrolux, 1944

Sixten Sason initially studied with Swedish artists living in Paris and later trained as a pilot. He subsequently worked for Husqvarna making motorcycles and small arms and later worked as a graphic designer and magazine illustrator. During the Second World War, he joined **Saab**, then a fledgling aircraft manufacturing company, and worked as a technical illustrator. Around 1945 the company decided to diversify, and Saab's chief engineer Gunnar Ljungström and his team began developing its first car, the Saab 92 (1947). Inspired by aeronautical design, Sason transformed this front-wheel-drive vehicle into a streamlined vision with integrated side panels, bonnet and boot – an innovation that was later adopted by **Pininfarina** and **Raymond Loewy**. The Saab 92 was Sweden's first small car and became the blueprint for later Saab models, such as the 95, 96 and 99 (all of which were designed by Sason). As one of the first European industrial design consultants, Sason also designed highly influential streamlined products for **Electrolux** and Hasselblad.

Design for *Model
248* vacuum cleaner
for Electrolux, 1943

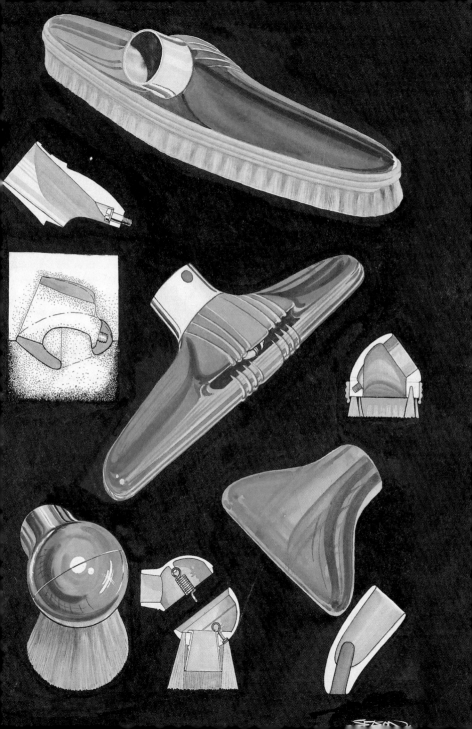

MARGARETE SCHÜTTE-LIHOTZKY

BORN 1897 VIENNA, AUSTRIA
DIED 2000 VIENNA, AUSTRIA

Margarete Schütte-Lihotzky was the first woman to graduate from the School of Applied Arts in Vienna. Influenced by the work of Heinrich Tessenow (1876–1950), she was dedicated to the improvement of workers' housing and initially collaborated with the Rote Wien group. In 1925 Ernst May (1886–1970), the director of town planning and building in Frankfurt, embarked on a large-scale mass-housing programme which focused on the concept of the "minimum-existence dwelling". For this, Mayer sought highly efficient solutions to housework and asked Schütte-Lihotsky to design the most rational kitchen possible. The resulting standardized *Frankfurt Kitchen* was equipped like a scientific laboratory, while its layout and size were designed in accordance with findings gleaned from time and motion studies. Strongly influenced by Christine Frederick's book, *Scientific Management of the Home* (1915 – translated into German in 1922), the kitchen was devised solely for food preparation and its space-saving built-in cupboards, storage units and worktops were similar to those in ship's galleys. The kitchen provided a well-lit and unadorned space and was equipped with a swivelling stool and an adjustable overhead lamp. Schütte-Lihotsky's prefabricated and standardized design was inexpensive to produce and was used for the new housing estates built in Frankfurt during the 1920s. The *Frankfurt Kitchen* played a significant role in the transformation of the art of cooking into the science of home economics.

The *Frankfurt Kitchen*, 1924

DOUGLAS SCOTT

BORN 1913 GREAT BRITAIN
DIED 1990 GREAT BRITAIN

Routemaster bus for
London Transport,
1954

Pay-on-answer coin
box for the General
Post Office, c. 1960

Douglas Scott trained as a silversmith at the Central
School of Art & Design in London and subsequently
worked in **Raymond Loewy**'s London office from 1936
to 1939. While there, he worked on designs for vacuum cleaners and refrigerators commissioned by **Electrolux**. After the Second World War, he set up
his own design office and was responsible for establishing the industrial design course at the Central School of Arts and Crafts in London. In 1949 he
co-founded the Scott-Ashford Associates design office with Fred Ashford.
He later designed the classic *Routemaster* bus (1954) for **London Transport**,
which is still in use in London today and became a widely recognized and
much-loved symbol of the city. The Routemaster (also known as the RM)
was the last of a series of buses designed specifically for London and intended to compliment and enhance the look of the
capital's streets. Having first entered service in 1956,
the *Routemaster* is still considered by many a highpoint in public transport design – a triumph of pragmatic engineering that is simple to maintain and operate. With its two-person crew and open back, the
Routemaster can pick up and set down passengers
more quickly than modern driver-only buses. Despite
these advantages, the *Routemaster* is unable to comply with recent European health and safety legislation
and is slowly being phased out.

Following this successful foray into transport design
and engineering, in 1955 Scott established his own
independent industrial design consultancy, Douglas
Scott Associates. He subsequently designed products
for companies such as **Marconi**, Ideal Standard and
ITT. Scott believed that there was no place for personal aesthetics in industrial design. His telephone
coin box (c. 1960) for the GPO (General Post Office),
which combines American **styling** with the traditional
restraint of British design, is highly representative of
his work.

Model K6 public telephone boxes, 1936 – also known as Jubilee Kiosks

Sir Giles Gilbert Scott was the grandson of the famous Gothic Revivalist architect, Sir George Gilbert Scott (1811–1878), who designed most notably the Albert Memorial in Kensington Gardens (1863–1872). Following in his grandfather's footsteps, Gilbert Scott trained as an architect and in 1898 articled with Temple Lushington Moore (1856–1920). In 1904, at the age of just 24, he won the competition for Liverpool Anglican Cathedral, defeating such strong contenders as Charles Rennie Mackintosh (1868–1928). His Gothic-style cathedral revealed the bold massing that came to characterize his later buildings, most notably Battersea Power Station (1932–1934) and the colossal Bankside Power Station (1947–1960), which is now the home

of the new Tate Modern art gallery. As well as buildings, Scott also designed two public telephone boxes for the GPO (General Post Office) in 1924 and 1936. The latter was the classic red *Model K6*, which is sometimes referred to as the *Jubilee Kiosk*. This design, which was inspired by the architect Sir John Soane's tomb in London, is as quintessentially British as **Douglas Scott**'s *Routemaster* bus and the famous black London cab. Like Gilbert Scott's buildings, the phone box is a successful synthesis of modernity and tradition. When British Telecom began replacing Gilbert Scott's well-loved phone box with new kiosks in 1988, it prompted a public outcry, with the result that many examples of this classic design can still be seen across Britain today.

Sundberg-Ferar,
automatic washing
machine and drier
manufactured by
Whirlpool for Sears,
Roebuck & Co.,
c. 1950

SEARS, ROEBUCK & CO.

FOUNDED 1893
CHICAGO, ILLINOIS, USA

Beyond design and manufacturing, the success of a
product has always hinged on distribution. The concept
of distribution by mail order was first pioneered in
America by the retailer Montgomery Ward in 1872. Fourteen years later, in
Redwood Falls, Minnesota, Richard Warren Sears (1863–1914) established
the R. W. Sears Watch Company as a mail-order company selling watches.
In 1887 he hired Alvah Roebuck as a watch repairer and moved the business
to Chicago. That same year he published his first catalogue, in which he
offered watches, diamonds and jewellery with a money-back guarantee.
Having become almost immediately successful, the company was sold for
$100,000 only two years later. Sears and Roebuck subsequently founded
another company, A. C. Roebuck & Co, which became Sears, Roebuck &
Company in 1893. The new venture published its first catalogue the same
year, running to 196 pages and offering a wide array of products from sewing
machines and saddles to bicycles and shoes. By 1894 the catalogue had
swollen to 507 pages. Sometimes known as the "Wish Book", the Sears catalogue was often the only book apart from the Bible in many rural North
American households. To those living in the American

Cover of Sears,
Roebuck & Co.
Consumers Guide
catalogue, No. 108,
1899

mid-west or on the Canadian prairies, the catalogue offered a powerful link to the cosmopolitan world and inspired many hours' dreaming, which the local general
store could not possibly match. Richard Warren Sears,
who also wrote the copy for the catalogues, would often
only attempt to locate a manufacturer for an item once
orders began pouring in – a remarkably early example
of "just-in-time" production. By 1914 consumers could
mail-order practically anything, from cutlery and furniture suites to complete houses. From 1924 Sears, Roebuck & Company was headed by General Robert Wood,
who recognized that the automobile was making retail
outlets in urban centres more accessible to consumers
in outlying suburbs and rural areas. To exploit this situation and the boom in car ownership, he opened the first
Sears retail store in Chicago in 1925. Many other stores

→Clarence Karstadt,
*Silvertone Model No.
6111* radio, 1938

were established over the next few years and by 1931, retail store sales had exceeded those of mail order. In 1934 Sears Roebuck launched the *Coldspot* refrigerator designed by **Raymond Loewy**, which was the first domestic appliance to be marketed primarily on its aesthetic appeal. The company also produced and retailed the futuristic white plastic *Silvertone* radio (1938) designed by Clarence Karstadt. During the 1950s, Sears' in-house design division was headed by Carl Bjorncrantz, who designed sleek, streamlined products ranging from household appliances to adding machines. For over 100 years, Sears has been one of the world's largest retailers of general merchandise. Today, its famous catalogues continue to offer a wide range of products, which like the early "Wish Books" provide small and remote communities access to quality goods that are difficult if not virtually impossible to obtain locally.

Raymond Loewy,
Coldspot refrigerator,
1934

SEYMOUR POWELL

FOUNDED 1983
LONDON, ENGLAND

Richard Seymour (b. 1953) trained as a graphic designer at the Royal College of Art, London, and subsequently worked in the advertising industry, which gave him a useful insight into the minds of consumers. Dick Powell (b. 1951) also graduated from the Royal College of Art, where he studied industrial design. After working as a freelance designer, Powell teamed up with Seymour in 1983. Three years later, Seymour Powell received widespread recognition for its *Freeline* kettle (1986) for Tefal – the first cordless model on the market. The team has also designed many other consumer products, but it is perhaps best known for its design of the *MZ Skorpion* bike (1993), which was created for the ailing former Eastern Block manufacturer, MuZ. Described by Dick Powell as "like a **BMW** that has had an affair with an Italian **Ducati**", the *MZ Skorpion* was designed to be "sustainable, cheap, simple and easy-to-maintain" and was styled to make an "emotional connection". Seymour Powell's ability to combine the technical with the aesthetic is also evident in the digital wristwatches it designed for **Casio**, including the *Tri Chrono* (1992) and *Overland* series (1993) as well as the highly successful *Baby G* series (1995), all of which were less aggressively masculine than Casio's previously launched *G-Shock* wristwatches. The *Vectis GX-4* waterproof camera (1995) for Minolta also reveals a similar **"Soft-Tech"** approach to design. Seymour Powell has been the subject of two English television series (*Designs On ...* 1998 and *Better by Design* 2000), which explored the duo's approach to redesigning a number of diverse products, from bras to electric cars.

Freeline cordless kettle for Tefal, 1986

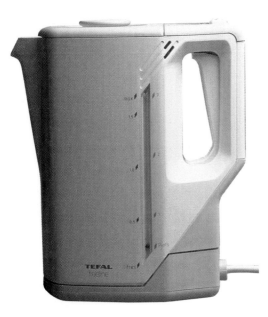

Vectis GX-4
waterproof camera
for Minolta, 1995

MZ Skorpion
motorbike for MuZ,
1993

Sharp *QT-12* stereo-radio cassette player, 1973

SHARP

FOUNDED 1912
TOKYO, JAPAN

Having founded a metalworks in 1912, three years later Tokuji Hayakawa designed the world's first mechanical pencil known as the *Ever-Sharp*, from which the company derives its name. Sharp Corporation began the production of crystal radio sets in 1925 and four years later introduced the *Sharp Dyne*, the first AC vacuum-tube radio set. In 1951 the company developed a prototype television and two years later, to coincide with the first Japanese television broadcasts, began mass-producing black-and-white television sets. These were followed by colour television sets in 1960 and microwave ovens in 1962. In 1964 Sharp produced the world's first all-transistor-diode electronic desktop calculator. During the 1970s, the company continued developing calculators and introduced the revolutionary *EL-805* calculator (1973), which was the first ever calculator to incorporate a liquid crystal display (LCD). Sharp went on to become the world leader in LCD technology. In 1973 it also launched the *QT-12* stereo-radio cassette player, which broke away from the convention of the black box by being offered in either silver or yellow finishes. Sharp introduced numerous consumer electronic products in the 1980s, from the world's first automatic front-loading VCR system (1980) to combination toaster microwave ovens (1986). The company also produced the world's first 4" and 5" colour LCD televisions in 1990 and later formed development partnerships with Intel and **Apple Computer** to produce advanced flash memory products and personal information equipment respectively. The company is dedicated to the promotion of "sharp products" – designs that "help individuals, families and corporate teams connect effortlessly, communicate clearly and unleash creativity."

MD-M11 mini disc player, 1995

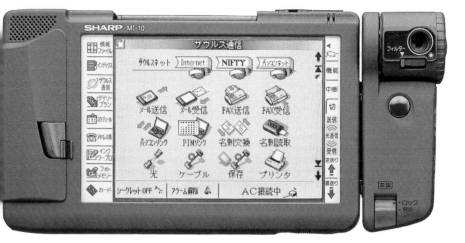

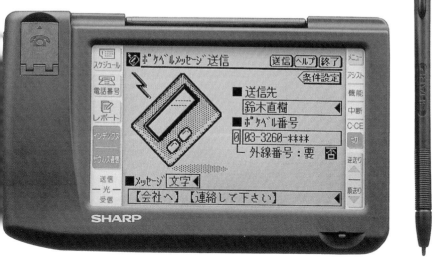

Siemens is a highly innovative electrical engineering and electronics company that manufactures approximately one million different products and aims to "benefit mankind, help protect the environment and utilize resources responsibly" within its six areas of operation – energy, industry, information and communications, healthcare, transportation and lighting. The company

T1200 telex,
1983–1986

was founded in 1847 by Werner von Siemens (1816–1892) and Johann Georg Halske (1814–1890) as the Siemens & Halske Telegraph Construction Company. The following year, the company began the construction of Europe's first long-distance electrical telegraph line between Frankfurt and Berlin. Between 1853 and 1855 Siemens constructed the Russian state telegraph network. In 1866 Werner Siemens discovered the dynamo-electric principle and invented the self-excited generator, which heralded the era of economical **electric power** that could be produced in large quantities. Siemens continued laying its telegraph cables, including the Indo-European line from London to Calcutta in 1870 and the first transatlantic line from Ireland to the United States in 1874. The company demonstrated the first electric rail-

Siemens Design,
Dental system
Sirona C1, 1994

way at the 1879 Berlin Trade Fair and two years later built the first electric tramway in Lichterfelde, near Berlin. Like **AEG**, Siemens understood early on the benefits that **standardization** could offer industrial production, and implemented an extensive program of standardized component interchangeability for its products, including its differential arc lamp produced from 1878 and its wide range of domestic appliances retailed from around the 1890s onwards. 1903 saw the establishment of a sister company, Siemens-Schuckertwerke GmbH, which took over the

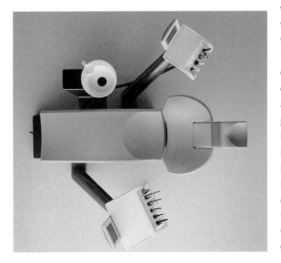

company's engineering activities and in 1905 developed an electrically powered car. That same year, Siemens & Halkse developed the tantalum lamp,
which was the first practical metal-filament incandescent light bulb. In 1919
the company, together with two other German incandescent lamp manufacturers, founded Osram GmbH KG for the large-scale mass production of
light bulbs. During the First World War, the Nonnendamm industrial quarter
of Berlin was officially named Siemensstadt (Siemens City), but by the end
of the conflict the company had lost approximately 40 % of its plant capacity. In 1932 another subsidiary was founded, Siemens-Reiniger-Werke AG,
which developed and manufactured diagnostic and therapeutic medical
equipment, including microscopes and X-ray equipment. After the Second
World War some 90% of the three companies' manufacturing facilities and
equipment was expropriated, but the firms survived and merged in 1966 to
form Siemens AG with headquarters in Munich. Since then, Siemens has
grown into a global company with over 49,000 employees actively engaged
in product research, design and development.

IGOR SIKORSKY

BORN 1889 KIEV, UKRAINE
DIED 1972 EASTON, CONNECTICUT, USA

Igor Sikorsky studied engineering in Paris and later in Kiev. In 1909 he moved back to Paris to learn as much as he could about this fledgling science, before returning to Russia to develop rotary-wing aircraft. Although his first helicopter design and its successor were complete failures, Sikorsky, undeterred, began designing fixed-wing aircraft, most notably the S-6-A. He subsequently worked for the aviation subsidiary of the Russian Baltic Railroad Car Works, where he designed the world's first multi-engined aeroplane, the S-21 (1913), whose numerous innovations included an enclosed cabin and a lavatory. In 1919 he emigrated to America and four years later founded the Sikorsky Aero Engineering Corp., which built a number of successful flying boats. Turning his attention again to rotary-wing craft, Sikorsky perfected the world's first practical helicopter, the VS-300 (1939). Following the success of this revolutionary design, United Aircraft (which had previously purchased his company) began full-scale manufacture of the first-ever production helicopter, the R-4 (1943). Sikorsky subsequently pioneered innumerable "firsts" in the helicopter industry. He was a man guided by a deep religious conviction, and often remarked: "the work of the individual still remains the spark which moves mankind ahead."

VS-300 helicopter, 1939 – the first practical helicopter

CLIVE SINCLAIR

BORN 1940
RICHMOND, SURREY, ENGLAND

Executive (centre) and
Sovereign electronic
calculators, 1972–1975

Clive Sinclair, an early pioneer of consumer electronics, established Sinclair Radionics in 1962 initially selling radios and amplifiers by mail order. This changed in 1972 when Sinclair launched his landmark *Executive* calculator, the world's first pocket calculator and a significant step forward in the **miniaturization** of electronics. Sinclair produced other pocket calculators during the early 1970s, including the slender *Sovereign*, *Scientific* and *Cambridge* models. In 1976 Sinclair designed a miniature television set, the *Microvision*, which won a **Design Council** Award in 1978. He also pioneered the development of personal computers with his inexpensive and highly successful *ZX 80* (1980) and the later *ZX Spectrum* (1984), which had 48K of RAM. During the mid-1980s these designs sold in their millions, rivalling the sales of both **Apple Computer** and **IBM**. By the end of the decade, however, Sinclair's machines had been surpassed technologically by American computers. In 1985 Sinclair produced the C5, an innovative single-person electric vehicle that became a celebrated commercial failure. The following year, Sinclair established a firm devoted solely to research.

ZX Spectrum
microcomputer,
1984

JOSEPH SINEL

BORN 1889 NEW ZEALAND
DIED 1975 USA

Detail of Joseph
Sinel's scales,
c. 1927

Model S scales,
*Height & Weight
Meter* for the
International Ticket
Scale Corporation,
c. 1927

Joseph Sinel is reputedly among the first to have used the term "industrial design", which he defined as the art and science involved in the creation of machine-made products. He initially worked as a freelance designer in New Zealand, but moved to Australia in 1913 prior to emigrating to the United States around 1918. In America in 1921, while working for an advertising agency, Sinel began undertaking "product improvement" jobs for his employer's clients. He eventually established his own industrial design consultancy some eight years before designers such as **Raymond Loewy**, **Henry Dreyfuss** and **Walter Dorwin Teague** followed suit, and this led him to describe himself in later life as America's first industrial designer. His *Model S* height and weight machine (c. 1927) for the International Ticket Scale Corporation of New York reflected the popularity of the contemporaneous Art Deco style, with its sky-scraper form and machine-polished stainless steel and chromed metal elements. Sinel also worked extensively as a commercial artist, designing trademarks and **packaging**, and in 1923 wrote *A Book of American Trademarks and Devices*. In 1930 he became a member of the American Union of Decorative Artists and Craftsmen. During the 1970s he was a professor emeritus at the California College of Arts & Crafts, and in 1975, the year of his death, he became a founding member of the American Society of Industrial Design (ASID) – a field of design in which he was an important pioneer.

Isaac Singer's first
sewing machine, 1851

Isaac Merit Singer (1811–1875) served an apprenticeship as a machinist and in 1839 patented a machine for drilling rock. While working in a Boston machine shop in 1851, he was asked to repair a Lerow & Blodgett sewing machine. Seeing a way in which the design could be improved, he set about developing his own model. Within eleven days he had built his first sewing machine. Unlike **Elias Howe**'s earlier design, Singer's machine could be operated by a treadle, allowing both hands to be kept free, and had a presser foot which kept the material in place. Both these features had been used previously by Barthélemy Thimonnier in France, but it was Singer who was the first to incorporate them successfully into a practical sewing machine. In 1851 Singer patented his design and established I. M. Singer & Company to manufacture the machine, which from 1855 was sold around the world. By 1860 Singer had become the world's largest producer of sewing machines, and three years later the firm was incorporated as the Singer Manufacturing Company. In 1885 the company unveiled its first elec-

Singer Manufacturing
Company's factory at
Kilbowie, Scotland,
1867

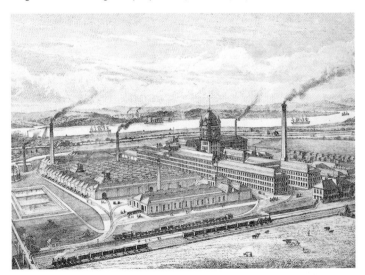

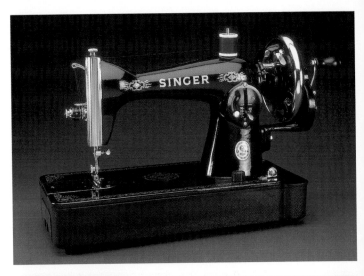

Singer sewing machine, c. 1900

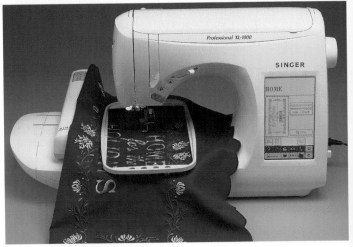

XL-1000 top-of-the-range sewing machine, late 1990s

trically powered sewing machine, although it would be another fifteen years before the first mass-produced electric sewing machines went on sale. Isaac Singer also pioneered the use of instalment credit plans, which had an enormous impact on consumer sales in modern society. In later years, the Singer Manufacturing Company diversified its product range to include furniture, power tools and floor-care products.

BR Trike for Kärcher, 1990s

HANS ERICH SLANY

BORN 1926
ESSLINGEN, GERMANY

Hans Erich Slany studied engineering in Eger (Hungary) and in Esslingen. Between 1948 and 1955 he developed products for Ritter Aluminium, before establishing his own design studio, Slany Design, in 1956. Around this time he collaborated with **Heinrich Löffelhardt**, co-designing the *Ikonette* compact camera for Zeiss-Ikon (1956). In 1959 Slany was one of the founders of the Verband Deutscher Industriedesigner (German Association of Industrial Designers), and that same year designed his first electric drill for **Bosch**. Utilizing **ergonomics** data and shockproof and heat-resistant thermo**plastics**, Slany went on to design numerous other power tools for Bosch, including an electric screwdriver (c. 1960), the *Combi-E* range of DIY tools (1966) and the first electro-pneumatic rotary hammer weighing less than 2.5 kg (1981). Since 1962 Slany has designed over 200 high-pressure cleaning products for Alfred Kärcher GmbH and has also designed a wide variety of other products, ranging from medical equipment to electronic rocketry components. He was made an honorary professor at the Berlin Hochschule der Künste in 1985. Over its 40-year history, Slany's office – which in 1997 renamed itself Teams Design – has won more than 900 national and international design awards, more than any other design consultancy in the world. Slany believes that "individual product personality" is essential for success in today's increasingly homogenous marketplace.

GBH 24 VRE drill for Bosch, c. 1992

Earliest socket
wrench set, 1920

Joseph Johnson &
William Seidemann
with a display of the
"original tool set"
on Snap-on's 25th
anniversay, 1945

Joseph Johnson worked for a company that manufactured traditional socket wrenches – tools that were constructed as one-piece units with the socket permanently attached to the handle. While there, Johnson questioned the logic of having to buy a handle in order to have the use of a single socket and wondered if it would be possible to have, for example, five handles and ten sockets that could "snap-on" to provide fifty different combinations. A colleague, William Seidemann, saw the potential of "interchangeable sockets" and together they developed a sample socket wrench set that was marketed with the slogan "Five Do the Work of Fifty". Johnson and Seideman established the Snap-on Wrench Company in 1920 and enlisted representatives to demonstrate the benefits of the Snap-on system directly to potential customers. In 1923 the first company catalogue featured over 50 designs, including open-ended wrenches, and by 1935 sales had exceeded $1 million. The Great Depression proved a difficult trading period, however, and so Snap-on introduced T. P. (time payment) selling, which allowed customers to purchase their "dream orders" and pay over time. This was an astute sales strategy that also bred much goodwill and customer loyalty. From 1941 Snap-on tools were retailed over-the-counter and the company continued developing new products, such as the patented *Flank Drive* wrenching system (1945) – a revolutionary design developed for the US Navy, which prevented the corners being sheared off aircraft fasteners under high torque conditions. The system was later incorporated into most Snap-on wrenches. In 1990 another innovative feature was pioneered, the *Flank Drive Plus* wrench, which has serrated jaws to increase the grip of open-ended wrenches. By 1993 Snap-on's sales had surpassed $1 billion and today the company remains a leader in its field.

Flank Drive
wrenching system,
1945

Flank Drive Plus
wrench, 1990

TR-55 transistor radio, 1955 – Japan's first transistor radio

→*TC50 compact cassette recorder, c. 1969*

SONY

FOUNDED 1946
TOKYO, JAPAN

In the period immediately following the Second World War, Japan saw a huge surge in demand for radios, fuelled by a populace desperate for news from around the world. In September 1945 the engineer Masaru Ibuka responded to this opportunity by opening a small electrical repair shop in Tokyo. Called the Tokyo Tsushin Kenkyujo (Tokyo Telecommunications Research Institute), the business repaired war-damaged radios and shortwave units and made its own shortwave adapters and converters that could turn short-wave radios into all-wave receivers. Ibuka was joined by his friend, the physicist Akio Morita (1921–1999), and together they founded the Tokyo Tsushin Kogyo Kabushiki Kaisha (Tokyo Telecommunications Engineering Corporation) in May 1946. Although the company's best-selling product was initially an electrically heated cushion, in 1950 – by now in larger premises – it introduced the first Japanese magnetic tape, the *Soni-Tape*, and also began marketing Japan's first reel-to-reel tape recorder, the *G-Type*. Later, the company developed the *H-type*, a less bulky and easier-to-use tape recorder in a case, which was more suitable for home use and was especially popular for educational purposes. In 1952 it devel-

TV8–301 television, launched in 1959 – the world first's portable television

oped a stereophonic audio system for the first Japanese stereo broadcast by NHK. That same year, Ibuka visited the United States and while there discovered that Western Electric was planning to release rights to its transistor patent to companies prepared to pay royalties – the transistor having been previously developed in 1948 by scientists at **Bell** Laboratories. Ibuka's company managed to obtain a licence to manufacture the transistors in 1954 and the following year introduced the first Japanese transistor radio, the *TR-55*. This was followed by the

Voyager Watchman, 1982

←Walkman, first launched in 1979

AIBO robotic dog, 1999

world's first pocket-sized transistor radio, the *TR-63* (1957). In 1958, aware of the need to appeal not just to Japanese consumers but to a global audience, the company changed its name to Sony Corporation, which sounded more Western. Around this time Ibuka noted: "The days of radio are over. The future lies in television", and in 1959 the company launched the first-ever transistor television, the *TV8–301*. Two years later, Sony entered into a contract with Paramount Pictures that involved the studio providing "technical assistance in the production of a chromatron tube and colour television utilizing it". This agreement led to the development of the small but revolutionary *Trinitron* colour televisions, which were first introduced in 1968. They included the *TV5–303*, the smallest and lightest micro-television in the world, which created a boom in America for micro-TVs. Sony's pioneering miniaturization of electronic technology continued with the introduction of the first *Walkman* personal stereo in 1979. With excellent and unwavering sound reproduction, the light and highly portable *Walkman* was an instant success and initially Sony could not keep up with consumer demand. Unlike conventional stereos with large speakers, the *Walkman* needed only a small amount of battery power because the sound was directed straight into the listener's ears. Sony has subsequently produced many versions of this landmark design, including models suitable for sporting activities. It went on to launch a flat-screened portable television, the *Watchman*, in 1982, and a portable compact disc player, the *Discman*, in 1990. More recently, Sony has been developing various digital technology products, including its miniature *Memory Stick*, which provides compact portable storage for digital data, and the sound-responsive *AIBO* robotic dog, which has infra-red eyes and can be trained to perform tricks. Sony predicts: "The Eighties was the age of the PC and the Nineties was the age of the Internet, the 2000s will be the age of the robot."

ETTORE SOTTSASS

BORN 1917
INNSBRUCK, AUSTRIA

Valentine typewriter
for Olivetti, 1969
(co-designed with
Perry King)

Ettore Sottsass studied architecture at the Politecnico di Torino and in 1958 began working as a design consultant to **Olivetti**. He designed some of the company's best-known products, including the *Logos 27* calculator (1963), the *Praxis 48* typewriter (1964), the *Valentine* typewriter (with Perry **King**, 1969) and the *Synthesis* office system (1973). His most remarkable project for Olivetti, however, was the design of the *Elea 9003* main-frame computer (1958), for which he was awarded a Compasso d'Oro in 1959. During the 1960s and 1970s, Sottsass became a prominent member of the Radical Design movement. In 1979 he collaborated with Studio Alchimia and in 1981 established the Memphis design group. His bold and colourful Memphis designs heralded a new direction in the decorative arts – Post-Modernism. As the most influential propagandist of Post-Modernism in the 1980s, Sottsass showed that it was possible and in many circumstances desirable to break away from the rational constraints traditionally imposed on the design of industrially manufactured products, in favour of a new language of design concerned with the articulation of creative expression and in particular, emotion.

Elea 9003 main-
frame computer for
Olivetti, 1959

Detail of *Model 1865* rifle

SPRINGFIELD ARMORY

FOUNDED 1794
SPRINGFIELD, MASSACHUSETTS, USA

In 1777, during the American Revolution, an arsenal was built in Springfield to produce cartridges and gun carriages. Later, the new Federal Government decided to begin manufacturing its own muskets so that it would no longer have to rely on imported arms. To this end, it established a new Armory in Springfield in 1794 and a year later the production of muskets began, with 40 workers producing 245 flintlock muskets per month. In 1819 Thomas Blanchard invented a special lathe which enabled rifle stocks to be mass-produced to a standardized design. In the period running up to the outbreak of the American Civil War, the Armory became a beacon of the **Industrial Revolution**. During the 1840s, the flintlock was replaced by a percussion cap and hammer, an innovation which led to better weather performance and faster firing. The development of rifling – cut spiral grooves in a barrel – to spin projectiles also ensured greater range and accuracy. The Union Army's subsequent victory at Gettysburg in 1863, and indeed in the Civil War as a whole, can to some extent be attributed to the *Springfield* rifle (1861) and the unprecedented scale of its mass production: between 1863 and 1864, the Springfield Armory boosted its productivity to a phenomenal 25,000 weapons per month using rationalized manufacturing techniques such as the division of labour, mechanization where possible and **standardization**. In testimony to the influence of the Springfield Armory and other pioneering small arms manufacturers such as **Eli Whitney**, mass production in America was initially referred to as "armory practice" when it was first employed by automobile manufacturers in De-

Springfield rifle with telescopic sight, 1903

troit. The Springfield Armory also produced the well-known *M1* rifle (1926), also called the *Garand* Rifle after its inventor John Garand, of which 4.5 million were made. Today, in recognition of its contribution to both American history and the evolution of industrial manufacture, the Springfield Armory has been designated a National Historic Site.

STANLEY

FOUNDED 1843
NEW BRITAIN, CONNECTICUT, USA

Frederick T. Stanley

In 1843 Frederick Stanley established The Stanley Works, a small company manufacturing door hardware, including bolts and hinges. His goal was to create a hardware company that would be unequalled in its product innovation anywhere in the world. With the early success of this mission and an increasing demand for his high-quality goods, Stanley began exporting his products in the 1870s. In the early years of the 20th century, the company opened the first of many plants outside the United States, thereby assisting in the realization of Frederick Stanley's global vision. Shortly afterwards, The Stanley Works acquired America's largest manufacturer of hand tools, Stanley Rule & Level, which had been established by a relative in 1857. By the 1920s, Stanley hardware and tools, including its famous high-quality planes, were available on every continent. In accordance with its founder's belief in expansion and diversification, Stanley today manufactures some 50,000 different products that are sold throughout the world. In recognition of its design innovation, the Industrial Designers Society of America (IDSA) awarded Stanley two silver Industrial Design Excellence Awards (IDEA) in 1999. That same year saw the launch of the *MaxGrip* line of hand tools, including ergonomically designed self-adjusting pliers and a locking wrench described as "another step forward in the evolution of the wrench", as it eliminates the tendency of the wrench to slip from the fastener. Today, Stanley works closely with the Ergonomic Technology Center at the University of Connecticut to ensure that all its products are designed to have an "ergonomic advantage."

Stanley plane, early 20th century

MaxGrip line of hand tools, 1999

PHILIPPE STARCK

BORN 1949
PARIS, FRANCE

Philippe Starck has the highest profile of any designer working today, not only because of his multitude of highly successful products but also because of his compelling showmanship, which has helped to promote the cause of design to new audiences. Starck studied at the École Nissim de Camondo and later became art director of the Pierre Cardin studio, where he produced 65 furniture designs. During the 1970s he worked as an independent designer, most notably designing nightclub interiors. In 1980 he established his own manufacturing and distribution company, Starck Products, to commercialize his earlier designs such as the *Easy Light* (1977) and *Von Vogelsang* sofa (1978) and the *Francesca Spanish* chair (1979–1980). During the 1980s, Starck became the leading "superstar of design" and worked prolifically on numerous projects. He designed elegant and sumptuous interiors, most notably for the Royalton Hotel, New York (1988) and Paramount Hotel,

Motó 6,5 motorbike for Aprilia, 1995

New York (1990), which incorporated his own designs for furniture, lighting, door handles, vases and other objects. He also became celebrated for his numerous furniture designs, from the three-legged *Costes* chair (1982) and the plastic injection-moulded *Dr. Glob* chair (1990) to the elegant *Lord Yo* tub chair (1994) and collapsible *Miss Trip* chair (1996), created for manufacturers such as Vitra, Disform, Driade, Baleri, XO and Idée. Like his furniture, Starck's lighting and product designs were also personalized with characterful names and sensual, appealing forms. Amongst the most commercially successful of these were the *Arà* table lamp for **Flos** (1988) and the *Juicy Salif* lemon squeezer (1990–1991), *Max le Chinois* colander (1990–1991) and *Hot Bertaa* kettle (1990–1991) for **Alessi**. During the 1990s, Starck designed consumer electronic products for Thomson, Saba and Telefunken that attempted to humanize technology. His *Jim Nature* television (1994) for Saba is particularly innovative with its high-density chipboard casing. Starck also designed the *Moto 6,5* motorcycle (1995) as well as the *Lama* scooter (prototype 1994) for **Aprilia** with seductive rounded forms. Today, Starck acknowledges that much of the design produced during the 1980s and early 1990s, including some of his own, was narcissistic "over-design" driven by novelty and fashion. He now promotes product durability (longevity) and believes that morality, honesty and objectivity must be an integral part of the design process. He has also stated that "the designer can and should participate in the search for meaning, in the construction of a civilized world" and predicts that "the 21st century will be immaterial and human."

STEELCASE

FOUNDED 1912
GRAND RAPIDS, MICHIGAN, USA

David Hunting (far
right) and colleagues
demonstrating the
strength of the Metal
Office Furniture
Company's filing
cabinet, c. 1919

→Steelcase adver-
tisement, 1935

Peter M. Wege and eleven other stockholders founded the Metal Office
Furniture Company in 1912. The company initially manufactured only "steel
cases" (ie. metal office safes) and it was not until 1914 that it began diversifying with the introduction of the *Victor* fireproof steel wastebasket. David
Hunting was recruited as sales and marketing vice-president that same year,
and he subsequently steered the firm into desk manufacturing. Having
learned that the Boston Customs House required non-flammable furniture,
it was obvious to him that steel was the ideal material for the job, so he bid
for the contract even though the company was not yet making desks. The
company managed to produce a prototype desk within 90 days and eventually won the contract to supply 200 fireproof metal desks. In 1919 the company introduced its first steel filing cabinets, which were advertised on the
merits of their superior strength. Around 1936, the company manufactured
one of the first Modern office systems, the revolutionary steel workstation
and office chair designed by Frank Lloyd Wright (1867–1959) for the S. C.
Johnson & Son Administration Building. In the late 1940s the company introduced the industry's first standard desk sizes, based on multiples of 15
inches, and in 1953 launched the first office furniture available in different

Steelcase *Co-
ordinated* files,
c. 1962

THING
LS LIKE
BALL"

For
**FASTER AND
BETTER FILING**
Finger-tip Action

STEELCASE
Business Equipment

**TRY THIS NEW AND DIFFERENT
PATENTED DRAWER SUSPENSION
IT WILL NEVER WEAR OUT**

Frank Lloyd Wright,
workstation and
office chair, 1936

Archival photograph
showing Frank Lloyd
Wright's office
system in use

colours (previously grey had been the only option). In 1954 the company
was renamed Steelcase Inc. and over the succeeding decades continued
developing advanced office systems. In 1974 Steelcase became the largest
furniture manufacturing company in the world – a title it retains to this
day thanks to its innovative product designs, such as the *Pathways* system,
which demonstrates the company's pursuit of product integration so as to
provide the most responsive office environments possible. With a strong
global culture, Steelcase is highly committed to continuous environmental
improvement – from the recycling of its workstations and the revamping of
certain manufacturing processes in order to reduce pollution, to the mini-
mization of waste wherever possible.

↗Office system,
1947

→ *Pathways* office
system, 1998

A/S Stelton was founded in 1960 to sell the stainless steel hollowware – sauceboats, platters, salad bowls etc. – manufactured by the Dansk Rustfrit company based in Fåre vejle, Stationby, a little town in Northwest Zealand. The products, which were of mediocre design, were intended by Stelton to form the basis of a range of contemporary housewares aimed primarily at the Danish market. Salesman Peter Holmblad was hired to market the line and in 1964, with a view to raising the company's profile, he convinced the well-known architect and designer **Arne Jacobsen** to design a range of hollowware for the firm. Jacobsen subsequently produced some sketches of cylindrical forms that were "terse, logical and functional", but the technology needed to translate his ideas into three-dimensional products did not yet exist – stainless steel being an extremely demanding material to work, as its surface can become easily distorted if it is not handled in exactly the right way. Stelton first had to develop new machines and welding techniques, therefore, before it could produce

Erik Magnussen,
No. 140 water kettle,
1988

→Erik Magnussen,
No. 460 cake knife,
1980, *No. 462* pizza
server, 1990 and
No. 463 fish server,
1990

Arne Jacobsen,
Cylinda-Line, 1967

Jacobsen's *Cylinda-Line*, which was eventually introduced in 1967. In recognition of the superior quality of its design, the *Cylinda-Line* was awarded an ID-prize by the Danish Society of Industrial Design in 1967 and the International Design Award by the American Institute of Interior Designers in 1968. The *Cylinda-Line* initially comprised around 18 pieces, but Jacobsen continued to add new designs up to 1971, while others were introduced posthumously from drawings. Today, the range comprises 34 pieces in total. Described as "products which live up to the demand for a beautiful, functional, and practical hollowware range in contemporary design", the success of the *Cylinda-Line* can be largely attributed to Jacobsen's ability to express the inherent qualities of stainless steel through highly rational yet aesthetically pleasing, timeless forms. From 1976, **Erik Magnussen** also worked for Stelton and designed a number of well-known table pieces in both stainless steel and plastic, including his classic three-piece vacuum jug set of 1977. The company's range today also includes designs by Peter Holmblad, such as his *No. 100–15* houseplant watering can, which has a similar rational aesthetic to the *Cylinda-Line*. Stelton's high-quality products exemplify Danish design and are represented in the permanent collections of museums around the world, including the Museum of Modern Art, New York, and the Victoria & Albert Museum, London.

GEORGE & ROBERT STEPHENSON

BORN 1781 WYLAM, NORTH-
UMBERLAND, ENGLAND
DIED 1848 CHESTERFIELD,
DERBYSHIRE, ENGLAND

BORN 1803 WILLINGTON QUAY,
NORTHUMBERLAND, ENGLAND
DIED 1859 LONDON, ENGLAND

George Stephenson's father was a mechanic who operated an atmospheric-steam engine, designed by Thomas Newcomen (1663–1729) and used to pump water out of coal mines. By the age of 19, the younger Stephenson was operating one of these revolutionary engines himself and the skill he developed led, in 1812, to his appointment as enginewright (chief mechanic) at the Killingworth High Pit colliery. A year later, he visited the neighbouring Middleton Colliery to appraise a "steam boiler on wheels" conceived for hauling coal trucks out of the mines. This double-cylinder locomotive had been devised by John Blenkinsop (1783–1831) and employed a patented system of toothed wheels that fitted into a racked rail. After seeing Blenkinsop's design, Stephenson believed he could produce an improved model and in 1814 built his first locomotive, the *Blucher*. This design did not meet with much success, however, until he added a funnel for the waste steam (as Richard Trevithick had done on his earlier engines), which effectively doubled the engine's power. The following year, Stephenson designed a miner's safety lamp (contemporaneously with **Humphrey Davy**'s better-known lamp) and built the *Killingworth* engine, which incorporated an innovative sprocket and chain drive. In 1817 the Duke of Portland commissioned Stephenson to construct one of his engines for use on the Kilmarnock & Troon line. In 1821 he was commissioned to build a proper steam locomotive for the new

The *Rocket* locomotive, 1829

Stockton to Darlington line, a project which culminated on 27 September 1825 with Stephenson's *Active* engine pulling the world's first public passenger train – an achievement that heralded the age of steam railways. The *Active* engine later became known as the *Locomotion* and could carry 450 passengers at a speed of 15 mph. Stephenson was subsequently asked to survey and construct a new line running from Liverpool to Manchester. In 1829, shortly before construction

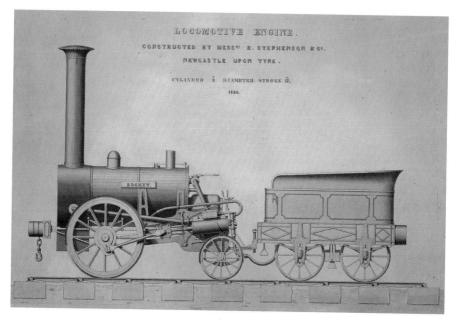

LOCOMOTIVE ENGINE.

CONSTRUCTED BY MESS.ʳˢ R. STEPHENSON & Cᵒ.

NEWCASTLE UPON TYNE.

CYLINDER 8 DIAMETER. STROKE 5.

1828.

Coloured engraving dating from 1836 showing the *Rocket* as rebuilt in 1831

of the line was completed, the famous Rainhill Trials were staged in Merseyside, with a prize of £500 being offered for the best locomotive. For the competition, Stephenson built a new engine – his now legendary *Rocket* – featuring single drive wheels on both sides which were coupled to a steam cylinder on either side of the firebox. It won with a top speed of 36 mph, and when the line was eventually opened in September 1830, eight modified locomotives, which had been built at Stephenson's works in Newcastle, were put into operation. Almost immediately, the *Rocket* came to symbolize the **Industrial Revolution**. Known as "the father of the railways", Stephenson greatly influenced the rapid transition from horse power to **steam power**. He went on to build other railroads in the Midlands, as well as acting as a consultant engineer on numerous railway projects both at home and abroad. His son, Robert Stephenson, was also a celebrated engineer and assisted his father during the 1830s with improved designs for locomotives. He too constructed railways and was acclaimed for his design of long-span rail bridges, such as the structure linking Anglesey to mainland Wales. The success of this unique tubular design led to his building of other innovative tubular bridges both in England and overseas.

Sunseeker *Predator*
75, 1999

SUNSEEKER

FOUNDED 1962
CHRISTCHURCH, DORSET, ENGLAND

Sunseeker International is the largest and most prestigious privately-owned motor yacht manufacturer in the world. The company evolved from humble beginnings when, in the early 1960s, John Macklin and Idris Braithwaite began selling small speedboats made by Owens from an automobile garage business in Christchurch. Braithwaite and his sons, Robert, a trained marine engineer and John, a designer, eventually moved the business to Poole. Converting an old yard for boat storage, they set up a chandlery shop and began selling boats by a number of other manufacturers. Robert and John Braithwaite then decided to build their own wooden craft, and their first design for the company (now called Poole Powerboats) was a revelation – the *Sovereign Sports 17*. With its high flared bow, this new design was extremely seaworthy, roomy and had a comfortable cruising speed of around 25 knots. Two of these boats were exhibited by the Braithwaite brothers at the 1970 Genoa Boat Show and sold immediately, with orders being placed for others. Also in 1970, Robert Braithwaite and John Macklin won the Royal Motor Yacht Club Trophy racing a *Sovereign* with a cutaway transom and mounted with twin 100hp outboard motors. By the end of 1970, *Sovereigns* were being sold through a network of dealers, the very first of whom was Henry Taylor in the south of France. Taylor introduced Robert and John Braithwaite to the racing boat designer Don Shead, and they subsequently commissioned him to produce a completely new type of hybrid boat – the super-fast cruiser. The ground-breaking results, which were co-designed with the Braithwaite brothers, were chris-

Sunseeker *XS 2000*,
designed by Sunseeker
in collaboration with
Fabio Buzzi, 2000

Sunseeker – XS2000

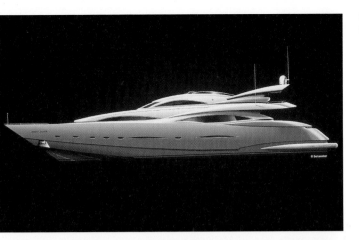

Sunseeker *105 Yacht*, 2000

Sunseeker *Superhawk 40*, 2000

tened Sunseekers and the first two 20-foot models, the *Rio* and the *Sports*, caused a sensation when unveiled at the Paris Boat Show in December 1971. Over the succeeding years, Sunseeker has produced increasingly larger and faster boats, including the *105 Yacht*, *Predator 75*, *Superhawk 40* and the awesome *XS 2000*, all with a distinctive aerodynamic form and unparalleled build quality. While not surprisingly the preserve of a very affluent minority, Sunseeker's exquisite high-performance craft represent "a triumph of technical mastery over design and materials".

Access, 1997/1998

SWATCH

FOUNDED 1983
BIEL-BIENNE, SWITZERLAND

Unable to compete with the plethora of cheap electronic watches being mass-produced in the Far East, and in particular Japan, by the mid-1970s the Swiss watch industry was deep in crisis. Against this background, Nicolas Hayek snr. (b. 1928) masterminded a four-year reorganization of the ASUAG and SSIH watch companies (founded in 1931 and 1930 respectively) which eventually merged in 1983 into the SMH (Swiss Corporation for Microelectronics and Watchmaking Industries). Hayek also made the strategic decision to develop a new watch brand that would rival the sales of Japanese models and rescue the embattled Swiss watch industry. The resulting slim plastic watch was based on a prototypical design by Ernst Thonke, Jacques Müller and Elmar Mock of Hayek Engineering AG. Comprising only 51 components (compared to the 90 or more components normally required in a conventional watch), it was very well suited to large-scale mass production. By combining high technology with affordability and artistic, emotional styling, the Swatch became a must-have fashion accessory following its

Range of early
Swatch watches,
1983

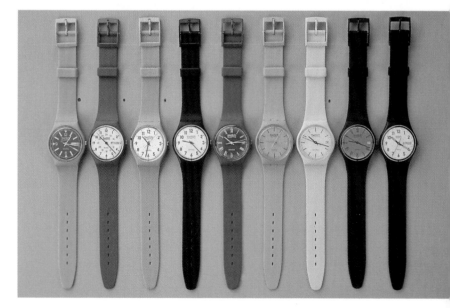

launch in 1983. Its success was boosted by the fact
that the watch was being marketed along the same
lines as *haute couture*, with new "collections" launched
every season. By producing limited editions for collectors as well as a range of "classic" watches in every possible style and colour, Swatch ensured that it had a design to suit almost every taste. Thanks to the remarkable success of the *Swatch*, the Swiss watch industry
regained its leading position in the sector in 1984. The
following year, Hayek and a number of Swiss investors
gained control of the Swatch Group, which included
other brands such as Blancpain, Omega, Longines,
Rado, Tissot, Certina, Mido, Hamilton, Balmain, Calvin
Klein, Lanco and the children's range Flik Flak. Today,
the company is the world's largest manufacturer of finished watches. Some 200 million *Swatch* watches have
been sold to date, ranging from the standard plastic
model to the metal cased *Irony*. The Swatch Group was
also highly instrumental in the development of the
diminutive *Smart* car in the late 1990s.

Micrograph stopwatch, first introduced in 1916

TAG HEUER

FOUNDED 1860
ST. IMIER, SWITZERLAND

In 1860 Edouard Heuer established a watchmaking workshop in a small town in the Swiss Jura mountains and focused on the design and manufacture of sports watches and chronographs. In 1880 he became the first watchmaker to serially produce chronographs, and nine years later his pocket chronograph was awarded a silver medal at the Exposition Universelle in Paris. In 1916 the company introduced the highly precise *Micrograph* stopwatch, which could accurately measure to 1/100th of a second – a landmark achievement that heralded the advent of modern sports competition. During the 1920s, Heuer's stopwatches were used for timekeeping at all three Olympic Games. In 1930 the company launched its first water-resistant case and over the next few decades concentrated its efforts on the development of superior performing wristwatches, such as the *Solunar* (1949), which was the first watch ever to measure ocean tides, the *Carrera* (1964), and the famous *Monaco* (1969), which was the first ever chronograph fitted with a self-wind-

Kirium watch, 1997

ing mechanism. In 1985 Heuer merged with TAG (Techniques d'Avant-Garde

to form TAG Heuer, a company dedicated to the application of cutting-edge technology. Two years later it launched its best-selling *S/el* watch, which is still regarded as a benchmark design for sports watches. TAG Heuer has been closely involved with Formula One racing for many years (TAG Heuer's parent company has owned the MacLaren team since 1986) and the company is the official timekeeper of many premier sports events. TAG Heuer's exquisite high-precision designs reflect its credo: "Technology determines function, function creates design".

ROGER TALLON

BORN 1929
PARIS, FRANCE

Roger Tallon is one of France's foremost industrial designers. He studied electrical engineering in Paris from 1947 to 1950, before joining the design consultancy Technès in 1953. While there, he designed numerous innovative products, including cameras for SEM (1957 and 1961), a typewriter for Japy (1960), a portable television for Téléavia (1963), the *Module 400* furniture range (1964), drinking glasses for Daum (1970) and the *Chronograph X* watch for LIP (1973). He also worked as a consultant, designing Frigidaire refrigerators for **General Motors** from 1957 to 1964. In 1973 he established the multi-disciplinary design consultancy, Design Programmes SA, and went on to gain an international reputation for his transportation design, which included the *Corail* locomotive (1977) for SNCF. In 1983 Tallon founded the design partnership ASDA + Partners with Pierre Paulin (b. 1927) and Michel Schreiber (b. 1950). He has continued designing trains, including the high-speed *TGV-Atlantique* (1988) for SNCF and the *Trans-Euro-Star* shuttle for Euro-Tunnel (1987). In 1973 he was elected an Honorary Royal Designer for Industry in London and in 1985 was awarded the National Grand Prix for industrial design.

Chronograph X
2230.194 wristwatch
for Lip-France, 1975

Flatliner comb
for Brian Drumm,
1995 – won the IF
International Design
Award in 1995

The industrial design consultancy Tangerine was founded in 1989 by Martin Derbyshire (b. 1961) and Clive
Grinyer (b. 1960). One of its earliest clients was **Apple Computer,** for which
it designed the Macintosh *SketchPad* (1990) and the *PowerBook* (1991).
Since then, Tangerine has forged a reputation for its designs for telecommunications and consumer electronic products and has also worked extensively
in the healthcare and transportation industries. Like many British design
consultancies, the majority of its briefs come from overseas, with around
70% of the work it undertakes being for companies based in Continental
Europe, North America, Japan and Korea. Its clientele includes British Airways, Hitachi, Waterford Wedgwood, Proctor & Gamble and LG Electronics.
Tangerine promotes a direct and personal dialogue between its partners
and its client's product development teams, in order to realize the most
creative solutions with the greatest efficiency. Tangerine states: "Our
strength is to be able to impart both spirit and structure to the design
process from the outset. It is here that creativity counts. It is here that the
important decisions are made that set not only the appearance of a product
but also its cost, longevity, environmental impact and many other factors."

Prototype television
for Goldstar, 1993

Cash register for the National Cash Register Company, 1937

WALTER DORWIN TEAGUE

BORN 1883 DECATUR, INDIANA, USA
DIED 1960 FLEMINGTON, NEW JERSEY, USA

Walter Dorwin Teague was one of the great pioneers of industrial design consulting in America. Having moved to New York in 1903, he subsequently attended evening classes at the renowned Art Students League. He later worked as an illustrator for a mail-order catalogue and also for the Hampton Advertising Agency. In 1912 he established his own studio in New York, working as a typographer and graphic designer, and in the mid-1920s also began designing **packaging**. In 1926 he founded Walter Dorwin Teague Associates, which was one of the earliest industrial design consultancies. Teague's first client was Eastman **Kodak**, for which he undertook a comprehensive design programme that included design research and product development. His first camera, the *Vanity Kodak* (1928), was designed specifically for the female market and was produced in various colours with matching silk-lined cases. Teague later received widespread acclaim for his *Baby Brownie* camera (1933), which was one of the first consumer prod-

Bantam Special camera for Kodak Company, 1933–1936

ucts made of plastic. His best-known camera, however, was the distinctively styled *Bantam Special* (1936), which was more compact and user-friendly than earlier Kodak models. In 1930 Teague designed the streamlined body of the Marmon *Model 16* car, which was among the most aerodynamically efficient automobiles of its time. He also designed other streamlined products, including glassware for **Corning**, kitchenware for **Pyrex**, pens and lighters for Scripto, lamps for **Polaroid**, mimeographs for A. B. Dick, radios for Sparton and the *Centennial* piano for Steinway. As well as consumer prod-

ucts, Teague also designed a plastic truck body for UPS, supermarkets for
Colonial Stores, interiors for the **Boeing** 707 airliner, United States pavilions
at various international trade fairs, exhibition interiors for **Ford**, service stations for Texaco and various exhibits at the 1939 **New York World's Fair**, including a gigantic cash register, which recorded visitor numbers and was
based on his earlier design for the National Cash Register Company. He also
wrote *Design This Day – The Technique of Order in The Machine Age* (1940),
which celebrated the potential of machines and the "new and thrilling style"
of the Modern era. Now headquartered in Redmond, Washington, with a regional office in California, Walter Dorwin Teague Associates Inc. employs a
staff of 200 professionals and specializes in transportation design, systems
engineering, human factors and **corporate identity.**

→ *Bluebird* radio
for the Sparton
Corporation,
1934–1936

Ruben Rausing

Dr. Ruben Rausing (1895–1983), the visionary founder of Tetra Pak, studied at the Stockholm School of Economics and later at Columbia University, New York. While in America, he was exposed for the first time to "self-service" stores and was convinced that the concept would eventually be adopted in European countries, which would in turn increase the demand for pre-packaged foodstuffs. On his return to Sweden, he founded, with Erik Wallenberg, the first Scandinavian specialist **packaging** company, based on Rausing's belief that "A package should save more than it costs". In 1943 the company began researching and developing a form of milk packaging which would require the minimum amount of material while providing the maximum hygiene. These early studies ascertained that the optimum shape for a milk carton was a tetrahedron. The following year saw the invention of new techniques for coating paper with **plastics** and for sealing packages. In 1951 Rausing and Wallenberg established AB Tetra Pak as a subsidiary of Åkerlund & Rausing, and the same year launched their new packaging system based on the tetrahedron-shaped carton. A year later, the Lundaortens Mejerifsrening dairy in Lund became the first to use a *Tetra Pak* machine to package cream into 100ml cartons. Several other Swedish dairies introduced the revolutionary packaging system in 1953, and that same year the company began using polyethylene as a coating for its paperboard. In 1954 the first *Tetra Pak* machine was exported to Germany and the half-litre packaging machine was introduced.

Detail of *Tetra Brik Aseptic ReCap* packaging, 1994

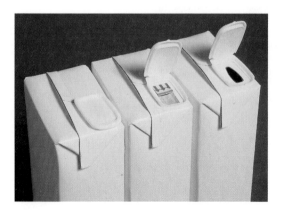

The commercial packaging of milk in 300ml cartons subsequently began the following year. The company eventually opened a much larger manufacturing plant in Lund and commenced the development of an aseptic packaging system (launched in 1961). By 1959 the factory was producing one billion cartons per annum and began working on a new brick-shaped design, the *Tetra Brik* (introduced in 1963).

Over the next few years, the business expanded its production operations
into Mexico, the United States, Lebanon and Italy. In 1965 Åkerlund &
Rausing was sold, but Ruben Rausing retained its subsidiary, AB Tetra Pak,
and with the funds from the sale was able to invest more resources into
the research and development of even better packaging designs, such as
the ubiquitous gable-top package known as *Tetra Rex* (1965). During the
1970s, the company exported its packaging machinery all over the world,
including to the USSR, Japan, China, Australia and Iran. By the beginning
of the 1980s, over 30 billion packages were being sold per annum for the
storage of milk, cream, fruit juice and wine. In 1991 AB Tetra Pak merged
with Alfa-Laval and subsequently diversified its product line to include PET
bottling equipment. Today, Tetra Pak remains the undisputed world leader
in the field of packaging, selling over 200 million products a day in some
150 countries. Its simple yet ingenious designs have revolutionized food
retailing and hygiene.

TEXAS INSTRUMENTS

FOUNDED 1951
DALLAS, TEXAS, USA

Jack St. Clair Kilby
holding up an early
integrated circuit,
c. 1959

Texas Instruments originated from a company, Geophysical Service, established in 1930 by Clarence "Doc" Karcher and Eugene McDermott. Geophysical Service specialized in the reflection seismograph method of exploration and, later, submarine-detection equipment and airborne radar systems. In 1951 it became a wholly-owned subsidiary of the newly-founded Texas Instruments. In 1952 Texas Instruments acquired a licence from the Western Electric Company to manufacture transistors, and in so doing entered the semiconductor industry. In 1958 Texas Instruments' in-house inventor, Jack St. Clair Kilby, demonstrated the first-ever integrated circuit (IC) and three years later the company delivered the first integrated-circuit computer to the US Air Force. 1964 saw the introduction of the first consumer product to incorporate an integrated circuit – a hearing aid – and the same year the company began mass-producing the first plastic-packaged ICs. Over the next decade, Texas Instruments was at the forefront of electronic technology, developing the first electronic hand-held calculator (1967) and the first laser-guided missile system (1969). In 1971 the firm pioneered the single-chip

The first electronic
hand-held calculator,
developed by Jack
St. Clair Kilby, Jerry D.
Merryman and James
Van Tassel, 1967

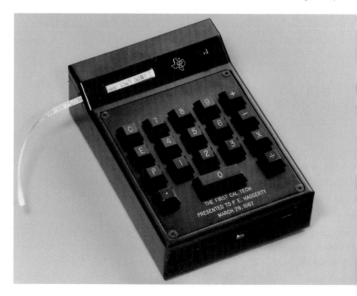

microprocessor and the single-chip microcomputer, and a year later entered the consumer market with its *Datamath* hand-held calculator. In 1974
it launched the *TMS1000* one-chip microcomputer and the following year
introduced three-dimensional data-processing technology. The first single-
chip speech synthesiser was unveiled in 1978, as was the first consumer
product to incorporate low-cost speech synthesis technology, the *Speak &*
Spell learning aid. In 1987 the company pioneered the first single-chip 32-bit
artificial-intelligence microprocessor, and four years later became the first
American manufacturer of semiconductors to establish a research and development facility in Japan. Over the years, Texas Instruments has won numerous awards for manufacturing excellence and its many inventions, which
have driven completely new industrial design typologies.

THERMOS
IN THE TROPICS.

THERMOS

FOUNDED 1904
BERLIN, GERMANY

Advertisement for
Thermos flask, 1909

→Thermos flask
Model 65, 1928

Advertisement for
Thermos flask, 1909

The philosopher and scientist, Sir James Dewar (1842–1923), was a pioneer of vacuum technology and in 1892 devised an innovative flask for the storage of liquefied gases that worked on the principle of temperature retention. This ingenious design comprised two glass bottles, sealed together one inside the other, with the air in the cavity between them pumped out so as to produce a vacuum. Later, Dewar silvered the glass with mercury to increase its insulating performance. Dewar managed to create a stronger, more resilient version of his flask with the assistance of a German glass blower, Reinhold Burger. While developing these flasks for scientific experiments, Dewar and Burger came up with the idea of making a flask with a protective metal casing for domestic use. The resulting "Thermos" was patented in Germany in 1903 and a year later the Thermos GmbH manufacturing company was subsequently established in Berlin. The flasks were retailed under the Thermos tradename from 1904 and by 1905 they were being marketed in the United States. Thermos Ltd. was founded in Britain two years later and a factory was opened in London shortly afterwards. The early Thermos flasks were time-consumingly handblown and cost an expensive one guinea. By 1911, however, production had become mechanized and the price of the vacuum bottles plummeted to 2s 6d. The Thermos flask became very popular with both workers and picnickers in the 1920s and 1930s, and during the Second World War Thermoses were used extensively by troops. With its double-walled design that isolates heat, the Thermos flask works equally well at keeping liquids hot or cold for many hours. Although other companies have manufactured insulating vacuum flasks, the name Thermos has become a generic term and the company continues to export its products throughout the world.

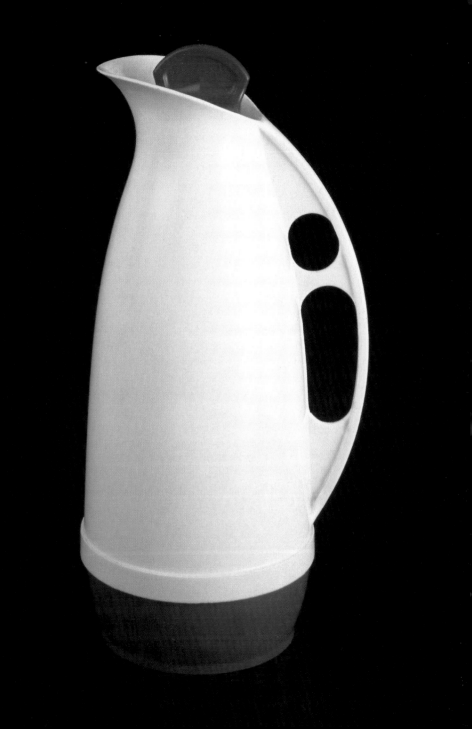

Michael Thonet, *Boppard Chair I,* 1836–1840

Michael Thonet established a furniture workshop in Boppard am Rhein in 1819. From around 1830, he began experimenting with laminated wood and produced a number of innovative chairs in the Biedermeier style. These designs prompted the Austrian chancellor, Prince Metternich, to invite Thonet to Vienna, where he was granted a patent for his new process for bending wood laminates in 1842. Having secured the necessary financial backing, Thonet moved to Vienna and in 1849 established a furniture workshop in Gumpendorf, a suburb of Vienna, with his sons, Franz, Michael, August and Joseph. For the next two years, the Thonet family concentrated on developing techniques for mass-producing furniture, including the steam bending of solid wood. In 1851 Thonet exhibited its new furniture designs at the **Great Exhibition** in London and won a gold medal. By 1853 Gebrüder Thonet, as the company was now known, had moved into larger premises and was fully

Page from Thonet catalogue, 1904

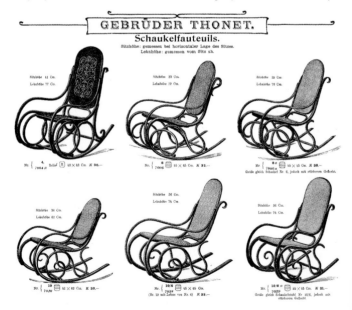

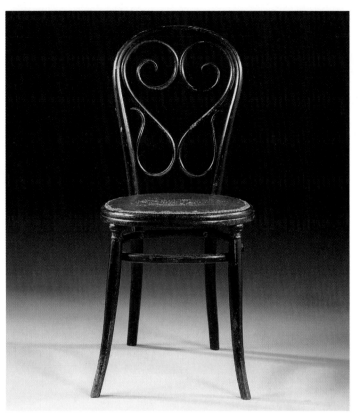

Model No. 4 side chair, c. 1850

mass-producing chairs distinguished by a reduction of elements and the elimination of extraneous ornament. In 1856 the company opened its first factory in Koritschan, Moravia. Gebrüder Thonet's remarkable success was due to its adherence to mechanized methods of production, which allowed it to sell its products at very competitive prices. In 1860, for example, the firm's best-known model, the *No. 14* chair, cost less than a bottle of wine, and by 1891 a staggering 7,300,000 of these ubiquitous café chairs had been sold. In the early 1900s several leading Viennese architects, including Josef Hoffmann (1870–1956), began designing furniture for Gebrüder Thonet in the Secessionist style. In 1929 a French subsidiary was established, Thonet Frères, which produced progressive **tubular metal** furniture designed by **Marcel Breuer** and others. The production of this furniture was later moved to Frankenberg, Germany. Thonet continues to operate and manufactures re-issues of its earlier seating as well as contemporary designs.

Osamu Kondo,
Libretto lap-top
computer, 1995

TOSHIBA

FOUNDED 1939
TOKYO, JAPAN

In 1875 Hisashige Tanaka (1799–1881) founded Japan's
first telegraphic equipment manufacturing company,
the Tanaka Seizo-sho (Tanaka Engineering Works). Later
renamed the Shibaura Seisaku-sho (Shibaura Engineer-
ing Works), his company became one of the largest Japanese manufacturers
of heavy electrical apparatus. Meanwhile, another company – Hakunetsu-
sha & Co. – was established in 1890 to manufacture the first Japanese electric incandescent lamps. Renamed Tokyo Denki (Tokyo Electric Company)
in 1909, this firm became a notable manufacturer of consumer electronic
products. Between them, the two companies were responsible for producing the first Japanese induction motors (1895), the first Japanese X-ray tubes
(1915) and the first Japanese washing machines and refrigerators (1930).
In 1939 the two companies merged to form the Tokyo Shibaura Electric Co.
(officially renamed Toshiba in 1978). A year later, this new venture produced
the first Japanese fluorescent lamps and in 1942 constructed Japan's first
radar equipment. In 1954 the company introduced the first Japanese digital
computer, *TAC*, which was designed specifically for the University of Tokyo,
launched the first practical electric rice cooker in 1955 and developed the
first Japanese transistor televisions and microwave ovens in 1959. Since
then, Toshiba has been a leading pioneer of electrical and electronic products, from computers and home appliances to medical diagnostic equipment, elevators and escalators. Toshiba's commitment to "universal design"
is powerfully expressed in the company's user-friendly products and its

↘Toshiyuki
Yamanouchi,
PDR-M4 digital
still camera, 1998

Toshiyuki
Yamanouchi,
TLP651 LCD data
projector, 1999

Yoshiharu Iwata,
ER-4 electric rice
cooker, 1958

pioneering **design for disability**, such as Braille operating panels on washing machines for the visually impaired. The company's design centre runs
a special programme in which its engineers and designers are made to
wear adapted goggles that lets them experience what it is like for a person
living with impaired sight. This type of "human-centric" design also involves
a systemized approach to **ergonomics**. Toshiba views "universal design" as
a means of "harmonizing social responsibility and business requirements."

TOYOTA

FOUNDED 1937
KARIYA, HONSHU, JAPAN

Japan's most famous inventor, Sakichi Toyoda (1867–1930), designed
the first Japanese power loom in 1896 and in 1924 devised the first-ever
automatic loom capable of replenishing the shuttle during continuous
operation. In 1926 he went on to found the Toyoda Automatic Loom Works,
which was one of the world's leading manufacturers of weaving machinery.
Three years later Toyoda signed a patent royalty agreement with a British
firm, Platt Brother & Co., for the manufacture of one of his weaving machines. This provided sufficient funds for Toyoda's eldest son, Kiichiro, to begin the research and development of an automobile. A steel manufacturing
division was also established to develop a "speciality steel" that could be
used for car engines and bodies, and in 1933 an automobile department
was formed. This produced its first *Type A* engine in 1934 and its first *Model
A1* car prototype in 1935. Serial production of the *Model AA* car commenced
in 1936 and the Toyota Motor Co. was officially founded the following year.
In 1947 Toyota launched its first small car, the *Model SA*, and from 1959 began building a vast network of plants outside Japan in the belief that "local-
izing operations provides customers with the products they need where they
need them". Today Toyota is the largest car manufacturer in Japan and the
third largest worldwide. In 1996 the company's Tokyo Design Research &
Laboratory division was centralized so that the entire design process, from

Toyota *AA*
automobile, 1936

Toyota *2000 GT*, 1966

Toyota *Yaris*, 1998

concept sketches to prototyping, could be undertaken at a single site. Since then, Toyota has produced award-winning cars such as the fuel-efficient *Yaris* (European Car of the Year 2000) and the world's first production hybrid car, the *Prius* (Japan's Car of the Year 1997–1998), which runs on both gasoline and electric power. With its innovative *RAV4* electric car, Toyota became the first automobile manufacturer to comply with the State of California's zero-emissions legislation.

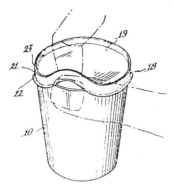

TUPPERWARE

FOUNDED 1938
LEOMINSTER, MASSACHUSETTS, USA

Earl Tupper (1907–1983) gained his love of invention
from his father, who used to construct labour-saving
devices on the family farm. In 1936 Tupper met Bernard
Doyle, who had developed a plastic known as Viscoloid,
and he subsequently took a job in the **plastics** manufacturing division of DuPont. Although he only stayed there
a year, Tupper later recalled that it was where "my education really began".
In 1938 Tupper established his own plastics manufacturing company in Leominster, Massachusetts. The company initially undertook subcontracted
work for DuPont, but subsequently won its own contracts for military products such as gas masks and signalling lamps. During the immediate post-
war years, Tupper turned his attention to the design of plastic consumer
products, such as sandwich picks, bathroom tumblers and cigarette cases,
which firms could give away to their customers as promotional sales tools.
Plastics in the 1940s still had numerous unpleasant properties, ranging
from brittleness and odour to greasiness. To overcome these problems,

Oyster, multi-purpose storage containers, 1990s

Tupper developed a technique for purifying black polyethylene slag (a by-product of the oil industry) into a strong, flexible, non-porous material that was translucent and non-greasy. Around the same time, he also patented his famous airtight and watertight Tupper seal, the design of which was based on the lids used for paint containers. Tupper combined these two innovations to create a range of airtight plastic containers, but found that the product line "fell flat on its face" in retail stores. He realised that his sealing system had to be demonstrated before consumers would buy into it. In the late 1940s **Stanley** was successfully selling its household products through local distributors, who arranged demonstration home parties, and in 1948 Tupper met several of these distributors. Amongst them was Brownie Wise, who expanded this novel retailing concept for Tupper so that his food storage containers were only available through such demonstrations. From 1951 to 1958 Wise was vice-president of the company, which was then called Tupperware Home Parties. As a skilful saleswoman and motivator of people, Wise understood the importance of her party hostesses, declaring: "If we build the people, they'll build the business." With her large sales force, she transformed the company's fortunes and Tupperware parties became as famous as Tupperware products. Today Tupperware produces a vast range of high-quality plastic housewares, from kitchen equipment and multi-purpose domestic storage containers to organizing products. A Tupperware demonstration party is held on average every two seconds somewhere in the world.

←Advertisement showing a Tupperware party scene, 1960s

GINO VALLE

BORN 1923
UDINE, ITALY

Cifra 3 clock for
Solari & C., 1966

Gino Valle trained at the Università di Architettura in Venice until 1948 and then worked in the architecture practice founded by his father, Provino Valle. He completed his studies at the Harvard Graduate School of Design in Cambridge, Massachusetts, graduating in 1951. Valle and his brother, Nani, subsequently took over their father's practice in Udine and designed numerous commercial and industrial buildings in northern Italy. In 1954 Lino **Zanussi** commissioned Valle to design a range of household appliances. Valle's resulting product system was based on the concept that appliances should be viewed as integral components of the kitchen. His rational style was also translated into his designs for automated schedule boards for train stations and airports, a large public-area clock (1966), and his well-known domestic *Cifra* range of clocks for Solari & C. (1956–1966), which incorporated a similar modular flap-down numeral system. He was awarded a Compasso d'Oro for his *Cifra 5* clock (co-designed with Nani Valle) in 1956 and another in 1962 for his indicator board, which is still widely used in railway stations and airports throughout the world.

Large metal public-area clock for Solari & C., 1963

HAROLD VAN DOREN

BORN 1895 CHICAGO, ILLINOIS, USA
DIED 1957 PHILADELPHIA, PENNSYLVANIA, USA

Harold van Doren was a pioneer of industrial design consulting and **stream-lining** who designed products with a clean, contemporary aesthetic. He first studied languages before moving to Paris, where he worked at the Louvre. After returning to the United States, he worked as an assistant to the dir-ector of the Minneapolis Institute of Arts, but resigned from this position when he was given the chance to work in the fledgling field of industrial de-sign. One of his first commissions came from the Toledo Scale Company, who asked him to design some commercial weighing scales. Van Doren's lightweight and innovative solution was one of the first products to incorp-orate a large-scale **plastics** moulding. Having based his practice in Phila-delphia, he went on to design many streamlined products for Maytag, Goodyear, Ergy, Swartzbaugh and DeVilbiss. Together with John Gordon Rideout, he designed the widely acclaimed green skyscraper-shaped plastic

First range
manufactured
by Philco, 1950

radio for Air-King (1930–1931) and a child's scooter (1936). In 1940 van Doren wrote *Industrial Design: A Practical Guide to Product Design and Development*, and nine years later published an article in *Design* magazine entitled "Streamlining Fad or Function?". Here he argued that the curved lines common to refrigerators at the time were the result of a manufacturing process involving a metal press known as a "bulldozer", and that the formal language of the product was thus "imposed on the designer by the necessity of obtaining low cost through high-speed production". Van Doren was particularly adept at re-styling products without altering their existing layout, as illustrated by his cooking range (1950) for the **Philco** Corporation. This streamlined design was a re-modelled version of an earlier range produced by Electromaster Inc. that "even experienced men in the industry believed to be entirely new". For Philco, van Doren also designed a streamlined refrigerator (c. 1950) whose door was emblazoned with the words "Advanced Design" – an early example of the use of design as a marketing tool.

Magnalite tea kettle for Wagner Ware, 1940 (co-designed with J. G. Rideout)

← Photograph showing Harold van Doren (left) and Hubert Bennett (right) with van Doren's redesigned scales for the Toledo Scale Company, 1930s

VICTORINOX

FOUNDED 1884
IBACH, SWITZERLAND

Karl Elsener

Karl Elsener (1860–1918) spent a year as a journeyman cutler before setting up a small business close to Schwyz in 1884. In order to boost the local industry, in 1891 he founded the Swiss Cutlery Guild (later re-named Victorinox) to produce soldiers' knives for the Swiss Army, which had hitherto sourced its knives in Germany. Elsener delivered his first batch of knives to the Swiss Army that same year, and soon after began manufacturing penknives such as the *Student Knife*, the *Cadet Knife* and the *Farmer's Knife*. While the *Soldier's Knife* was extremely robust, its corkscrew, punch, can opener, screwdriver and second blade also made it quite heavy, prompting Elsener to design the lighter and more elegant *Officer's Knife*. This hugely successful model, which incorporated just two springs for its six multi-purpose tools, was registered in 1897. Over the succeeding years, numerous other tools were added to the knife, and today this classic design is sold all over the world. Victorinox currently produces 77,000 Swiss Army knives a day, amongst them its top-of-the-range *Swiss-Champ*, which although weighing only 185 grams has 31 different features.

Swiss Army knife,
original model
designed c. 1891

MASSIMO VIGNELLI

BORN 1931
MILAN, ITALY

Massimo Vignelli studied architecture at the Politecnico di Milano from 1950 to 1953 and later trained at the Università di Architettura, Venice. From 1956 he designed glassware for Venini and from 1958 to 1960 taught at the Institute of Design, Chicago, while his wife, Lella Vignelli (b. 1936), worked for the architects Skidmore, Owings and Merrill. In 1960 the couple returned to Italy and opened a studio in Milan. Four years later Massimo Vignelli began working for the Container Corporation of America, Chicago, and designed its logo. In 1965, together with Bob Noorda (b. 1927), Jay Doblin (b. 1920) and **Reinhold Weiss**, Vignelli founded the design consultancy Unimark International in Milan. The Vignellis moved permanently to America that same year and in 1966 Unimark opened a New York office, specializing in **corporate identity** work. In 1971 the couple established Vignelli Associates and subsequently designed: corporate identity programmes for Knoll, American Airlines, Bloomingdale's, **Xerox**, Cinzano, **Ford** and Lancia; **signage** for the subway systems in Washington DC and New York City; furniture for Poltronova, Sunar, **Rosenthal** and Morphos; glassware for Venini, Steuben and Sasaki; melamine tableware for Heller; and showrooms for **Artemide** and Hauserman.

➤Stacking cups for Heller, 1970

Stacking dinnerware for Heller, 1967

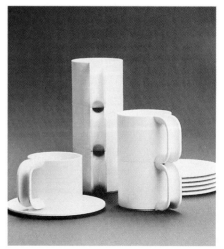

JEAN-PIERRE VITRAC

BORN 1944
BERGERAC, FRANCE

Courier motorbike
for Bunny Courses,
1992

Jean-Pierre Vitrac trained at the École des Arts Appliqués, Paris and in 1968 started to work as a designer for Lancôme. He later established his own company, specializing in the design of **packaging** and in-store advertising. In 1970 he designed the floor lamp *Fleur* for Verre et Lumière which featured detachable, chromed metal segment-like elements and which predicted the elegant yet innovative forms of his later work. In 1974 he founded Jean-Pierre Vitrac Design and over the succeeding years the office became increasingly involved in product design. One of his earliest forays into this field was his design of the brightly coloured, disposable and inexpensive *Plack* picnic set (1977) for Co & Co., which cleverly incorporated a plate, cup, knife, fork and spoon in a single thermoformed sheet of polystyrene. He later designed the strong but supple *Superior* range of suitcases for Superior SA (1981–1989), which were the first luggage articles constructed of a thermopressed synthetic foam. Other products designed by Vitrac include *Les Pros* kitchen knives for Sedasco (1985), packaging for automobile bulbs for Norma/**Philips** (1987), the *Hide Series* of desk accessories for Arco (1994),

Plack picnic set for
Co. & Co., 1977

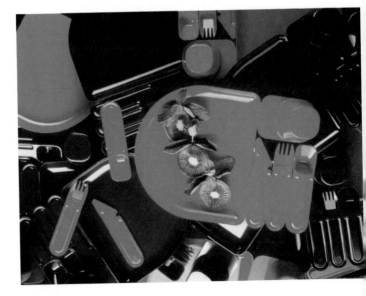

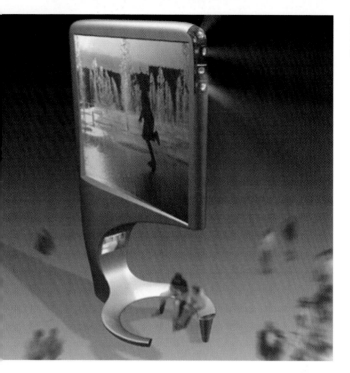

Proposal for urban furniture for advertising for Arcomat, late 1998

...acuum cleaner for Panasonic (1992), a desk for Mobilier International ...91), a dentist's chair for Trophy (1986), a television for National (1992), ...*Mairie Mobile* (a concept Mobile Town Hall) for **JC Decaux** (1993), and ...*Welcome* display board for FA Technology (1994). During the 1980s, Vit...wrote a text entitled *Comment gagner de nouveaux marchés par le design* ...*ustriel* (How to win new markets through industrial design), and today ...office, now called Vitrac (Pool), has affiliations with other leading Euro...n consultancies, most notably **Minale Tattersfield** in Britain and Windy ...derlich in Germany. Vitrac's transportation designs, such as the courier ...torcycles for Bunny Courses (1992) and his concept bicycle for Diam ...95), exemplify his belief that design should be used for the "conception ...adically new products".

← *Moonkey* foldable sunglasses for Alpha Cubic, 1989

Ferdinand Porsche, original Volkswagen *Beetle*, 1938 – first designed in 1934

VOLKSWAGEN

FOUNDED 1938
WOLFSBURG, GERMANY

In January 1934 the automotive designer Professor Ferdinand **Porsche** submitted a design proposal to the new German Reich government for a revolutionary "car for the masses", called the *Volkswagen* (People's Car). Soon afterwards he signed a contract with the Reichsverband der Automobilindustrie (Automobile Industry Association), which provided him with a development budget on the understanding that a prototype of the car would be completed within ten months. In his pursuit of technological excellence, Porsche introduced innovations into the chassis, engine and transmission, regardless of cost; indeed, despite the availability of cheaper alternatives, he selected the more expensive air-cooled, horizontally-opposed engine for the car because of its full-throttle endurance. The first *Volkswagen* prototype was unveiled in October 1935, and by the following spring successive prototypes were undergoing extensive testing. The earliest cars, with their modern all-steel bodywork, were later improved with the addition of bumpers and running boards. **Erwin Komenda**, the acclaimed aerodynamicist, designed a distinctive streamlined body for the car, which, although compact, could easily accommodate five passengers. In 1938, in advance of the full-scale mass-production of the *Volkswagen*, the Gesellschaft zur

Volkswagen-Werk GmbH brochure, c. 1937, designed by Thomas Abeking

Volkswagen *Beetle*,
model of the late
1970s
(first re-designed
in 1945)

Volkswagen
New Beetle,
1998

Vorbereitung des Volkswagens (Company for the Preparation of the People's Car) was established, later renamed Volkswagen-Werk GmbH. When a prototype of the *People's Car* was shown to the press later the same year, it was immediately dubbed the "Beetle" by the *New York Times*. The German government, on the other hand, preferred the propagandist "Kraft durch Freude" (Strength through Joy) name for the new car, which was initially intended to be sold through a savings stamp scheme. Work began on the construction of the world's largest car plant, where the cars were to be manufactured, and a town for workers was also built (the Town of Strength through Joy,

Volkswagen
Transporter, model
of the 1970s
(first re-designed
in 1950)

today Wolfsburg). In 1940 the still incomplete factory started producing armaments, and *Volkswagen* production took a back seat while Porsche converted the car into two military jeep-style vehicles called the *Kübelwagen* and the *Schwimmwagen*. A year later, serial production of the *Volkswagen* was initiated with the manufacture of 41 cars, which were mainly used by the Nazi Party for propaganda purposes. After the war, *Volkswagen* production accelerated considerably, with the British Military Authorities placing an order for 20,000 vehicles in September 1946. Some 100,000 *Volkswagens* had been manufactured by 1950, the year in which another "classic" design was launched – the VW *Transporter*. This famous camper van was developed from a rough sketch by a Dutch VW importer who was inspired by the "flat vehicles" used in factories. Functioning as either a mini-bus or a commercial

vehicle, the *Transporter* was hugely successful, and by 1951 Volkswagen was producing some 12,000 units per annum. The *Karmann Ghia Coupé* was launched in 1955 and two years later a sophisticated cabriolet version was unveiled at the Frankfurt International Motor Show. By 1962 Volkswagen was producing 1,000,000 cars per annum. In 1964 the company built a huge research and development complex in Wolfsburg, which included a state-of-the-art wind tunnel opened in 1965. With this facility, VW was able to increasingly use technological advances as the basis for its future designs. In 1972 the *Beetle* became the world's most produced car. Volkswagen launched the *Passat* in 1973, followed in 1974 by the *Golf* designed by **Giorgetto Giugiaro**. Both cars served as blueprints for subsequent generations of Volkswagen models. The *Golf* – the intended successor of the *Beetle* – proved one of Volkswagen's greatest successes: one million of these "classless quality cars" were manufactured within the first 31 months and in 1977 the *Golf GLS* received a "Gute Form" award sponsored by the Federal Ministry of Economic Affairs. By 1988 an impressive ten million *Golfs* had been produced. Volkswagen's highly successful *New Beetle* was launched in 1998 and immediately received numerous awards and accolades, including being named "The Best Design of 1998" by *Time* Magazine. Based on the new *Golf* chassis, this superb design – like all other Volkswagens – "expresses the German engineering passion for designing and building cars". The Volkswagen Group now owns **Audi**, **Bugatti**, **Lamborghini**, Seat and Skoda and is one of the largest car producers in the world.

Volvo was the first car manufacturer in the world to offer safety belts as standard in 1959

VOLVO

FOUNDED 1927
GOTENBURG, SWEDEN

Sweden in the 1920s was heavily reliant on imported cars, and so Gustaf Larson and Assar Gabrielson decided to produce their own automobiles. In 1926 they came up with the idea of designing high quality car components that would be manufactured to their exacting specifications by selected suppliers, and which they would then assemble with the assistance of experienced car builders. Their first series-manufactured car, the *OV4*, left their factory in Gotenburg in April 1927 – marking a turning point in Swedish industrial design history. In 1929 Volvo introduced its first six-cylinder model, the *PV651*, and in 1935 the company launched the streamlined *PV36* designed by Ivan Örnberg, who had spent the whole of his career working in the US car industry. With its integrated headlamps, this sleek design was undoubtedly inspired by **Carl Breer**'s *Airflow* of 1934. During the Second World War, Volvo produced several military vehicles, including a jeep, and also developed a number of civilian prototypes. 1944 saw the launch of the *PV444*, which combined American **styling** with an unconventional unitized all-steel body. Developed by Helmer Petterson, the *PV444* was the first Volvo that "could be afforded by the average-man-in-the-street" In 1956 the *120* (known as the *Amazon* in Scandinavia) was introduced, featuring numerous innovative safety features including the world's first three point safety belt and a padded dashboard. With its emphasis on safety, in 1959 Volvo became the first manufacturer in the world to equip its cars with

Volvo *Amazon 120*, 1956

safety belts as standard. During the 1960s and 1970s, models such as the *144* sedan (1966), *145* estate (1967) and *240/260 series* (1974) became best-sellers while pioneering other innovative safety features such as disc brakes all round, split steering columns and outer rear-seat three-point safety belts. In 197 Volvo presented the world's first catalytic exhaust emission control

and by 1988 70% of its new cars were fitted with catalytic converters. The
highly successful *700 Series* was launched in 1982, with the subsequent *740/*
760 estate models (1985) totally dominating their market segment. During
the 1990s the company attempted to shake off its conservative image with
the introduction of a number of new cars, such as the *S80* (1998), which
featured smoother styling yet retained a strong Volvo identity. As a pioneer
of revolutionary safety features, such as the safety belt tensioner, the integrated child booster seat, the self-adjusting belt reel and the integrated side
impact protection system including side airbags, Volvo has had a significant
impact on car design around the world.

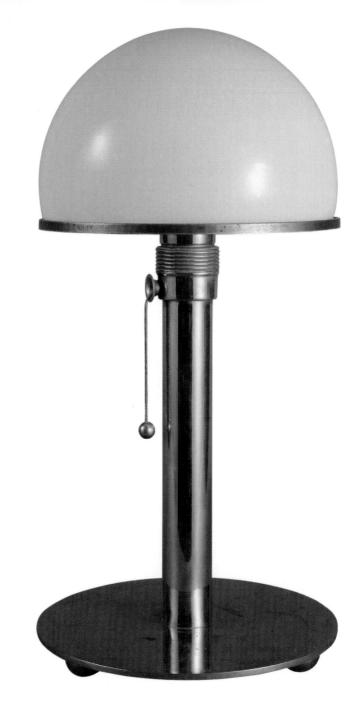

WILHELM WAGENFELD

BORN 1900 BREMEN, GERMANY
DIED 1990 STUTTGART, GERMANY

Wilhelm Wagenfeld trained with the silverware manufacturer Koch & Bergfeld and attended classes at the Kunstgewerbeschule in Bremen from 1916 to 1918. From 1919 to 1922 he trained at the Zeichenakademie (Drawing Academy) in Hanau, where he executed expressionist woodcuts. In 1923 he enrolled on the preliminary course at the Weimar **Bauhaus** and subsequently apprenticed under László Moholy-Nagy (1895–1946) in the school's metal workshop. While there, Wagenfeld co-designed with Karl J. Jucker (1902–1997) the famous "Bauhaus" table lamp (1923–1924), which was put into serial production by the workshop. After completing his journeyman's exam, he remained in Weimar and became an assistant to Richard Winkelmayer in the metal workshop of the Staatliche Bauhochschule. In 1927 he began independently producing designs for industrial production, including metal household items for Walter & Wagner, Schliez and door handles for S. A. Loewy, Berlin. A year later he took over the directorship of Bauhochschule's metal workshop, where he continued designing functional objects such as his *M15* tea caddy, which incorporated geometric forms that were less severe than those of his earlier Bauhaus designs. Although strongly connected with

Table lamp for the
Dessau Bauhaus,
1924

Heat-resistant glass
tea set for Schott &
Gen. Jenaer Glaswerke, 1930–1934

Kubus containers for Vereinigte Lausitzer Glaswerke AG, 1938

the Bauhaus, Wagenfeld rejected the institution's design doctrine as being too theoretical and self-centred. He shared its belief, however, that function was a prerequisite of good design and that industrial design practice necessitated close cooperation between designer and manufacturer. From 1929 Wagenfeld also worked as a freelance designer for the Schott & Gen. glassworks in Jena and in 1930 established his own design studio. He redesigned Schott & Gen.'s domestic glassware range, which included his famous glass tea set (c. 1930). Unlike so many of his former Bauhaus colleagues, Wagenfeld did not emigrate from Germany to escape Nazi persecution – his unquestionable skill in designing for industrial production being put to use during this period. Having become a professor at the Staatliche Kunsthochschule Berlin in 1931, he gave up teaching in 1935 when he was appointed artistic director of the Vereinigte Lausitzer Glaswerke, for whom he designed glass products that were highly suited to industrial mass-production, including utilitarian pressed-glass ranges of restaurant wine and beer glasses, as well as commercial-use bottles and jars. His best-known design from this period was his famous *Kubus* range (1938) – a modular system of stacking storage containers for kitchen use. During the 1930s Wagenfeld produced

Max und Moritz salt
and pepper shakers
for WMF, 1952–1953

designs for **ceramics**, most notably a porcelain service (1934) for Fürsten-
berg and the *Daphne* service (1938) for **Rosenthal**. Wagenfeld's refusal to
join the Nazi party eventually led to him being sent to the Russian Front,
where he was captured and made a prisoner-of-war. After the war and his
release, he outlined his functionalist approach to design in articles published in several journals, including *Die Form*, between 1947 and 1949.
During this period, he was also professor of industrial design at the Hochschule für Bildende Künste in Berlin and continued working as an independent designer, most notably designing cutlery for the Württemburgische
Metallwarenfabrik and electric light fittings, including graduated-sized plastic globe lamps, for Lindner of Bamberg. In 1954 he established the Wagenfeld Workshop in Stuttgart to develop products for industrial manufacture,
including metalware for WMF and a melamine in-flight meal tray for Lufthansa (1955). Wagenfeld's teachings stressed the moral, social and political
obligations of designers and were enormously influential upon the younger
generation of design practitioners in Germany. Today, as a celebrated Bauhaus figure, Wagenfeld's functional and aesthetically restrained industrial
designs are seen to epitomize the aspirations of the institution.

Waterman's Ideal Fountain Pen

WATERMAN

FOUNDED 1887
NEW YORK, USA

Early advertisement for Waterman's *Ideal* fountain pen

Patent drawing for Lewis Waterman's *Regular* fountain pen, c. 1884

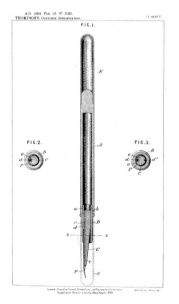

Lewis Edson Waterman (1837–1901) worked as an insurance broker in New York. According to legend, he was on the verge of selling a valuable policy when he loaned the client a newly-purchased fountain pen to sign the document. The pen failed to work and, to make matters worse, it made a large blot on the paper. While Waterman hurried back to his office to get another copy of the policy, a rival broker signed up the client. Furious at this turn of events, Waterman decided to dedicate himself to the development of an improved fountain pen. Whether this story is apocryphal or not, Waterman did invent the multi-channel feed in the 1870s. His first pen design to incorporate this feature was the *Regular*, patented in 1884. This forerunner of the modern fountain pen used a capillary system of small grooves under the nib to ensure a smooth flow of ink, which was contained in a reservoir filled using an eyedropper. Waterman subsequently established a small workshop and began manufacturing his innovative *Ideal* fountain pen, which was cased in hard **rubber**. Guaranteed for five years, the *Ideal* was the first truly reliable and leak-proof fountain pen and as such was the first pen to be a large-scale commercial success – Waterman produced around 500 of these pens during his first year, and in 1900 opened a factory in Montreal. After Waterman's death in 1901, his nephew headed the company and in 1905 introduced a revolutionary feature – the pen clip. Around 1914, Jules Fagard obtained the French manufacturing licence and later established Jif-Waterman, which produced exquisitely decorated pens. In 1954 Waterman's American operation ceased trading, after which Jif-Waterman became the largest Waterman concern. Now a subsidiary of the **Gillette** Group, Waterman is the largest manufacturer of fountain pens in Europe and the second largest worldwide.

JAMES WATT

BORN 1736 GREENOCK, RENFREWSHIRE, SCOTLAND
DIED 1819 HEATHFIELD HALL, NEAR BIRMINGHAM,
ENGLAND

At the age of 17, James Watt trained as a mathematical instrument maker
and studied under a master in London for one year before ill-health forced
him to return to Scotland. In 1757 he subsequently established a shop within the precincts of Glasgow University (then one of the epicentres of the
scientific community in Britain) and produced various mathematical instruments, such as quadrants, compasses and scales. While there, Watt met
many leading scientists, including the professor of chemistry, Joseph Black
(1728–1799), whose investigations into latent heat formed the basis of modern thermal science and inspired Watt's later improvements to the steam
engine. In 1764 Watt was asked to repair a model of the atmospheric steam
engines designed by Thomas Newcomen (1663–1729). In the course of his
work on the engine, Watt realised that it wasted steam, and therefore power.

Patent drawing for
James Watt's steam
engine, 1781

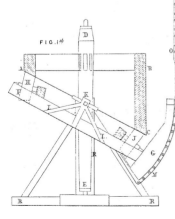

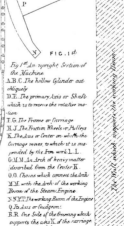

He also identified the biggest design flaw in Newcomen's engine, namely its loss of latent heat (the heat used to alter the state of a substance), a problem which Watt calculated could be solved by adding a separate condenser connected to the cylinder by a pipe. Soon after this discovery, Watt entered into partnership with Dr. John Roebuck (1718–1794), who owned a coal mine in Linlithgowshire that could not be drained with one of Newcomen's engine-driven pumps. With Roebuck's financial assistance, Watt constructed an (unsuccessful) experimental engine in 1768 and the following year patented his famous "New Invented Method of Lessening the Consumption of Steam and Fuel in Fire Engines". Unable to find financing to develop his steam engine further, Watt worked as a surveyor for several years before entering into another partnership in 1775 with Matthew Boulton (1728–1809), who owned the Soho Engineering Works in Birmingham. With the necessary backing secured, Watt was able to improve his engine and an example was built to power a section of Boulton's works. By 1776 two Boulton & Watt engines were in commercial operation – one pumping water at a Staffordshire colliery, the other blowing air into the furnaces at John Wilkinson's ironworks. Between 1776 and 1781 Watt installed a number of his pumping engines in Cornish tin and copper mines. Boulton recognized the enormous potential of **steam power** for milling and persuaded Watt to design a rotative engine to replace his earlier model, which used a reciprocating action. The

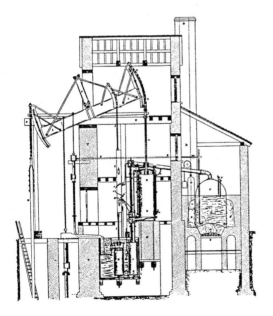

resulting design of c. 1780 incorporated a crank driven by two pistons, but this element had already been patented by James Pickard. Watt got around this problem in 1781 by devising the novel "Sun-and-Planet" wheel, and a year later harnessed the power derived from the expansiveness of steam to create a double-stroke engine. As well as continuing to improve upon his steam engines for powering machinery, Watt also independently discovered the composition of water (H_2O). As an important pioneer of the practical steam engine, Watt made an immense contribution to the **Industrial Revolution**.

Drawing of single-action Watt engine showing condenser, 1788

Sir Joshua Reynolds,
Josiah Wedgwood,
1783

The youngest of twelve children, Josiah Wedgwood came from a family whose members had been potters since the 17th century. After the death of his father in 1739, he worked for the family pottery, which had been inherited by his eldest brother, Thomas. After a five-year apprenticeship there, in 1754 Wedgwood entered into partnership with an established potter, Thomas Whieldon. Around this time, he began carefully recording his experiments, which included his highly successful formula for green glaze. In 1759 he established his own business at Ivy House Works, where he produced "a species of earthenware for the table, quite new in appearance, covered with a rich and brilliant glaze". A tea and coffee service of this warm cream-coloured earthenware was ordered by Queen Charlotte and in 1765 Wedgwood was given permission to re-christen it *Queen's ware* and to call himself "Potter to Her Majesty". Well finished and clean in appearance, *Queen's ware* was extremely versatile, in that it could be left plain, incised with decoration, hand-painted or transfer printed. Its durability and its serviceable

Wedgwood's Etruria
factory opened on
13 June 1769

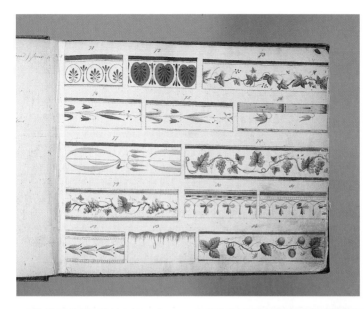

Page from Josiah
Wedgwood's first
pattern book –
border designs
from c. 1775–1780

Large *Queen's ware*
(Royal Shape) dish
with *Pink Antique*
border, c. 1775

forms also made it highly practical, and it became the standard domestic pottery selling successfully around the world. The success of *Queen's ware* was not only the result of the increasing demand for pottery fuelled by the growing popularity of tea-drinking, but also due to its simple forms, which were utterly in tune with the emerging Neo-Classical style. To meet the demand for his high quality wares, in 1764 Wedgwood moved his business to the larger Brick House Works in Burslem. Here he pioneered the logical division of labour, employed rationalized production methods and used innovative marketing techniques, while the forms of his cream-ware became less and less ornate so as to be better suited to mass-production. Wedgwood also discovered that cream-coloured earthenware could be successfully coloured with oxides to imitate stones, most notably agate, porphyry, granite and Blue John. In 1768 he entered into partnership with Thomas Bentley, a merchant from Liverpool, to manufacture ornamental wares using this type of coloured earthenware in the Neo-Classical style. A year later, they established the famous canal-side Etruria factory to produce these

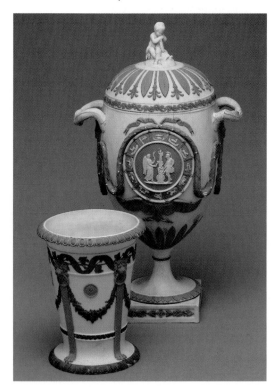

Pair of white *Jasper ware* vases with green and lilac decoration, 1862 (large) and 1871 (small)

Keith Murray, coffee
service, c. 1934

new moulded wares. The pieces were so popular that the company's London agent was, according to Wedgwood, "mad as a March hare for Etruscan vases". After literally thousands of experiments, in 1774 Josiah Wedgwood finally perfected his cameo-like *Jasper ware*. Many leading artists, including George Stubbs (1724–1806) and John Flaxman (1755–1826), designed relief decorations for *Jasper ware*, which was unashamedly Neo-Classical in style. Wedgwood explored every kind of shape and function possible in **ceramics**. He also investigated new industrial methods of production and in 1782 his Etruria factory became the first to install a steam-powered engine for the mass production of his moulded wares. During the 19th century, the Wedgwood factory continued manufacturing both domestic and ornamental wares. In the 1930s and 1940s, the factory began producing Modern wares designed by Keith Murray (1892–1981) and Eric Ravilious (1903–1942), among others. Alongside its more traditional wares, Wedgwood today continues to produce designs that combine simple forms with modern decoration.

HE 1 electric kettle
for Braun, 1962

REINHOLD WEISS

BORN 1934
STUTTGART, GERMANY

Reinhold Weiss studied at the Hochschule für Gestaltung in Ulm, where he trained under the industrial
designer **Hans Gugelot**. In 1959 Weiss became an in-
house designer for **Braun** and three years later was
appointed deputy director of its design department.
While at Braun, Weiss was also executive designer
for the appliance division and designed numerous household products, including the well-known *HL 1* desk fan (1961), which was awarded a "Gute
Form" Bundespreis in 1970. This relatively austere design exemplified the
highly rational and systematic approach to design that was advocated at
Ulm and promoted by Braun. Weiss's *HT 1* toaster (1961), *HE 1* electric kettle (1962) and *KMS 1/11* coffee grinder (1967) similarly articulated this design vocabulary. In 1967 Weiss emigrated to the United States and together
with **Massimo Vignelli** opened the New York office of Unimark, a design
consultancy specializing in **corporate identity**. Three years later Weiss established his own design office in Chicago, which concentrated on graphics
and product design. In 1972 Weiss received an award from the American
Institute of Graphic Arts for a **packaging** design and in 1982 was awarded
a gold medal at the Industry Fair in Brno. Reinhold Weiss currently lives
in Evanston, Illinois, and continues to run a design office.

KMS 1/11 coffee
grinder, 1967

ROBERT WELCH

BORN 1929
HEREFORD, ENGLAND

Cast-iron nutcracker,
1964

Cast-iron candle-
stick, 1962

Robert Welch studied at the Malvern School of Art, where he won the Charlotte Jacob Prize for Silversmithing in 1950. Although he went on to train as
a silversmith under Ralph Baxendale and Cyril Shiner at Birmingham College
of Art, his prospects of a career looked bleak since Britain during this period
was still imposing a 100% purchase tax on silver. Welch therefore decided to
learn how to design for industrial production and enrolled on the industrial
design course at the Royal College of Art in London.
During the mid-1950s, the influence of Scandinavian design was immense and Welch was inspired by its simplicity, function and understated beauty. Throughout
his career Welch has emulated these virtues in his work
both as a silversmith and as an industrial designer. His
designs for industry include the *Merlin* alarm clock for
Westclox; an enamelled steel kettle for Carl Prinz; plastic
handled scissors and a nylon knife sharpener for Harrison Fisher; a range of eight acrylic lamps for Lumitron;
bathroom fittings for Adie & Nephew; *Kitchen Devil*
knives for Wilkinson Sword; and most notably stainless
steel tableware for Old Hall, including serving dishes,
a coffee service, a toast rack, cutlery, a nutcracker, can-
dlesticks and a cruet set. Welch has also designed a
range of utilitarian cast-iron products, including candle-
sticks, salt and pepper mills, fruit bowls, trivets, a tape
dispenser, a pestle and mortar, weighing scales and a
range of cookware, which are sold from his studio-shop
in Chipping Campden (where he has chosen to work
since 1955). Throughout his career, Welch has balanced
"hand and machine" in his design of domestic products
which possess rich, craft-like qualities and yet are vol-
ume-produced using industrial techniques.

Revere Ware sauce-pans for the Revere Copper & Brass Company, 1938

W. ARCHIBALD WELDEN

AMERICAN, ACTIVE 1920S–1950S

The cooking pot is one of humankind's oldest designed artefacts, with an ancestry that can be traced back to prehistoric times. One of the most innovative designs along its evolutionary path was W. Archibald Welden's copper-bottomed stainless steel *Revere Ware* cooking pans (1938) produced by the Revere Copper & Brass Company. Founded in 1892, the Revere Copper & Brass Company had over many decades attempted to develop "a pot that would combine the thermal efficiency of copper with easier cleaning and maintenance". While stainless steel appeared to offer the solution, it had poor thermal conductivity. It was not until 1936 that a manager of the company, J. M. Kennedy, came up with the idea of depositing a copper coating on the outer base of a stainless steel pan so as to unite the heat conductivity of copper with the durability and imperviousness of stainless steel. After literally hundreds of tests and experiments, the company eventually arrived at a revolutionary low-cost process for the electrodeposition of a heavy, hard and adherent copper coating on stainless steel. In the meantime Welden, who was working as a design con-

Revere Ware sauce-pan for the Revere Copper & Brass Company, 1938

Drawings of detailed
elements from *Revere
Ware*

sultant to the company, determined that the fundamental criteria for a cooking pot were heating efficiency, ease of carrying and handling, especially with
liquid contents, and ease of cleaning. His resulting designs were entirely
logic-driven – the pots were carefully dimensioned in accordance with the
design and size of gas, coal, oil and electric range burners, a non-geometric
curve was developed for the inside base of the pans that conformed to the
shape of the fingers so as to facilitate easier cleaning, while the form of the
moulded plastic handles was guided by numerous ergonomic considerations. Revere eventually began retailing Welden's *Revere Ware* in 1939 and in
testimony to the timeless functional simplicity of the design, the range remains in production today with only a few minor modifications. For Revere,
Welden also designed the first successful stainless-steel pressure cooker
and later an institutional range of pans (1954).

WESTINGHOUSE

FOUNDED 1886
PITTSBURGH, PENNSYLVANIA, USA

In 1865 George Westinghouse (1846–1914) filed his first patent for a rotative steam engine. Although this invention proved impractical, he later applied the engineering principles on which it was based to the successful design of a water meter. Also in 1865, he designed a device for placing derailed freight carriages back onto their tracks. His first major invention, however, was the air brake, which he patented in 1869. Westinghouse became an early advocate of **standardization** when he realised that there were significant advantages to standardizing the design of his air brakes, so that they would be compatible between different rail cars on different lines. Apart from his important contribution to rail transport safety, Westinghouse was also responsible for the widespread acceptance of electrical alternating current in America, which during the 1880s was still employing the direct current system.

Westinghouse advertisement for washing machine, 1950s

Europe, on the other hand, was developing a number of alternating current systems utilizing the transformers designed by Lucien Gaulard and John Gibbs. In 1885 Westinghouse imported a Gaulard-Gibbs transformer and began adapting its design. With this device, Westinghouse was able to "transform" the voltage of alternating current so that electricity could be carried over considerable distances at high voltage, before it was stepped down to the proper voltage for its intended use. In March 1886 the redesigned transformer was demonstrated in Great Barrington, Massachusetts, where people marvelled at the electrically powered lights "that produce no odour, heat or danger of fire". This first Westinghouse transformer heralded the golden age of electric-

... of course, it's electric!

Just Plug It In!

SAVE INSTALLATION COST WITH NEW WESTINGHOUSE

EASY LOADING ... NO STOOPING
No lifting or carrying of heavy wet clothes. No lines to string or tighten.

"No defrosting anywhere..not even here!"

Yes, the Westinghouse FROST-FREE
Is the _First_ and _Only_ Completely Automatic Refrigerator

See all the famous Westinghouse Appliances . . . at your retailer's . . . today!

YOU CAN BE SURE..IF IT'S Westinghouse

ity and its basic features have since remained virtually unchanged. Having founded the Westinghouse Electric Company the same year, by 1890 the company had constructed over 300 power stations and achieved an annual sales figure of $4 million. As well as being a major power provider, the company also became a leading manufacturer of electrical appliances, introducing the first fully automatic electric cooking range in 1917 and the first practical automatic electric iron in 1924. During the Second World War, the company designed the first long-range ground warning radar system (1939) and built the first American-designed and American-built jet engine (1941). In the 1950s Westinghouse also became a pioneer of nuclear power technology. Although its large range of domestic appliances made it a trusted household name, it was never able to rival the **General Electric** Company and in 1975 ceased production of all household appliances, apart from light bulbs. Today, Westinghouse remains a leading developer and manufacturer of defence electronics and a major player in the commercial nuclear power industry.

ELI WHITNEY

BORN 1765 WESTBORO, MASSACHUSETTS, USA
DIED 1825 NEW HAVEN, CONNECTICUT, USA

One of the great pioneers of the **Industrial Revolution** in America, Eli Whitney studied science and technology at Yale College (now Yale University). After graduating in 1792, he accepted a teaching post in South Carolina, but was shipwrecked on his way there and ended up in New York, where he met Phineas Miller and Catherine Greene. Travelling south with them, he stayed at Greene's plantation, Mulberry Grove, where he learned of the need for a new machine that would help remove seeds from the short-fibred cotton that grew in the South, and which unlike long-fibred cotton could not be easily cleaned. Within six months, Whitney had constructed a working model of his revolutionary cotton gin. "This machine may be turned by water or with a horse, with the greatest of ease", he declared, "and one man and a horse will do more than fifty men with the old machines". The device pulled the cotton through hundreds of closely set teeth mounted on a revolving cylinder that combed the seeds out. The fibre was then passed through slots in an iron breastwork that were narrow enough not to let seeds through. Whitney's cotton gin had an enormous impact on the South's economy, which became almost completely cotton-based. This simple yet ingenious design was patented in 1794. Whitney subsequently

Engraving of Eli
Whitney's cotton
gin, designed in 1793

went into business with Miller to begin mass-producing the cotton gin
using purpose-built machinery. The patent rights to the device, however,
were widely infringed by planters who made their own copies, which resulted in Miller & Whitney going out of business in 1797. Having learned from
this experience, Whitney decided to turn his attention to the manufacture of
small arms, as the US Government, fearing the outbreak of war with France,
was attempting to solicit 40,000 muskets from private contractors. Muskets
at this stage were still made almost entirely by hand and individually, so that
each one differed slightly from the next. Whitney resolved to produce small
arms using specially-designed machinery allowing for the mass-manufacture
of interchangeable parts. This novel and far-reaching concept of **standardization** was completely revolutionary and had enormous implications for
later industrially produced designs. Standardized interchangeable parts enabled faster assembly and much easier replacement of broken elements.
Funded by the US government, Whitney established an armoury in New
Haven to produce his famous muskets. There he almost single-handedly
pioneered what became known as the "American System" of mass-production, which by the late 1900s was being used for the large-scale manufacture
of numerous products, from sewing machines and clocks to automobiles.

WILBUR & ORVILLE WRIGHT

BORN 1867 NEAR MILLVILLE,   BORN 1871 DAYTON,
INDIANA, USA   OHIO, USA
DIED 1912 DAYTON,   DIED 1948 DAYTON,
OHIO, USA   OHIO, USA

Wilbur (top) &
Orville Wright

Drawing of Wright
brothers' basic
patent, 1904

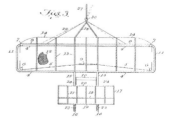

Having grown up in a home where "there was always much encouragement to children to pursue intellectual interests; to investigate whatever aroused curiosity", the Wright brothers initially established a printing shop where they designed and built their own presses. In 1892 they opened a shop for the sale and repair of bicycles and four years later began constructing bicycles to their own designs. They used the profits from these two ventures to fund aeronautical experiments, whereby the experience gained from the design and construction of the presses and bicycles, and their subsequent understanding of the inherent characteristics of materials such as **tubular metal**, wood and wire, was of enormous benefit to their research. After corresponding with the French civil engineer and leading authority on aviation, Octave Chanute, in 1900, the Wright brothers realized that a successful flying machine would need wings to generate lift, a propulsion system and a control system. Of these, control posed the greatest problem. Having determined that the best means of control would be through the precise manipulation of the centre of pressure on the wings, they devised a method of mechanically inducing a helical twist across the wings in either direction. This provided lift on one side and decreased lift on the other so that the pilot could raise and lower either wing tip at will. This concept of "wing-warping" was used to control the Wright's first glider of 1900, which was mainly flown as a kite without a pilot on board. Having constructed and flown another glider in 1901, they subsequently built a small wind tunnel in which they tested the performance of over 100 wing designs. Their researches led to the very successful 1902 glider with an improved control system comprising a forward monoplane elevator and a moveable rudder linked to the wing-warping system. First tested in late 1902 at Kill Devil Hills, near the village of Kitty Hawk, North Carolina, the 1902 Wright glider demonstrated that the brothers had overcome the key problems of heavier-than-air flight. So important was this aircraft that the Wright's later basic patent featured a glider

design with the control system of the 1902 model, rather than a powered
aircraft. Having made considerable technical progress with their gliders,
the brothers moved on to powered machines and built their 1903 *Flyer*.
This aircraft was powered by a 12.5 horsepower four-cylinder engine of the
Wright's own design, which was linked through a chain-drive transmission
to twin counter-rotating pusher propellers. On 17 December 1903 at Kill
Devil Hills, with Orville as pilot, this aircraft achieved the first ever controlled, powered flight, flying 120 feet in 12 seconds. Over the next couple
of years the Wrights refined their designs so that by 1905 their *Flyer* could
remain in the air for 39 minutes. Three years later, the Wrights were contracted by the US government to supply a military aeroplane capable of
flying for more than one hour at an average speed of 40 mph. In 1909 the
US Army conducted trials of the resulting *Model A*, which was the world's
first military aeroplane. That same year, the Wright Company was incorporated, an aeroplane factory was established in Dayton and a flying school
was founded at Huffman Prairie. In tribute to the Wright brothers' historic
achievements, the label beside their famous 1903 *Flyer* on display in the
Smithsonian Institute reads: "By original scientific research, the Wright
brothers discovered the principles of human flight. As inventors, builders
and flyers, they further developed the aeroplane, taught man to fly, and
opened the era of aviation."

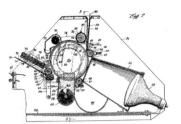

XEROX

FOUNDED 1906
ROCHESTER, NEW YORK, USA

British patent
drawing, 1952 –
the American
patent was
originally filed
in 1940

→Xerox copier,
mid-1970s

Chester Carlson (1906–1968) briefly worked for the **Bell Telephone Company**, prior to joining the New York-based electronics firm P. R. Mallory Company. While working there as a patent attorney, Carlson became increasingly frustrated by the difficulties he experienced in obtaining copies of patent drawings and specifications. From 1934, he began exploring techniques for copying both text and drawings using electrostatic methods rather than photographic or chemical methods, which were already being researched by a number of large corporations. Having set up a makeshift laboratory in Queens, New York, he developed a prototype photocopier that produced the first xerographic image in 1939. Carlson then spent several years trying to sell his invention to various large corporations, including **General Electric** and **IBM**, but was met with what he described as "an enthusiastic lack of interest". The reasons for this were manifold – Carlson's prototype copier was both bulky and messy to use, for one, and many business people felt that carbon copy paper remained perfectly adequate for their needs. Eventually the Battelle Memorial Institute in Colum-

Chester F. Carlson
with early photo-
copier

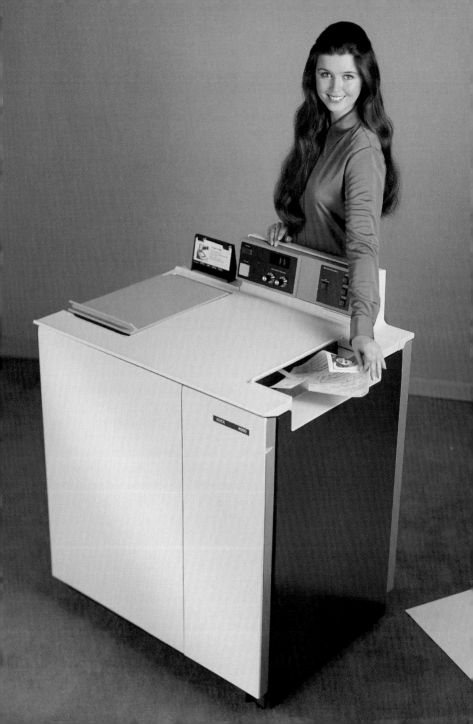

Telecopier 400
facsimile machine
manufactured by
Rank Xerox in
England, 1974

bus, Ohio, contracted Carlson to refine his "electrophotography" process in 1944. The Haloid Corporation, a photographic paper manufacturer and retailer founded in 1906, subsequently purchased the rights to manufacture and market a copying machine based on Carlson's invention. Later, the company obtained all the rights pertaining to Carlson's pioneering technology and, with him, agreed that a new and shorter title for the process was needed. The name "Xerography" was eventually coined and the word "Xerox" was trademarked in 1948. The following year saw the introduction of the first xerographic copier, the Model A. In 1958 the Haloid Company changed its name to Haloid Xerox Inc. (becoming the Xerox Corporation in 1961). The first automatic plain-paper office copier, the Xerox 914, was launched in 1959 and remained in production until 1976. This design was so successful and influential that the company had to battle to prevent the name "Xerox" from becoming a generic term. Since then, Xerox has been a world leader in the photocopying field and has diversified into other product areas, including word-processors, facsimile machines and the highly successful office communications network Ethernet (1979). Since the early 1990s, Xerox has been committed to the design of "waste-free" and "sustainable" products that "optimize resource utilization and minimize environmental impact".

→ Model 1075 photo-
copiers, late 1980s/
early 1990s

Global G-3 carving knife & G-13 fork, 1985

Komin Yamada studied industrial design at the Tokyo Ikuei High School, graduating in 1968. He subsequently worked as a studio assistant in the industrial design department at Chiba University prior to establishing the Y & N Design Office in 1972. In 1985 he founded his own independent industrial design consultancy, the ZAC Design Office. Around this time, he was commissioned by the Master Cutlery Corporation of Japan to design a range of revolutionary kitchen knives harnessing the best available materials and the latest technology. Developed with an almost limitless budget, the resulting *Global* knives (1985) had a single form construction made of the finest stainless steel. The ice-tempered blades utilized molybdenum/vanadium stainless steel that was hardened to Rockwell C 56–58 for superior edge retention. Yamada's knives are very carefully weighted: like Samurai swords, they have perfect balance in the hand and as a consequence are especially favoured by Sushi chefs. Their smooth contour and seamless construction eliminates dirt traps, thereby making them extremely hygienic. Sold in over 65 countries, the *Global* range now embraces over 50 different knives for every conceivable type of food preparation. In 1990 *Global* knives received a Japanese Good Design award and in 1991 were selected as the "Best Cooks' Knives" in the Benelux countries.

Global GS-3 cooking knives, 1985

YAMAHA

FOUNDED 1897
HAMAMATSU CITY, JAPAN

GK Design Group,
-vot audio device,
987

Torakusa Yamaha produced his first organ in 1887 and ten years later founded the Nippon Gakki Company (renamed the Yamaha Corporation in 1987 to mark the company's centenary). In 1900 the company began manufacturing pianos and concentrated on the production of keyboard instruments until 1954, when it began diversifying and developed its first hi-fi player and first motorcycle, the 125cc YA-1. The subsidiary Yamaha Motor Company was founded in 1955 and in 1960 began production of fibreglass powerboats and outboard boat motors. Since the 1960s, Yamaha has developed a wide range of mainly sports vehicles that utilize the company's expertise in engine and fibreglass technology. These include the SL350 snowmobile (1968), Yamaha's first golf cart (1975), and the MT500T WaveRunner personal watercraft (1986). Yamaha's motorcycle business proved particularly successful – in 1977 motorcycle production reached the ten million mark. Many Yamaha Motor Company products have been designed by the **GK Design Group**, including the company's highly successful all-terrain vehicles (ATVs). Yamaha continues to design and manufacture high-quality audio equipment and keyboard instruments, and like many other Japanese companies is also actively involved in the design of robots for automated production lines.

GK Design Group,
Morph II motorcycle,
991

MARCO ZANUSO

BORN 1916
MILAN, ITALY

One of Italy's leading postwar industrial designers, Marco Zanuso is celebrated for his innovative applications of state-of-the-art materials and technology in the design of forward-looking office equipment, consumer electronics and furniture. Having studied architecture at the Politecnico di Milano,
graduating in 1939, in 1945 he established his own Milan-based practice
and executed product and furniture designs as well as commissions in the
fields of architecture and town planning. From 1946 to 1947 Zanuso edited
the journal *Domus* jointly with Ernesto Rogers (1909–1969), and also edited
Casabella magazine from 1947 to 1949. In 1948 he was commissioned by
Pirelli to explore the potential of latex foam as an upholstering material. His
subsequent *Antropus* chair (1949) was the first chair to exploit this material
and was produced by Arflex, a manufacturing company set up by Pirelli. It
was followed by several other latex foam upholstered seating designs, including the *Lady* chair (1951) and the *Triennale* sofa (1951). These two pieces

Duna cutlery for
Alessi, 1995

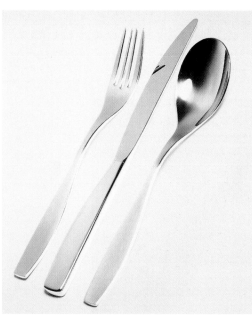

were first exhibited at the IX Milan
Triennale of 1951, where Zanuso
was awarded a Grand Prix and two
gold medals. In 1956 he was awarded a Compasso d'Oro for his *Model
1100/2* sewing machine for Borletti
– a design that epitomised his work
through its rational yet sculptural
form. Zanuso also undertook several architectural commissions,
including the **Olivetti** manufacturing plants in San Paolo (1955) and
Buenos Aires (1955–1957) and the
Necchi factory in Pavia (1961–1962)
which revealed his interest in **product architecture** and prefabricated
structures. From 1958 to 1977 Zanuso collaborated with **Richard Sapper**
on numerous landmark product
and furniture designs, including

the enamelled light-weight stamped steel *Lambda* chair for Gavina (1959–1964); the *No. 4999/5* stacking child's chair for **Kartell** (1961–1964), which was the first seating design to be produced in injection-moulded polyethylene; the *TS 502* radio (1964) and *Doney 14* (1964) and *Black 201* (1969) televisions for **Brionvega**, which were remarkable for their **miniaturization** of technology and synthetic Pop aesthetic; the *Grillo* telephone for Italtel/ **Siemens** (1966) with its innovative folding form; and the *2000* kitchen scales for Terraillon (1970). In 1956 Zanuso became a member of CIAM (Congrès Internationaux d'Architecture Moderne) and the INU (Instituto Nazionale Urbanista). During the same year, he also co-founded the Associazione di Disegno Industriale (ADI). He was a city councillor in Milan from 1956 to 1960 and in 1961 became a member of the city's planning commission. Having experimented with new materials, manufacturing methods and types of construction throughout his career, Zanuso often produced product designs that redefined the formal potential of existing types. From 1966 to 1968 he was a professor of materials applications at the Politecnico di Milano, becoming Professor of Industrial Design there in 1976.

Stacking child's chair *No.4999/5* for Kartell, 1961–1964

Gastone Zanello &
Zanussi Design
Center, *Series 700*
mini cooker, 1961

In 1916 Antonio Zanussi opened a small workshop to make stoves and
wood-burning ovens. Four years later, the Officina Fumistera Antonio Zanussi launched its first export design – a wood-burning stove with a cast-
iron hotplate. In 1946 Zanussi's son Lino took over the company and began
a programme of modernization by adopting efficient industrial production
processes. He was also responsible for establishing the Zanussi Design
Center in 1954 and for commissioning **Gino Valle** to design household appliances for the firm. With their crisp modern lines, Valle's product designs
helped Zanussi gain a reputation for innovation. From 1958 to 1981 the company's design department was headed by Gastone Zanello, who introduced
the concepts of modularity and dimensional co-ordination, thus heralding
the advent of built-in models. Between 1976 and 1989, the freelance designer Andries Van Onck also produced designs for the company that were
based on the concept of "product families", which served to aesthetically
differentiate the various brands within the group. In 1982 Roberto Pezzetta

A. Van Onck &
Zanussi Design
Center, "A" washing
machine, 1980

succeeded Zanello as director of the Zanussi Design Center and two years later the company became part of the **Electrolux** group. This provided it with the necessary funding to re-establish its reputation for ground-breaking designs, such as the *Jetsystem* washing machine (1985), which automatically adjusted its energy consumption to the size of the load. As part of the Electrolux Design Families Strategy, in 1987 Pezzetta designed the post-modern *Wizard's Collection*, which was considered "a cornerstone in household appliances design history". During the 1990s, Pezzetta also designed several experimental prototypes for "a new generation" of "bio-design" appliances that combine "emotion, style, ergonomy", including the award-winning *Oz* refrigerator and the *Zoe* washing machine. Today, around five million Zanussi appliances are sold every year.

Roberto Pezzetta, *Colours* oven, 1998

Roberto Pezzetta & Zanussi Design Center, *Colours* refrigerator, 1999

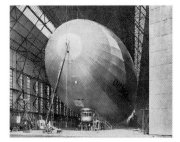

LZ 126 airship being repaired, c. 1924

FERDINAND GRAF VON ZEPPELIN

BORN 1838 KONSTANZ, BADEN, GERMANY
DIED 1917 CHARLOTTENBURG, BERLIN, GERMANY

In 1863 Ferdinand Graf von Zeppelin made his first balloon ascent in St. Paul, Minnesota, while acting as a volunteer military observer for the Union Army during the American Civil War. He became fascinated by balloon technology and after retiring from the military in 1890 dedicated his energies to the design of rigid dirigible airships. After ten years of research, Zeppelin's *LZ-1* airship made its first test flight. 128 metres long, it had a lightweight **aluminium** frame and could achieve speeds of up to 32 kph. The *LZ-1* was the first of many lighter-than-air airships which Zeppelin developed, and which performed better than Germany's fledging aeroplanes. After a Zeppelin airship made a 24-hour flight in 1906, the German government placed an order for an entire fleet. An airship passenger service was launched four years later, by the Deutsche-Luftschiffahrts AG (German Airship Travel Company). Over one hundred Zeppelin airships were used for military purposes during the First World War, after which commercial production resumed under the supervision of Hugo Eckner (1868–1954), who piloted the legendary *Graf*

The *Hindenburg* disaster – the airship crashed and exploded while landing at Lakehurst, New Jersey, on 6 May 1937

Zeppelin on its 21-day around-the-world flight in 1929. When the Nazi
Party came to power in 1933, the company was compelled to design an
even bigger and better airship that was intended to reflect the superiority
of the Third Reich. The result was the *LZ-129*, which made its first flight
in 1936. Better known as the *Hindenburg*, this airship famously crashed
and exploded while landing at Lakehurst, New Jersey, in 1937. Following
this tragedy, airships were deemed too dangerous and all but disappeared
from the skies.

THEMES AND MATERIALS

Bruno Sacco's
Mercedes-Benz 190
in wind tunnel, 1983

Aerodynamics is a branch of physics concerned with the study of air and liquid motion and the forces that act on a body, such as an aeroplane, car or ship, when it passes through such media. It is based on the fundamental principle that the less resistance, the faster and more efficiently a body can move.

The fact that air resists rather than impels the movement of a projectile was already recognised in the late 15th century by such pioneering engineers as Leonardo da Vinci (1452–1519). Astronomer and mathematician Galileo Galilei (1564–1642) subsequently demonstrated that resistance was proportional to the velocity of an object. In the late 17th century, the Dutch physicist Christiaan Huygens (1629–1695) and the British mathematician and physicist Sir Isaac Newton (1643–1727) established that the resistance acting on the motion of a body was proportional to the square of its velocity. These and other subsequent discoveries into the nature of aerodynamics provided the scientific basis for the eventual practical application of aerodynamics to the design of ships, trains, automobiles, aircraft and rockets. Modern aerodynamics emerged around the time of the **Wright** brothers' first powered flight in 1903. The German physicist Ludwig Prandtl (1875–1953) is generally acknowledged as the father of this modern science; it was his discovery, in 1904, of the boundary layer which adjoins the surface of a body moving in air or water, which led to a greater understanding of drag forces.

Alfa Romeo 40–60
HP, 1913

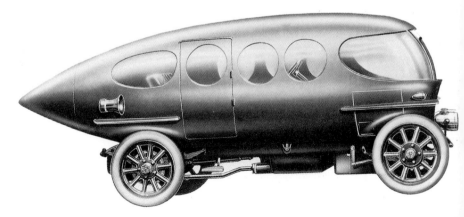

Prandtl and the British engineer Frederick Lanchester (1868–1946) later independently worked on wing theory, which explained the principles of airflow and the concept of lift. By the start of the 1920s, industrial designers such as **Paul Jaray** were employing wind tunnels to develop better and more efficient aerodynamic designs. In demonstrating that the reduction of drag through aerodynamic **streamlining** increased speed, reduced fuel consumption and improved stability, Jaray's findings had a profound impact upon automobile design, as evident particularly in the case of cars produced by Tatra, **Audi** and **Porsche**. Many of the streamlined forms of the 1930s, were driven purely by stylistic concerns, however, rather than by scientific data or – in the case of such domestic appliances as vacuum cleaners and refrigerators – by any real need for aerodynamic performance at all. An exception was **Chrysler**'s *Airflow* automobile designed by **Carl Breer** in 1934, which although not a commercial success, was a remarkably advanced and influential design. Other prominent pioneers of aerodynamics include the Hungarian-born engineer Theodor von Kármán (1881–1963), whose research led to major advances in turbulence theory and supersonic flight.

Rodney Kinsman,
Seville bench for
OMK, 1991

Aluminium is the most abundantly found metallic element in the Earth's crust (forming up to 8 % by weight) and is the world's most widely used non-ferrous metal. Although it never occurs in its metallic form in nature, its compounds are present to some extent in nearly all rocks. The principle ore from which aluminium is extracted is bauxite, a mixture of hydrated aluminium oxides. Although unsuccessful in his attempt, in 1807, to extract the metal, it was **Humphrey Davy** who gave it the name "aluminum" – a name

←Aluminium kettles and saucepans being mass-produced in a factory near Stratford-upon-Avon, 1945

Ernest Race, *BA* chair for Race Furniture, 1945 – made from re-smelted aluminium scrap

Yoshinori Ive,
Ultegra group set
6500 bicycle
components
for Shimano,
1997–1998

that was retained in Canada and the United States but modified to "aluminium" in England and many other countries. A crude form of aluminium was eventually isolated in 1825 by the Danish chemist Hans Christian Ørsted (1777–1851). Introduced to the public in 1855 at the Paris Exposition, aluminium was initially only available in small quantities and was very expensive. In 1886, however, following the advent of relatively plentiful and cheap **electric power**, the commercial method of producing aluminium was simultaneously discovered by the American Charles Martin Hall (1863–1914) and the Frenchman Paul Louis Toussaint Héroult (1863–1914). Their electrolytic process made low-priced aluminium available for the first time, and annual production rose dramatically – from around only 17 tonnes in 1886 to a massive 7,200 tonnes by 1900. In the 1930s aluminium became the material of choice for many industrial designers, especially in America, as its inherent properties suited the prevalent taste for smooth streamlined surfaces. Lightweight, ductile, highly malleable, non-toxic, corrosion-resistant and an excellent conductor of heat and electricity, applications of aluminium ranged from kitchenware to aircraft. Aluminium alloys can be cast, pressed, machined, rolled into a thin foil, spun and even extruded into tubes and other hollow forms. Aluminium can also be mixed with materials such as manganese to produce a metal alloy that is equal in strength to steel. In terms of industrial production, aluminium remains the most favoured metal; with increasingly sophisticated casting and extrusion technology, it continues to find new and highly innovative applications from furniture to highly engineered mechanical components.

Fig.2.

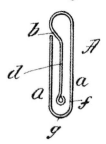

Early design for a paperclip

Six anonymous designs – from the humble screw to the handy safety pin

Many product designs that we take for granted in our daily lives have evolved anonymously over decades or even centuries, through a process of "natural selection" that is driven by practical need rather than by aesthetic concern. Objects such as these are often superlatively functional, as their performance has been honed by successive generations of designers, craftsmen and manufacturers. The designer **Gio Ponti** described these types of objects as "designs without adjectives" as they belong to no particular style or movement. As early as the mid-19th century, anonymous designs were being praised for their honesty, integrity and beauty – at a meeting of the Freemasons of the Church in London in 1849, William Smith Williams stated that: "The adaptation of the thing for its purpose, so far from producing ugliness, tends to beauty. ... In the commonest, rudest, and oldest implements of husbandry – the plough, the scythe, the sickle – we have

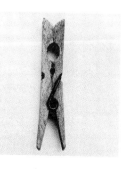

examples of beautiful curves. The most elementary and simple of forms, if well-proportioned and of graceful contour, are the most pleasing." Similar sentiments were later taken up by the **Deutscher Werkbund**, which illustrated industrially produced and highly functional anonymous products in its catalogue *Form ohne Ornament* (Form without Ornament) of 1924. As identified earlier by Williams, one of the best examples of anonymous design is the common plough – the single most important agricultural tool in history. Its evolution can be traced to prehistoric digging sticks, but it was during the Roman period, over 2000 years ago, that recognizable ploughs with iron blades or shares first appeared. The 18th century saw the development of ploughs with cast-iron mouldboards that turned the furrow which had just been cut by the plowshare. This design was considerably improved in the 19th century, by **John Deere**, who developed a plough with an all-steel one-piece mouldboard, similar examples of which remain in use today.

The classic English teapot sometimes known as the "Brown Betty" is another example of a remarkably functional anonymous design. Having evolved over hundreds of years, this teapot pours well, brews tea beautifully and does not stain – few designs can claim superior fitness for purpose. The

R. Farrier, Etching showing a labourer using a breast plough, 1882

study of anonymous designs, such as the paperclip, the clothes peg, the champagne cork, the wine bottle, the zip fastener, the thumb tack, the safety pin, the padlock, tweezers and the wood screw, has highlighted the fact that an evolutionary rather than revolutionary approach to design more often than not produces the best solutions.

"Brown Betty" teapot, England – a design type that has been in existence for centuries

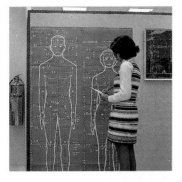

Anthropometric
chart, c. 1960s

Anthropometrics is the systematic collection and correlation of measurements of the human body. Having
originated in the late 19th century with social scientists
evaluating the physical differences between racial groups
and attempting to establish evidence of criminal physi-
cal types, anthropometrics did not become a factor in design until the 1920s,
when pioneering Scandinavian designers such as Kaare Klint (1888–1954)
began relating the dimensional aspects of the human form to the design of
everyday objects. Also known in the United States as "human engineering",
the application of anthropometric data became more common after the
Second World War, when wartime research was made public. The American
industrial designer **Henry Dreyfuss** was one the most important proponents
of anthropometrics and the closely related field of **ergonomics** as essential
tools for designers. His seminal *Designing for People* (1955) features scale
drawings of "Mr & Mrs Average" (whom Dreyfuss christened "Joe" and
"Josephine"), which illustrate numerous average measurements between

"Joe" – the average
man, from Henry
Dreyfuss' book,
Designing for People,
1955

physical "landmarks", such as the distance between
the wrist and elbow, and the knee and ankle. Dreyfuss
described how he used maquettes of Joe and Josephine
to determine the optimum layouts of product designs
ranging from tractors seats to telephone control con-
soles. Dreyfuss elaborated on his extensive anthropometric researches in his later book, *The Measure of
Man: Human Factors in Design* (1960), which helped
establish the application of anthropometric data as
standard practice in the design community. Dreyfuss'
research into anthropometrics was later developed by
Scandinavian design groups such as **Ergonomi Design
Gruppen**, which concentrated on **design for disability**.
The application of anthropometric data in design planning has in recent years been greatly assisted by the
use of advanced computer software and is now widely
used in most areas of design, from transportation and
furniture to communications and clothing.

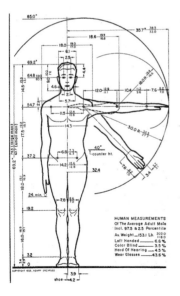

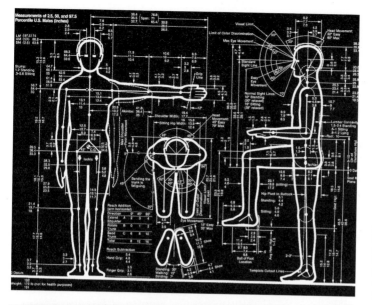

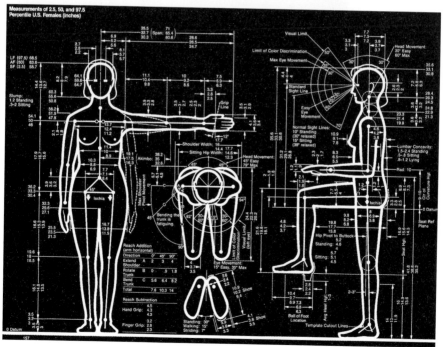

Diagrams showing percentile measurements of males and females in the United States from Henry Dreyfuss Associates' *Humanscale* publication

Isamu Noguchi,
Radio Nurse nursery
monitor for Zenith
Radio Corporation,
1937 – with Bakelite
housing

Leo Baekeland in his
laboratory

The Belgian-born chemist and entrepreneur Leo Baekeland (1863–1944) emigrated to America in 1889. His
first invention was Velox, a photographic paper that
could be developed in artificial light rather than in just
natural light, which made its processing much more
convenient for photographers. The rights to Velox were
purchased for the then astronomical sum of $1,000,000
by George Eastman, who went on to rename his company the Eastman
Kodak Company in 1902. The now wealthy Baekeland moved his family to
the sumptuous Snug Rock estate near Yonkers, where he converted a barn
into a laboratory. Around this time, the price of shellac – a resinous natural
plastic, derived from the secretion of the *Laccifer Lacca* beetle in Southern
Asia and used as an electrical insulator – began to increase dramatically
as demand far outstripped supply. This prompted Baekeland and other
chemists to begin searching for a synthetic alternative. Some 30 years earlier, in 1872, the German research chemist Adolf von Baeyer (1835–1917)
had succeeded in producing a synthetic horn-like resin
from a reaction between phenol and formaldehyde.
Baekeland now attempted to develop a more sophisticated synthetic insulating material, one that could be
dissolved in solvents to produce a varnish, but that
could also be moulded just like rubber. From 1904 he
experimented tirelessly until he had developed a substance that he referred to as Bakelite – a synthetic resin
formed from the chemical combination of phenols and
formaldehydes. This early thermoset plastic was manufactured using a heated iron vessel known as a "bake-
lizer", which accurately controlled the phenol-formaldehyde reaction. After patenting this remarkable material,
Baekeland began the commercial production of Bakelite
– the world's first completely synthetic plastic – in 1909.
A hard, infusible and chemically resistant plastic, Bakelite soon became known as "the material of a thousand
uses". It was moulded into casings for telephones, cam-

eras and pens, and into single-form products such as buttons, ashtrays
and mixing bowls. As an excellent non-conductor of electricity, Bakelite
was particularly useful for electrical appliances, especially radios and fans.
Significantly, the advent of Bakelite made possible, for the first time, the
large-scale mass production of many different types of consumer products.
Thanks to this and its suitability to the moulding process, Bakelite proceeded to change the aesthetic of many industrially manufactured products. The
streamlined and sculptural plastic forms of many of the housewares manufactured in the 1930s would have been inconceivable without the development of Bakelite. Phenol-formaldehyde resins, including Bakelite, continue
to be used as adhesives and paint additives in many industrial applications,
and are indispensable in the manufacture of chemical equipment, machine
and instrument housings, bottle closures, electrical components and insula-
tors. Leo Baekeland's revolutionary material not only heralded the "Age of
Plastics" but also ensured that **plastics** became *the* materials of the 20th
century.

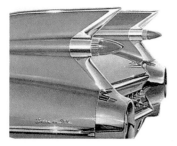

The tail fins of a
1959 Cadillac

→Seeburg jukebox
HF100G, 1953

BORAX

Borax is a derogative term that refers to the American
practice during the 1930s, '40s and '50s of adding surface details to products in an attempt to enhance their
consumer appeal rather than their function. The term was derived from the
famous Borax soap company and its well-known promotional strategy of
give-away special offers. The Borax aesthetic had much in common with
Art Deco **styling** and 1930s **streamlining**, in that it was highly expressive
and frequently ostentatious. Typically relying on the liberal use of gleaming
chromium detailing, Borax was extremely popular, but was also the very an-
tithesis of Good Design. In 1948 it was the subject of a damning critique by
Edgar Kaufmann in *Architectural Review*, entitled "Borax, or the chromium-
plated calf". Common to cheap furniture and other domestic products, during the 1950s Borax found its most extreme expression in automotive styling
and in particular the exaggerated tail fins on cars such as **Harley Earl**'s 1959
Eldorado and *Fleetwood* Cadillacs. In Britain, the term Borax was also used
during the post-war years to refer to extravagant streamlining and over-em-
bellished surface treatments, such as those found on juke boxes. By the late-
1950s, however, consumers in both the USA and Britain had become more
design-literate and the surface glamour of Borax began to tarnish.

Pontiac *Firebird*
concept car, c. 1959

Branded doll from
Sunny Jim flour

Pages from a
Cadbury's booklet
showing branded
gifts available with
Bournville Cocoa
coupons, 1935

Branding is the process through which meaning and value are added to products. At its simplest, a brand is a guarantee of authenticity and replicability, a badge of trustworthiness and a promise of performance. Thus, a brand exists as a collection of notions in the consumer's mind. Branding, however, can actually affect our perception of a product's physical characteristics and thus positively colour our experience of using the product. To this extent, branding is used widely by manufacturers as a cost-effective means of adding value to their products. On average, branding is responsible for over 80% of a product's added value, yet accounts for only approximately 20% of its cost. Manufacturers establish brand identity by various means, including naming, **packaging**, advertising and marketing. The idea of "brand personality" is also becoming an ever more important factor in the marketplace, as manufacturers attempt to differentiate their products from those of their competitors. Branding issues are thus increasingly a consideration

CHOCOLATE GIFTS

CHOCOLATE GIFTS

Gift No. 700—½ lb. Cadbury's Milk Chocolate, 2 oz. Bournville Fruit and Nut Chocolate, 2 oz. Milk, Fruit and Nut Chocolate, 2 oz. Bournville Chocolate.   15 Coupons.

Gift No. 701.—½ lb. Drum Cadbury's Chocolate Table Biscuits.   14 coupons.

Gift No. 702.—Selection of Cadbury's 2d. Chocolate Varieties.   15 coupons.

Gift No. 703.—Large Cocoa Jug, Whisk and Measure.   35 Coupons.

Gift No. 704.—Child's Money Box containing an assortment of Cadbury's Chocolates.   13 coupons.

Gift No. 705.—Child's Mug containing Cadbury's Milk Chocolate Drops.   14 coupons.

Gift No. 706.—Small Selection Box of Cadbury's Chocolates.   15 coupons.

Gift No. 707.—Large Selection Box of Cadbury's Chocolates, (Two layers.)   28 coupons.

How to obtain your gift.
SEE PAGE 8.

2

3

McDonald's restaurant at Des Plaines, Illinois, 1955 – the "golden arches" not only created a powerful brand image but came to symbolise post-war American lifestyle as well

for industrial designers working within the framework of an established brand identity, or required to incorporate brand names or logos in their overall design schemes.

The importance of the relationship between **corporate identity** and branding is illuminated by the success of **IBM**, **General Electric** and **Ford**. These companies rank among the top five brands worldwide, a position they have achieved not least by investing enormous resources into integrated corporate design programmes that encompass brand strategies. The implementation of a house style for both product and packaging is a key to the establishment of brand identity. Thus the functionalist aesthetic of **Braun** products and the high-tech vocabulary of **Bang & Olufsen** audio-visual equipment make them instantly recognizable. Brand logos, too, speak a visual, trans-cultural language in which McDonald's "golden arches" and **Nike**'s "tick" mean the same to consumers whether in Germany or in Thailand. Although in recent years it seemed that branded products were going to lose ground to cheaper no-name or own-brand products, the opposite has occurred, with brand identity carrying more weight than ever in today's highly competitive global marketplace. An outstanding example is Coca-Cola, ranked the world's most powerful brand according to a survey conducted in 1999, with a brand value of nearly $90 billion (Pepsi is estimated to be

Symbol of Friendship

Coca-Cola advertisement, "Symbol of Friendship", 1950s

worth only $3.75 billion). Where a branding programme is successful, it should produce a differential and sustainable advantage over competitive products or services. Today, product differentiation through branding is especially critical in sectors where there is increasingly little to distinguish the performance and/or technological advantages of one product over another. Beyond price, the decision leading to the purchase of a particular product is increasingly governed by the consumer's perception of and identification with the brand.

Classic Coca-Cola
bottle, c. 1950s
(originally designed
in 1916 by Alexander
Samuelson)

Mike Burrows, *Mono* superbike, 1992

Carbon fibre is an advanced composite material in which a reinforcing element of woven graphite fibres is embedded in an epoxy matrix. The material is highly versatile and has two outstanding properties, strength and lightness, which makes it especially suitable for high-performance products that require minimum weight with high tensile resiliency. Carbon fibres have extremely high elastic modulus values (up to five times that of steel) and so make excellent reinforcement. Usually loosely woven into a type of cloth, the fibres are laid in the matrix according to the strength requirements of the design – areas where the fibres are most closely woven will be those of greatest stress. While carbon fibre is still a relatively new material, one of its most celebrated applications was in **Mike Burrow**'s *Mono* superbike (1992) which was used by the British cyclist Chris Boardman to win a gold medal at the 1992 Barcelona Olympic Games. Carbon fibre is employed in state-of-the-art sports equipment, from golf clubs to kayaks, the aerospace industry and racing car components, all of which are fields of design where weight considerably affects performance. Carbon fibre has also been used experimentally in furniture, as in **Alberto Meda**'s *Light Light* chair (1987). Con-

Carbon Racing 2.0 ski helmet for Carrera, 1999

structed of a moulded carbon fibre skin surrounding a Nomex honeycomb core, this chair is remarkable for its strength yet extreme lightness, weighing only about 1 kg. Although the high cost of carbon fibre precludes its widespread use in mainstream consumer products, one of the most successful recent applications of the material has been in the serial production of ski helmets, with most manufacturers now using carbon fibre for their top-of-the-range-models.

Josiah Wedgwood,
*The Apotheosis of
Homer* vase, c. 1786

Ceramic materials are derived from naturally-occurring inorganic materials such as clay minerals and quartz sand. Through production processes that have been refined over centuries, ceramics have been rendered into a wide variety of products including china tableware, bricks, tiles, industrial abrasives, refractory linings and even cement. The inherent malleability of clay makes ceramics ideally suited to moulding processes and, thereby, mass production. Ceramics may be said to represent one of the very first processed materials used in the service of design, clay having been fashioned into ceramic pots and vessels since the earliest times. In Ancient Rome, oil lamps were replicated using decorated plaster moulds – an early example of serial manufacture. It was not until the mid-18th century, however, that the English potter **Josiah Wedgwood** became the first to truly industrialize the manufacture of ceramic wares. Due to their inherent resistance to chemical attack, ceramics can also be used for **packaging**, a practice especially widespread in the days before **plastics**. The last 20 years have seen the development of a new breed of high-tech engineering ceramics, with numerous industrial applications ranging from combustion engine

Ceramic dispensing
pots, England, 19th
century

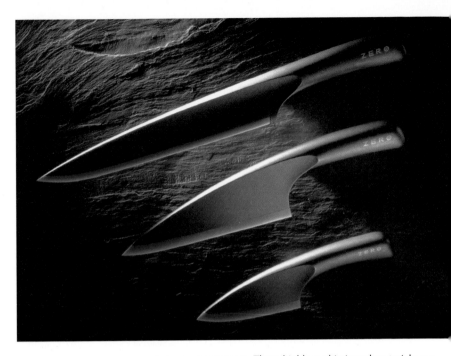

Seymour Powell,
Zero knives (internal
study project), 1998
– using zirconia
ceramics

components to oxygen sensors. These highly sophisticated materials are completely different from traditional ceramics in that they include in their composition metallic powders such as alumina, titania, yttria and zirconia. This lends them significantly different properties, most notably exceptional wear and corrosion resistance, high hardness and strength, and makes them suitable for – among many other things – cutting blades. Today, material scientists are pioneering other advanced ceramics that can be injection-moulded and finished to remarkably accurate tolerances. "Bio ceramics" have also been developed for various medical implants including hip replacements – the material's bio-inert properties, strength and toughness making it ideal for such applications. Ceramics have not only played an important role historically in the evolution of industrial design, but are considered by many to be the engineering materials of the future.

Oldsmobile *88* for
General Motors,
c. 1955

Chromium is a hard metallic element that can be polished to produce a
gleaming reflective surface. It is often used as an alloy to strengthen other
metals, especially steel, and to increase their resistance to oxidation and
corrosion. While a relatively abundant element, chromium always occurs as
a compound, most commonly as chromite. Although the element was first
discovered in 1797 by the French chemist Louis-Nicolas Vauquelin (1763–
1829), it was not until the First World War that it found a useful application
as a coating material for shell casings. In 1925 it became widely available
and was subsequently used decoratively for Art Deco furnishings and as a
corrosion-resistant coating for Modernist **tubular metal** furniture. During
the 1930s, American designers made extensive use of chromium plating,
as the brilliance of its reflectivity accentuated the seductive streamlined

George Scharfen-
berg, *Model T-9*
toaster for Sunbeam,
1937

forms of their product designs. The material had its true heyday in the 1950s,
however, when each year's new car models were laden with ever more chro-
mium-plated elements, which were intended to exude a sense of luxury.

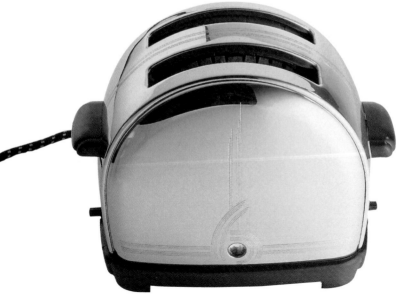

Kenneth Grange,
Protector razor for
Wilkinson Sword,
1992

The colour of a product is very often determined not by the designer but by their client, who sometimes employs an independent colour consultant to predict future trends in preference. This type of fashion-driven colour selection, however, belies the importance of colour in product design. Colour can radically alter the visual perception of a product and can dramatically highlight its form – the classic example being the red used for **Ferrari** automobiles. Colour can also be used to make a product more user-friendly – the limited use of colour in **Braun** alarm clocks, for example, highlights the product's various functions. Colour can furthermore be used to visually update an existing product so as to buy time for the manufacturer to develop its successor. While there are many conventions relating to the use of colour, a number of these have been famously overthrown with spectacular results. **Jonathan Ive**'s *iMac* computer (1998), for example, brilliantly contradicted the notion that all computers must be grey, while **Kenneth Grange**'s *Protector* razor (1992) broke the long-held rule that red should

Electrolux *Oxygen*
vacuum cleaner in
assorted colours,
1999

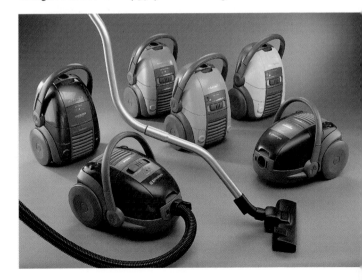

Alessandro Mendini,
Anna G. corkscrews
for Alessi, 1994

never be used for shaving products because of the colour's association with blood. Designers and manufacturers also have to be mindful of different cultural associations with colour. White, for example, is recognized as a symbol of purity in Western cultures yet is associated with mourning in China. There are also many cultural variations in colour preference – southern Europeans tend to prefer red or white cars, while northern Europeans tend to prefer black or silver. Colour is an important aspect of industrial design as it can have strong psychological and physiological effects on product users. Colour selection, therefore, should be treated as an integral part of the design process rather than as an afterthought motivated by the whims of fashion.

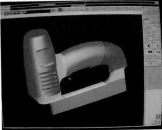

Ergonomi Design
Gruppen, computer-
generated design for
a tool, 1990s

Computer-Aided Design (CAD) was first developed at
the renowned Massachusetts Institute of Technology in
the 1950s. During the early days of computers, however,
CAD software was relatively primitive and was rarely used outside the aerospace and automotive industries. As hardware became increasingly more
powerful in terms of both memory and processing speed, and at the same
time less expensive, so CAD began to gain a foothold in the design community. Today CAD is a powerful and essential tool with which designers can
prepare drawings and translate them into three-dimensional models (either
in solid form or made up of structural lines) that can be rotated, scaled,
zoomed, panned or cross-sectioned. The information derived from CAD
programmes can then be translated by Computer-Aided Manufacture (CAM)

IBM 5080 graphics
system, which can
pan, zoom and
rotate two-dimen-
sional and three-
dimensional models,
late 1990s

software into useable data that assists in the control of the machines (such
as lathes and cutting tools) and robots used in the production process. Primary amongst the many benefits of employing a CAD/CAM combination
are thus greater engineering precision and increased productivity. CAD/
CAM programmes are not only invaluable in ensuring the **standardization**

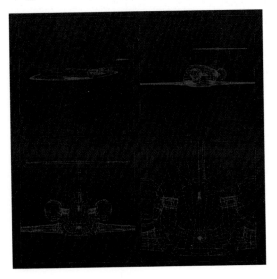

of components but also permit
greater flexibility in the manufacturing process, allowing for lower-
volume or irregular, batch-type production. Significantly, CAD/CAM
can also dramatically reduce the
time-lapse from initial concept to
working prototype, thereby making
the design process more efficient
and enabling manufacturers to get
new products to the marketplace
even quicker. The use of computers
has brought about a revolution in
the design industry and has led directly to the development of safer
and better performing products.

Teams Design, Computer rendering of hedge-clippers for Bosch, 1998

Teams Design, Computer rendering of the *Columbus* thermos flask for Leifheit, 1994

Raymond Loewy, redesigned bottle and drinks dispenser for Coca-Cola, c. 1948

→Selection of corporate logos designed by Raymond Loewy Associates

Corporate identity design, which is strongly related to **packaging** design, is a means by which companies and/ or brands can give their products or services a visually unified character that will differentiate them from others in the marketplace. Central to corporate identity is the company logo, which is normally used on all corporate projections from stationery to advertising. Some mainly design-led companies and brands, such as **Braun**, adopt a holistic approach to corporate identity, implementing a rigorously-managed design regime which impacts not only on the nature of their products but on the design of their offices and factories as well. **Peter Behrens** was the first designer to put such a programme into action when he became the artistic adviser to **AEG** in 1907. He applied an integrated language of design not just to the company's products and graphics, but to housing for its workers and even one of its factories – all of which was instrumental in forging AEG's easily recognizable identity. In view of the progressive globalization of today's markets, commercial organizations are increasingly embracing the universal language of corporate identity design in an effort to compete more effectively.

Peter Behrens, logos for AEG, 1908–1914

Banks

Carib

Andrex

U.S.MAIL

PANEM

ALPHA CREDIT BANK

GÜMÜŞSUYU

ELECTRICITY AUTHORITY OF CYPRUS

NABISCO

RISTORANTE

ROYALE
BELGE

McLaren

CHUBB

De Dietrich

Random Techno-
logies, *Kid's Phone*,
1998 (prototype)

→*Galaxy* system for
Kompan – won the
ID prize 1999

Products for children are designed within a completely different set of parameters than those for adults, and fall into two main categories – equipment and toys.

The equipment category embraces everything from feeding cups to pushchairs (strollers) and is primarily governed by function, although changing social trends also play a part. Thus the prams or "baby carriages" which date back to the 18th century and were the norm in Western society until the 1960s have now given way to pushchairs such as the lightweight folding Maclaren baby buggy, which when introduced in 1967 quite literally transformed the transportation of infants. Generally speaking, well-designed children's equipment will be ergonomically resolved for ease of handling, strongly constructed, characterized by smoothed surfaces for hygiene and the prevention of injury, and brightly coloured so as to attract the child's attention. All of these features are found in the design of the *Anyway* cup. A new prototypical design that also fulfils these criteria is the *Kid's Phone* produced by Random Technologies. This product features a simple layout with a "home" button that can be used by a child to call home even if they cannot remember their telephone number. The children's toy category is governed by similar criteria to equipment, with the most successful designs, such as **Lego** and **Meccano**, providing children of all ages with hours of stimulating creative play. The majority of toys, however, are poorly designed, being driven more by marketing strategies and profit margins than

Anyway cup for V&A
Marketing, c. 1999

any real desire for quality. Among the worst examples of this type of consumerist design are the "free" give-away toys that are common to fast-food restaurants. The play content of these toys is more often than not as short-lived as the meal that accompanies them. As future consumers, this is surely not the way to instil in children the importance of Good Design and of responsible consumption.

A&E Design, *Handy extension handles for Etac, 1977*

A&E Design, *Clean shower and toilet chair, 1999*

Design for disability falls into the category of "virtuous design", but could equally be called "design for ability" because it helps to empower users. The history of this design discipline can be traced to the development of sedan chairs and the later invalid carriages of the 18th and 19th centuries. It was not until the 1950s, however, that leading designers became involved in the design of products for the disabled – one of the earliest examples being an amputee's hook (c. 1950) designed by **Henry Dreyfuss** in collaboration with the Army Prosthetics Research Laboratory and the Sierra Engineering Company. This innovative device was designed so that with a slight flexing of the shoulder muscles the hook could be opened and closed, thus enabling the user to handle relatively small objects including coins and matches. Until the 1960s, equipment for the disabled was designed from a medical perspective and little if any thought was given to their aesthetic appearance. It was around this time, however, that an international disability symbol was devised and public environments began to be adapted or designed for greater accessibility. In 1969 *Design* magazine devoted a whole issue to the subject of design for disability, and two years later Victor Papanek's highly influential *Design for a Real World* was published. Highlighting the need for better design solutions, Papanek wrote: "Cerebral palsy, poliomyelitis, myasthenia gravis, Mongoloid cretinism, and many other crippling diseases and accidents affect one-tenth of the American public and their families (20 million people) and approximately 400 million people around the world. Yet the design of prosthetic devices, wheelchairs, and other invalid gear is by and large still on a Stone Age level." During the 1970s, the greatest advances in the design of enabling products for handicapped people were made in Sweden, most notably by **A&E Design** and **Ergonomi Design Gruppen**. The latter designed numerous products including the well-known *Eat and Drink* combination cutlery, drinking vessels and plates (1980) for RSFU Rehab. The design excellence of these products led one British theorist, James Woud-

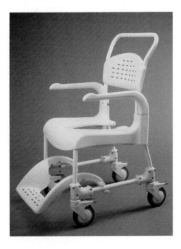

Smart Design, *Good Grip* vegetable peeler for OXO, 1990s

ᐱ David Farrage of Smart Design, *Good Grip* scrub brushes for OXO, 1997

huysen, to ask, in the "Svensk Form" exhibition catalogue (1981), whether it was a handicap not to be handicapped in Sweden. Other life-enhancing products developed over the 1980s and 1990s included London Innovation Limited's *Neater Eater* device, which enabled people suffering from severe tremors to feed themselves, and **Motivation**'s wheelchairs, developed predominantly for use in third world countries using locally available components. Over recent years, the Design for Ability organization based at Central St Martin's College of Art & Design in London, has conducted extensive market research into the needs of disabled people and made the findings available to the design community, resulting in better-performing and better-looking products such as **Tangerine**'s *Activ* walking frame, which bears little relation to its predecessor – the scaffolding-like *Zimmer*. The *Good Grips* range of kitchen equipment, which includes numerous easy-to-use products from dustpans and brushes to potato mashers, was developed by Sam Farber in response to a paper written by Mary Reader for the *Journal of the Institute of Home Economics*, in which she called for the development of "transgenerational" kitchen tools. Having noted that most kitchen tools were "at best indifferent and at worst hostile", especially to those suffering from arthritis, Faber and his wife Betsey commissioned the New York-based consultancy, Smart Design, to design a new range of tools without losing "sight of the final user". The subsequent introduction in 1989 of the *Good Grips* collection of everyday domestic tools with soft rubber ergonomic handles signalled a completely new approach to design for disability – inclusive design for most members of society regardless of age or physical ability.

Chubb fire extinguisher *Model WS9*, 1979

⌄Highmask Manufacturing & Co., Anti-stab vest, 2000 – selected as a Millennium Product by Design Council

Protector Technology, *Tornado* respirator, 2000

The issue of safety is a relatively recent phenomenon within the history of industrial design. It first came to widespread public attention when the young lawyer and consumer advocate, Ralph Nader (b. 1934), wrote a damning critique of the American automobile industry, *Unsafe at any Speed: the Designed-In Dangers of the American Automobile* (1965), which focused in particular on the design flaws of the Chevrolet *Corvair* (1960). This unusual looking mid-engine car was a potential killer because of its tendency to roll over when cornering sharply. While the management of **General Motors** was aware of this defect, it cynically decided to put profit before safety and continued with production. At the time, it was estimated that General Motors spent about $700 on **styling** per car but only around 23 cents on safety features. Nader's landmark legal triumph over this automotive giant heralded not only the beginning of the product liability industry but also the widespread awareness of safety issues within both the design and manufacturing industries. Design for safety falls into two main categories – the develop-

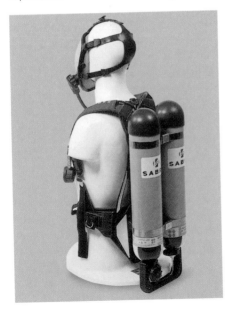

Crash test dummies
developed by Volvo

ment of products that are safer to use, and the design of safety equipment such as fire extinguishers, seat belts, airbags, smoke alarms, breathing apparatus etc. Some car companies such as **Volvo** are renowned for their remarkably long record of developing innovative safety "firsts", while others will only go as far as safety legislation compels them. The design of safer products has been assisted considerably, however, by the introduction of safety standards. These are effectively guidelines, drawn up by committees composed of representatives of relevant industries and then submitted to a process of public consultation. The resulting standards are in many cases incorporated into legislation governing product design and manufacture (e. g. the banning of certain furniture upholstering foams in Britain). In the sphere of toy manufacture, for example, toys complying with European Directives are allowed to bear the CE mark, enabling them to be imported and exported from one member state to another. Although such legislation is increasingly effective in improving product safety, designed products continue to cause injuries and fatalities. In some cases, increasing competitiveness leads design consultancies to just "give the client what they want", without any real in-depth consideration of potential safety issues.

Sycamore Origination, *SpinGrip Outsole* football boot for Umbro, c. 1999 – selected as a Millennium Product by Design Council

Sports equipment is one of the most interesting areas of design practice as it frequently involves pushing materials and technology to new heights of performance. In competition sports, how well a piece of equipment is designed can not merely make the difference between winning and losing, but can redefine the parameters of the sport itself. In recent years, for example, Atomic *Beta Race* skis, whose parabolic shape and highly innovative Beta titanium construction facilitate more precise and aggressive carving, have dominated World Cup skiing to such an extent that they have affected the design of race courses. Because sports equipment relies heavily on research and development, many companies have their own in-house design teams and only rarely commission an independent designer to produce new products for them. Sports equipment is often initially developed using state-of-the-art **Computer-Aided Design** (CAD) technology in accordance with the latest **ergonomics** data, and is then extensively field-tested by company-sponsored athletes so that the designs can be further honed. The knowledge gained from these trials is then incorporated into designs for mass production. The best-performing sports equipment, whether for competition or leisure use, is generally that which is designed to act like body prosthetics, which respond to every move that is made. Although the design of sports equipment is often evolutionary, it can also be revolutionary, as in the case of the *Windcheetah* recumbent high-performance vehicle (1992) designed by **Mike Burrows**, and the *Sea-Doo* personal watercraft invented by **Bombardier** in 1968. The manufacture and retail of sports equipment is now big business, but it is nonetheless subject to the vagaries of fashion. Sometimes a product that is heralded as a significant innovation performs only minimally better than its predecessors and therefore of only limited benefit to the amateur user. It is human nature, however, to believe that a new design – whether a golf club or a pair of football boots – will dramatically improve one's

Big Bertha metal wood driver for Callaway Golf, 1991

Alien roadcycling
helmet for Carrera,
1999

Cressi-Sub,
swimming fin,
c. 1998

game or performance, and perhaps this produces at least a placebo effect. Frequently, new lightweight yet robust materials find their first mass applications in the field of sports design, as with the introduction of **carbon fibre** in the construction of ski helmets. Over recent years, the design of sportswear and sports equipment has had an enormous influence on the mainstream fashion industry.

↘*Bug* sports chronograph for Animal, 1999

Naoto Fukasawa & Peter Spreenberg (IDEO), *Vertech* sports monitor for Avocet, 1993

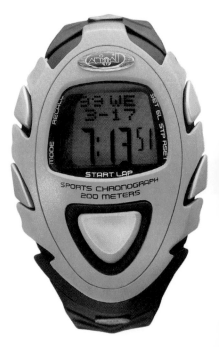

←Continuum of
Milan, Atomic *Beta
Race* skis, 1997

Giugiaro Design,
in-line skate for
Tecnica, 1999

↓Tomohiko Nishi-
mura, *MTB SH-
M320* cycling shoe
for Shimano, 1998

Motivation, *Mekong wheelchair*, 1993

Behind the concept of design for the Third World lies the goal of empowering developing nations to meet their own needs in ways which make economic and environmental sense. Culturally appropriate design can not only dramatically enhance the lives of those most in need, but – vitally, in the longer term – can also provide some of the key foundations upon which regional economies can be built. Examples of this type of design include the wheelchairs designed by **Motivation** for landmine victims, which are built from locally available components, the pedal-powered washing machines developed by the Industrial Design Laboratory of the University of Paraiba in Brazil, and the *Freeplay* self-powered radios and torches designed by **Trevor Baylis**, which are now manufactured in South Africa. Unfortunately, few Western designers have felt compelled to work within this extremely worthy area of design because it is generally regarded as financially unprofitable. The solution to this problem lies in education, as mooted by Victor Papanek in his seminal book, *Design for a Real World* (1971). Here he put forward the idea of Western designers travelling from one developing country to another, in order to train people from "the indigenous population of the country [in order] to create a group of designers firmly committed to their own cultural heritage, their own lifestyle, and their own needs."

WOBO (WOrld BOttle) house study and prototype bottles commissioned by A. H. Heineken, early 1960s – an unrealized proposal for shelters constructed from specially designed beer bottles

Michael Faraday (1791–1867) – one of the greatest pioneers of electromagnetism and inventor of the first dynamo

The transition from the Age of Steam to the Age of Electricity marked a massive turning-point in the history of mankind. It is difficult today to appreciate just how life-changing the introduction of electricity really was. The study of electrical phenomena can be traced right back to Classical Greece, when it was discovered that rubbing amber with a feather created a static charge. The word "electricity" in fact derives from the Greek word for amber, *elektron*. It was not until the 16th century that the British scientist William Gilbert (1544–1603) undertook the first serious scientific research into magnetic bodies and electrical attraction. Generally considered to be the father of electrical science, it was he who coined the terms "electric attraction", "electric force" and "magnetic pole". In 1752 Benjamin Franklin (1706–1790) contributed to the field with his famous kite experiment, which proved the electrical nature of lightning and led to his invention of the lightning rod. Franklin was also responsible for the adoption of the words "negative" and "positive" to describe different charges. Research into the nature of electricity accelerated during the Age of Enlightenment, bringing many new discoveries, including the fact that an electrical charge could be stored in an insulated conducting body. Pioneers in the field of electrostatics from this era included Joseph Priestley (1773–1804), Henry Cavendish (1731–1810), Charles Augustin de Coulomb (1736–1806) and Siméon Denis Poisson (1781–1840).

Early Edison light bulb, c. 1880

The new science only found its first practical applications with the dawn of the 19th century, however. In 1800 the Italian physicist Count Alessandro Volta (1745–1827), working in Como, invented the electric battery, which was the very first source of a continuous electric current. In tribute to this achievement, the "volt" (a unit of electric potential) was named after him. In 1808 the renowned British inventor **Humphrey Davy** showed that an arc of electricity could be produced when two charge-carrying charcoal electrodes were separated, thus demonstrating that electricity could potentially be used as a source of both heat and light. But it was Davy's assistant, Michael Faraday (1791–1867), who contributed most to the understanding of electromagnetism. In 1821 he discovered the scientific principles relating to the electric motor and the same year constructed a primitive electric generator (the first-ever dynamo). He was utterly convinced that there was a connection between magnetism and electricity and his later researches led

him to discover, in 1831, that an electric current could be produced when magnetic intensity was altered – this became known as Faraday's law of induction. Two years later, Faraday stated the laws of electrolysis and a unit of electricity was subsequently named after him. In 1869 the French electrical engineer Zénobe Théophile Gramme (1826–1901) invented a continuous-current dynamo that produced significantly higher voltages than earlier models. After demonstrating his invention in 1871, he began manufacturing it. He exhibited his design in 1873 at the "Welt-Ausstellung" (World Exhibition) in Vienna. While on show, it was discovered that the Gramme dynamo was reversible and could therefore potentially be used as an electric motor. The same year, Gramme demonstrated that overhead conductors could carry electricity. Around 1879 **Thomas Alva Edison** changed the course of history by designing the first incandescent electric light – a design that had eluded numerous inventors over a period of some 50 years. Unlike electric arc lamps that were connected to a series circuit, Edison connected his lights in a parallel circuit so that the failure of one bulb would not affect

Page from The Story of the Lamp published by the General Electric Company Ltd, c. 1907

the rest of the circuit. By 1881 Edison had installed the first operative lighting system on the *Columbia* steamship, and that same year he established the world's first commercially-operated power station in Manhattan. In 1886 George **Westinghouse** demonstrated the first alternating current system in the United States, and his Westinghouse Electric Company went on to build over 300 power stations within 4 years. The use of electric power spread rapidly in both domestic and commercial environments throughout most Western societies, and was used to power numerous landmark inventions from the telephone to the electric streetcar. During the 1890s, the English physicist Joseph John Thomson (1856–1940) discovered the electron that led to John Ambrose Fleming's (1849–1945) invention of the diode in 1904, heralding the birth of the modern science of electronics. The 20th century was fundamentally shaped by electric power, and its application resulted in the development of a multitude of life-changing design typologies, from re-frigerators to DNA sequencing machines. With potentially more advanced and environmentally sustainable means of large-scale energy generation on the horizon, including wind, solar, geothermal and possibly even nuclear fusion, electric power looks set to shape the 21st century as well.

Neo Ball fluorescent light bulb developed by Toshiba, 1998 – has six times the life of an ordinary bulb and uses a quarter of the energy

Environmental design is primarily concerned with minimizing waste and reducing the throughput of energy and materials in our society to sustainable levels. It was most famously pioneered by **Richard Buckminster Fuller,** who in the 1920s promoted a "design science" that was based on the concept of "providing the most with the least". It was he who, in the 1950s, coined the term "Spaceship Earth", which led people to think of the planet in a more holistic way. Other writers and theorists who have contributed to a greater understanding of environmental design include Vance Packard, whose book *The Waste Makers* (1961) was particularly damning of the practice of **planned obsolescence,** and Victor Papanek, author of *Design for a Real World* (1971), who related ecological awareness to the design process and urged for radical design solutions that were mindful of the environment. Views such as these were given greater currency in the early 1970s, when the oil crisis increased concern about the finite aspect of the world's natural resources. By the 1980s, several man-made environmental disasters, plus the growing realization that industrialization was contributing to global warming, had underlined the urgent need for environmental design. Environmental design – also known as "green design" – takes into consideration a product's whole life-cycle: the extraction of raw materials and the ecological impact of their processing; the energy consumed in the manufacturing process, together with any negative by-products; the energy required for and the impact of the distribution system; the length of a product's service life; component recovery and the efficiency of recyclability; and the ultimate effects of disposal on the environment through, for instance, landfill or incineration. Although

Pencil made of a recycled vending cup by Remarkable Pencil Limited, 1998

recycling can reduce energy consumption, it does not minimize it and in
some ways can be seen to actually perpetuate the throwaway culture. Increased product durability, on the other hand, minimizes waste and energy
consumption – by doubling the useful life of a product, its environmental
impact can be halved.

Zdenek Kovar,
ergonomic design
for scissors, 1952

Ergonomics, or "human factors" as it is more commonly known in the United States, is the systematic study of the characteristics of human users and their relationship with products, systems and environments. Closely related to **anthropometrics** (the systematic collection and correlation of human body measurements), ergonomics is concerned with anatomical, physiological and psychological factors in conjunction with human behaviour, capabilities and limitations. Through the scientific application of this data, ergonomics permits the design of better-performing, safer and more user-friendly solutions that are also easier to maintain and understand. In the workplace, therefore, ergonomics leads to greater efficiency and productivity. Because an ergonomic product is designed to work in harmony with the human body, it is also very often more comfortable to use, be it a pair of kitchen scissors or a chair. In terms of office seat furniture, where the provision of continuous flexible support is essential for healthful sitting, it is true to say that the majority of the ergonomic principles which now inform the design of chairs are based on human weaknesses, rather than on strengths. The remarkable success of **Emilio Ambasz**'s *Vertebra* chair (1977), which was the first office chair to re-

Peter Opsvik, *Balans
Variable* chair for
Stokke, 1979

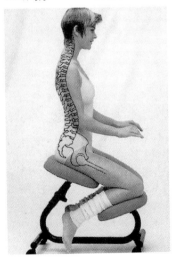

spond automatically to the body's movement, was almost entirely due to the fact that its design was based on ergonomic principles. Today's health and safety legislation, especially that governing the work environment, has ensured that ergonomic factors are increasingly taken into account by designers and manufacturers. As the International Centre of Ergonomics states: "Whether at work, on the road or in the home, ergonomics provides solutions that maximize convenience and effectiveness whilst minimizing the risk of accidents and injuries." The emergence of **Computer-Aided Design** (CAD) has further helped facilitate the application of ergonomic data to the design of products, systems and environments that are more functionally unified with the people who use them.

Emilio Ambasz,
Vertebra chair for
Castelli, 1977

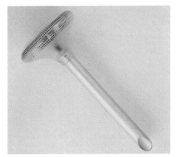

Ross Lovegrove,
Solar Bud eco-light
for Luceplan,
1996–1997

Essentialism is an approach to design concerned with
the logical arrangement of only those elements which
are absolutely necessary for the accomplishment of
a particular purpose. To this extent, essentialism is
based on the Modern concept of getting the most from the least and is
quite closely linked to **environmental design**. The origins of essentialism
can be traced to the Dymaxion (dynamic + maximum efficiency) design
science pioneered by **Richard Buckminster Fuller** in the 1920s, and Fuller's
subsequent attempts to bring about innovative design solutions using a
minimum of energy and materials. An essentialist approach to design frequently relies on a technologically-driven logic of construction and thus
has much in common with functionalist and rationalist design traditions.
Essentialism has characterized the work of many of the 20th century's leading industrial designers, yet the formal vocabulary of their designs has often
varied considerably. Seen as an evolution of Modernism, essentialism can
assume either geometric or organic form. There is no one defining style –
a design by **Dieter Rams**, for example, does not have the same aesthetic
qualities as a design by **Charles Eames**, yet both can be deemed Modern

Ross Lovegrove,
Oasis chair for
Driade, 1997

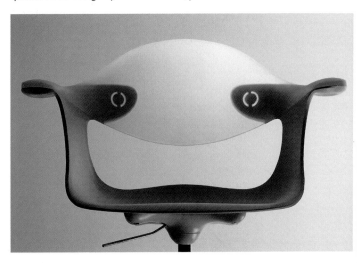

and utterly essentialist. Contemporary industrial designers such as **Jasper Morrison** and Konstanin Grcic (b. 1965) take an almost utilitarian approach to design in order to create essentialist products that have an intense aesthetic purity. Other contemporary designers such as Harri Koskinen (b. 1970) look to inherently essentialist vernacular design paradigms in order to reinterpret them in a modern idiom. Essentialism is increasingly partnering organic design and a new naturalism of form, as exemplified by the work of **Ross Lovegrove**. Commenting on the confusion that often exists between the pared-down formal vocabulary of essentialism, which is also known as "dematerialism", and minimalist **styling**, Lovegrove states: "I'm suspicious of minimalism because it doesn't really exist in nature ... I'm suspicious of it because I think life is not minimalist, generally it is quite complicated and detailed. Dematerialism – or essentialism – is another thing and relates more to the physicality of artefacts. Dematerialism means placing less emphasis on weight, density and thickness. It's the idea that things could be built in the future perhaps more organically, less constructed physically, growing rather than constructed ... Dematerialism is an absolute objective." While essentialism requires a great deal of understanding of structure, the nature of materials and industrial techniques, it is clearly the most appropriate approach to design for the 21st century.

Testing of *Model T* starting device, 1914

Fordism is a term used to describe mass production – a method of manufacture that has dominated the economies of most developed countries since the early 20th century. Named after Henry **Ford**, who pioneered an assembly-line system of production for his *Model T* car, Fordism revolutionized not only the structure of the work process but also the way in which products were conceived and designed. As a result of the overwhelming demand in North America for the *Model T*, which was launched in 1908, Ford turned to the problem of producing the car in large volumes and at a low unit cost. His solution of a moving assembly line, which involved the optimal arrangement of machines, equipment and workers for the continuous flow of workpieces, was inspired by the meat-packing industry in Cincinnati and Chicago. Here, meat was processed industrially on a very large scale with animal carcasses being moved past workers at a steady pace via a system of electrically-powered overhead trolleys. Stationary workers concentrated on one task, performing it at a pace dictated by the mechanized line, minimizing unnecessary movement and greatly increasing productivity. Drawing upon his observations of these techniques and the theories of Frederick Winslow Taylor – whose seminal book, *The Principles of Scientific Management* (1911), outlined the precepts of what became known as **Taylorism** – in 1913 Henry Ford implemented his first assembly-line production of magneto flywheels. That same

Flywheel production for *Model T*, 1914

year he developed a chassis-building system in which the chassis were pulled by rope past stockpiles of components, with the whole of the manufacturing process being compartmentalized into individual repetitive tasks. When the system was further improved with electrically-powered chain-drive movement, assembly time for the *Model T* was reduced from 12 hours and 8 minutes under the old system (in which parts were carried to a

Gas tank desk for
Model T, 1914

stationary assembly point) to just 1 hour and 33 minutes. Whereas *Model T* production figures for 1910 had stood at 20,000 units at a cost of $850 each, by 1916 this total had trebled to 60,000 units, at a cost of $360 each, clearly demonstrating the efficacy of the moving assembly line technique. When production of the *Model T* ceased in 1927, 15 million had rolled off Ford's assembly line and the company was now producing half of all the motor vehicles in the world.

While the success of assembly-line mass production has always rested upon elaborate planning and synchronization, the meticulous design and **standardization** of components, and the efficiency of the overall design of the product, it has also depended heavily on large-scale investment in the plants and tooling required. Generally, only very large companies can afford to make such investments. Thus the rise of Fordism has precipitated the ascendancy of a small number of increasingly global corporations – from **Boeing** to **IBM** – who dominate their particular markets. More recently, demands for increased productivity have prompted the introduction of automated systems of mass production, with robots being increasingly employed for manufacturing processes that are repetitive, unpleasant and potentially injurious to human health, such as welding and spray painting, or which require handling of heavy and awkward workpieces or tools. Meanwhile, automation has been carried to a new level in the shape of computer-

Body desk for
Model T, 1914

integrated manufacturing (CIM), which has evolved out of the use of **computer-aided design** and **computer-aided manufacturing** (CAD/CAM) systems, and which now goes well beyond design and production to include most of a firm's business functions.

While the automation of Fordist systems of production has resulted in the elimination of many unskilled jobs, it has increased demand for knowledgeable technicians to oversee the operation of automated devices. Against many expectations, automation has led to the reappearance of the skilled worker. It has also improved efficiency and expanded production while relieving the drudgery and increasing the earnings of the worker – exactly the aims of Frederick Winslow Taylor at the turn of the 20th century.

Josiah Wedgwood, *Queen's Ware* vase, c. 1765

The term "Industrial Revolution" was originally popularized by the English economic historian Arnold Toynbee (1852–1883), who used it to describe the rapid industrialization which swept Great Britain between 1760 and 1840 and which brought about far-reaching technological, socio-economic and cultural changes. Today, however, the term is used more broadly to describe industrialization in general over the last 250 years. There are several reasons why the Industrial Revolution emerged in Great Britain first. Since the 1600s, the dwindling influence of her monarchy and church had been paralleled by the rise of her ever more powerful merchant classes. Britain's free press also allowed a culture of free thinking to flourish. By 1760 the country had experienced internal peace for over 150 years and was politically, legally and economically more advanced than its European neighbours. Britain also enjoyed a natural abundance of coal, iron ore and minerals necessary for the production of iron and steel. These factors led to the conditions that were essential for the birth of the Industrial Revolution – a surplus of capital, a surfeit of labour and a new and growing market for goods both at home and abroad.

In the early days of industrial manufacture, however, both quality and production were low. They only increased with the emergence of inventions that improved on existing tools or eliminated the need for human power through the use of water power, and later **steam power**. The first successful invention of the Industrial Revolution was **James Hargreaves'** *Spinning Jenny* (c. 1764). This landmark design was followed in 1769 by **Richard Arkwright**'s water-powered spinning machine, which was itself surpassed by later, better-performing spinning machines designed by Samuel Crompton (1753–1827) in 1779 and Edmund Cartwright (1743–1823) in 1785. The same year as Cartwright's invention, **James Watt**

Print showing a Jacquard loom designed by Joseph Marie Jacquard in 1805

Bowstring Girder
Bridge, c. 1870

and Matthew Boulton (1728–1809) installed a steam engine at a Nottinghamshire cotton mill and harnessed steam power for the first time to a manufacturing process. The first successful steam engine had been designed as early as 1693 by Thomas Savery (c. 1650–1715) and was later improved by Thomas Newcomen (1663–1729) in 1705. Watt's subsequent improvements to Newcomen's "fire engine" (which had only been used for pumping water out of mines) resulted in a design that could also be made to drive machinery. The advent of motive power brought in its wake a multitude of labour-saving machines, which not only increased productivity but also enabled a greater degree of replication. The rapid industrialization that took place in Great Britain from the 1760s onwards led to an unprecedented urbanization of the work force. The phenomenal growth in density and size of Britain's industrial centres was accompanied by increasing commercialization, as mechanization meant the production of cheaper goods that could be afforded by more people. The skilled craftsmen and women of the past gave way to an unskilled workforce which mass-produced a wide variety of goods from boot buttons to domestic labour-saving devices such as knife cleaners. One of the greatest forces driving the Industrial Revolution was the production of iron and steel. As early as 1709, Abraham Darby (c. 1678–1717) had successfully smelted iron ore to produce marketable iron at **Coalbrookdale**. In c. 1740, the clock and instrument maker Benjamin Huntsman (1704–1776) became the first to develop a method of casting steel ingots in crucibles. The steel produced using Huntsman's process was more uniform in composition and free of impurities than any previous form of steel. Despite being perfect materials from which to construct the new industrial machinery, it was not until the 1780s that steel and iron saw a dramatic increase in production. Soon other countries followed Britain's example and began to industrialize. By the late 1840s France had established itself as an important industrial power, with a flourishing mechanized textile industry. Germany only began its industrial expansion in the 1870s, after becoming a unified state, but by the early 1900s had already overtaken Britain in the production of steel and chemicals. Some of the first inventions of the American Industrial Revolution were designed by **Eli Whitney** – who patented his famous cotton gin in 1794 and developed a musket with interchangeable components in 1800. Small arms manufacturers such as the **Springfield Armory** and **Remington** were among the first in America to successfully master mass production on a large scale, using modern manufacturing systems including **standardization** and the division of labour. "Armory practice", as it became known, was subsequently applied to the manufacture of other products – most notably cars by **Henry Ford** – with such success that by the early 1900s America had become a leading industrial power.

Since the 1760s, industrialization has brought untold material benefits to society – from **George Stephenson**'s *Rocket* locomotive to the "information superhighway". However, as traditional heavy industry is progressively shifted from Western economies to developing nations keen to benefit from their own latter-day industrial revolutions, it should be remembered that, while technology can be used as a force for good, its selfish and unrestrained exploitation can lead to much social discord and ecological destruction. The key to sustainable progress is a responsible and enlightened use of technology which acknowledges the critical interconnection between humankind and our shared environment.

◤ Boot buttons and box, c. 1900

Knife cleaner manufactured by Spong, c. 1900

Inhaler, British, late 19th century

←An artificial arm manufactured by P & K Artificial Limb Company, Belfast, 1925

Medical design falls into three broad categories: prosthetics, products designed for the administration of drugs, and equipment created for diagnostic purposes or medical therapies. Throughout the history of medicine, new designs have been devised to assist in the treatment of the sick and the disabled. It was not until the early 18th century, however, that accurate anatomical studies began to directly inform medical design. At this stage, the universities in Edinburgh and Glasgow led the world in anatomical research and it was a Scotsman, William Smellie (1697–1763), who designed the first modern obstetric forceps. Once in position, the two sections of this revolutionary design locked together so as to effectively cradle the baby's head. It is a tribute to Smellie's invention that the forceps used in today's hospitals differ little from his original design. During the 18th and 19th centuries, industrialization fuelled the growth of urban populations and with them the rise of infectious diseases such as tuberculosis, typhoid and diphtheria. Respiratory diseases were often made worse by airborne pollution from domestic fires and coal-burning factories. The thick fogs common to so many cities during Victorian times were disastrous to the health of inhabitants, and most households would have owned a ceramic inhaler as it offered one of the best forms of medical treatment prior to the advent of antibiotics. The greatest benefits to health, however, were not advanced by medical design or new drugs but by the implementation of proper sanitation systems from the mid-1800s onwards.

Obstetrical forceps, c. 1820 – based on a revolutionary design created by William Smellie in 1752

This period also saw the advent of modern industrialized warfare, which resulted in an unprecedented need for **design for disability** and in particular prosthetics. Since the beginning of the 20th century, technological advances have often found life-saving applications within the field of medicine, as evidenced by medical X-ray equipment, for example. When a new medical typology

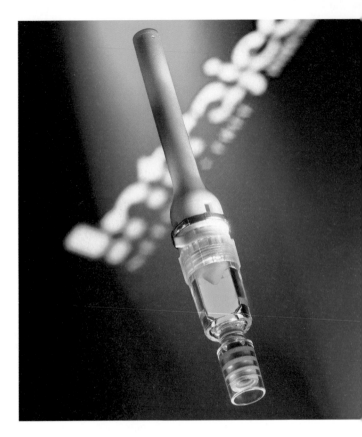

Western Medical,
Intrajet pen, 1990s

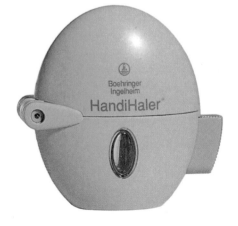

Kinneir Dufort,
HandiHaler asthma
inhaler for Boehringer,
1995

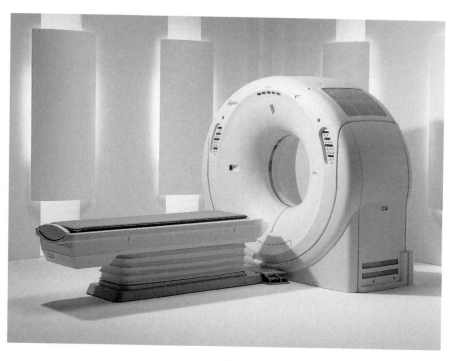

Masahiko Kitayama,
Aquilion CT scanner
for Toshiba, 1997

is introduced, its design may appear relatively unrefined because it is informed almost entirely by technology. As the typology evolves, however,
its design will become user-friendlier and less threatening in appearance.
Thus the brain scanner first introduced in 1970–1971 has today metamor-
phosed into the **soft-tech** *Aquilion* CT scanner (1997) designed by Masahiko
Kitayama for **Toshiba**. Increasingly, designers are also helping to develop
better ways of administering drugs, such as Weston Medical's needle-free
injector or the *HandiHaler* asthma inhaler (1995) designed by Kinneir Dufort
for Boehringer. With the emphasis today firmly on human-centric design, a
host of bio-compatible designs are being implanted into the human body to
replace worn-out body parts – from pace-makers to high-tech ceramic hip
replacements. Although medical design is by necessity driven by function
rather than aesthetics, designers and manufacturers are increasingly realising that the appearance of medical equipment can have a psychological effect upon patients and can thereby influence – and potentially enhance –
the therapeutic value of a design solution.

Ludwig Hohlwein, poster for Motorenfabrik Oberusel, Germany, 1910s

While the effectiveness of military weaponry and hardware is perpetually improved as a result of technological progress, the primary criteria for military design have remained the same for centuries, namely durability, strength, fitness for purpose, functionality, transportability, ease of maintenance and repair, and rationalized/standardized construction for ease of manufacture.

How well a weapon functions is often a decisive factor in war, and so in order to gain the advantage over an enemy it is essential to have the latest equipment available. This factor has led to the culture of continuous development within military design – a realm of many specializations that can be broken down into five main categories: offensive weapons, defensive weapons, transportation, communications and detection systems. Although advanced technologies such as railways and telegraphy were used for military purposes in the 19th century, very few technological innovations were incorporated into the design of weapons during this period. Many modern manufacturing techniques, however, found their first application in the manufacture of small arms. In 1800 **Eli Whitney** developed a musket with interchangeable parts which heralded the advent of industrial **standardization**, while other small arms manufacturers in the United States, most notably the **Springfield Armory**, mass-produced rifles using mechanization

Major Wilson and Sir William Tritton, *Big Willie* tank, 1915

Interior of a shell-
filling factory, Great
Britain, c. 1915

and the industrial division of labour. The mass production of weapons
changed the face of military conflict forever, with the American Civil War
being effectively the first example of modern (i.e. industrialized) warfare.
By the outbreak of the First World War, weaponry had become relatively
sophisticated and infinitely more deadly. The First World War saw the introduction of many "firsts", including the use of chemical weapons (such as
mustard gas), aeroplanes for bombing, submarines and tanks. This war,
however, was first and foremost an artillery conflict, and the quick-firing
Howitzer field gun, with its high angle of fire, significantly outperformed
older designs with flatter firing trajectories. WWI also saw the introduction of trench mortars and the development of primitive radio communications.

During the Second World War, military designs that had been introduced
in the earlier conflict were considerably improved. Battlefield communications became increasingly sophisticated and the design of field telephones
enabled much better command and control of troops. The aircraft carrier
also came of age and brought a completely new dimension to naval warfare, while the design of aircraft became increasingly specialized. Daylight
bombers such as the **Boeing** *B-17 Flying Fortress* (1934–1935), for example,
required the development of long-range escort fighters such as the **North
American** *P-51 Mustang* (1940). Tanks also played an important role in providing armies with greater manoeuvrability and the possibility of punching
holes through enemy lines – with its superior protection and firepower, the
Russian *T34* tank (1940) was among the most decisive weapons of WWII.

15-inch Howitzer,
1916

Browning machine
gun, 1918

M4 Sherman tank, 1942

The development of atomic weapons and the subsequent onset of the Cold War spawned a phenomenal proliferation of weapons and weapons systems, each progressively more sophisticated and deadly while at the same time more costly. This escalation was largely driven in the West by the American military-industrial complex – the powerful economic and political conjunction of the US military establishment, government officials and the defence industry. Since the end of the Cold War in the 1990s, however, there has been significant consolidation in the US arms industry, which has to some extent diminished the power and influence of the military-industrial complex. Global economic integration, combined with a rise in democracy all over the world, has finally made war and the build-up of its associated matériel unprofitable, unwise and unnecessary.

Gerhard Fuchs, *Titan Minimal Art Model 7373* eyewear for Silhouette, 1999

As technology has evolved throughout the Modern era there has been an increasing trend towards miniaturization. The benefits of reducing the scale of industrially manufactured products are clear – weight, volume, unit costs to the manufacturer, costs to the user/consumer and the amount of waste generated can all be significantly decreased. Not surprisingly, the aerospace and automobile industries were among the first to explore the potential of miniaturization. Cars such as the **Volkswagen** *Beetle* (1934), **Fiat** *500* (1936), Fiat *Nuova 500* (1957) and **Alec Issigonis'** *Mini* (1959) were not only among the smallest cars produced in their time, but were also the least expensive and the most popular. The extraordinary advances made in the field of semiconductors during the mid-20th century led to the radical miniaturization of consumer electronics. 1947 saw the development of the first transistor by John Bardeen, Walter Brattain and William Shockley at the **Bell Telephone** Laboratories, heralding what many regarded as the "second industrial revolution". By the late 1950s the transistor had been perfected to such an extent that it began rapidly replacing the use of electron tubes – thereby allowing industrial designers the possibility of developing more compact electronic equipment; such as the world's first pocket-sized transistor radio, which was introduced by **Sony** in 1957.

Ian Sinclair Design, *Eon* flashlight, 1990s

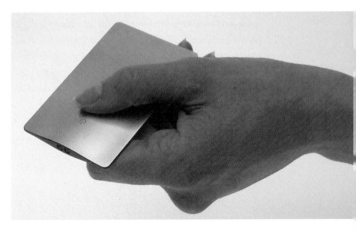

DaimlerChrysler MCC, *Smart Car*, introduced 1998

Another turning-point came with the development of the first integrated circuit by Jack St. Clair Kirby at **Texas Instruments** in 1958. Such revolutionary microchip technology enabled appliances to become more compact, and easier to handle, transport and store. The bulky desktop calculator was soon replaced by pocket models, while the room-filling mainframe computers of the late 1950s and early 1960s eventually gave way to desktop personal computers in the 1980s and laptop computers in the 1990s. Today, we are surrounded by mini-versions of every imaginable product, from micro-vehicles such as the *Smart Car* (1998) to the digital cameras (2000) by **Casio** which can be worn on the wrist. Minimal designs are also being made possible by new materials, such as the strong yet lightweight titanium employed in the wire-like eyewear produced by Silhouette. With the rapid development of "molecular electronics" and nanotechnology, the reduction in size of electronic devices will open the door to whole new classes of ultra-miniaturized products and information systems. It has been predicted that, in the 21st century, computing will be an automatic constituent of everyday objects across a wide range of product types.

Apothecary jars, 19th
century

Günter Kupetz, mineral water bottle,
1969 – this classic German design
(like the British milk bottle) has been
effectively recycled over a number
of decades and is used not only for
mineral water but also orangeade
and lemonade. It received a Gute
Form design prize in 1973.

Since the earliest commercial ventures, packaging has
been used as a means of preparing goods for efficient
transport, storage and sale. In the second half of the
19th century, however, the importance of packaging
increased dramatically when it began to be used by
Western manufacturers as a means of **branding** their
products. During this period, glass and ceramic jars,
pots, bottles and metal tins were emblazoned with
either moulded or transfer-printed names and logos,
together with claims that often exaggerated the properties of the products they contained. As packaging
became more sophisticated in the early 20th century,
many companies started commissioning leading
graphic designers to produce eye-catching designs,
often as part of a larger overall **corporate identity** programme. Lucian Bernhard's brightly coloured packaging for **Bosch** spark plugs, and Alfred Runge & Eduard
Scotland's bold graphics for Kaffee Hag coffee tins, for
example, helped the respective companies to powerfully
differentiate their products and achieve much brand
recognition. Since the onset of modern industrial food
production and mass retailing, packaging has become
increasingly necessary to protect goods from the hazards
of handling and environmental conditions, to provide
a convenient unit of the product for the manufacturer,
distributor and consumer, and to identify the product in
an appealing way to the consumer. Packaging today also
needs to be easy to manufacture and inexpensive relative to the cost of the final packaged product, to be
clearly tamperproof, and to comply with environmental
standards. As consumers have become more brand
conscious, the graphic design of packaging has become
evermore systemized, with particular graphic styles
being used to promote the idea of product families.

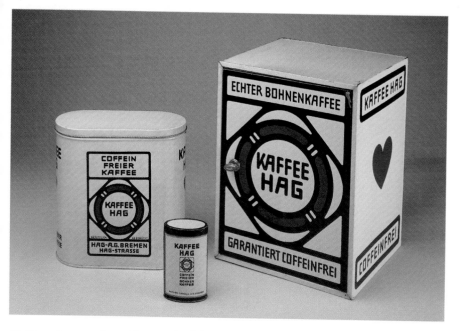

Alfred Runge &
Eduard Scotland,
Kaffee Hag coffee
tins, c. 1910

A notable example of this is Lewis Moberley's design of packaging for Boots the Chemist. The graphic clarity of this completely integrated programme gave items as diverse as low-calorie foodstuffs and laundry care products a strong visual coherence and brand identity. Now that supermarkets stock a greater selection of products every year and competition between brands is more closely fought, "on-shelf impact" has become an essential packaging criteria. Colour, typography and branding are just some of the tools used to grab consumers' attention when shopping. More important than what is printed on the packaging is, however, the actual design of the container. Food packaging in particular must be designed to retard spoilage and prevent physical damage and exposure to the elements. Especially critical is the design of closures, which must adequately seal the container in a sanitary and mechanically safe way. One of the greatest pioneers in this field of packaging was Dr. Ruben Rausing (1895–1983), who established **Tetra Pak** in 1951. He believed that packaging should use a minimal amount of material while providing the greatest degree of hygiene. This he achieved by developing a tetrahedron-shaped carton made of polyethylene-coated paperboard. During the 1950s and 1960s, **plastics** became widely available and began replacing **ceramics** and glass as packaging materials. Since then, a variety of plastics have been used in packaging applications, including poly-

Lewis Moberley,
packaging design
for Boots Laundry
range, c. 1995

styrene (PS), polypropylene (PP), low density polyethylene (LDPE), polyvinyl chloride (PVC), high density polyethylene (HDPE), and polyethylene terephthalate (PET). Owing to their high durability, light weight, flexibility and insulation qualities, plastics are used extensively for packaging liquids, perishable foodstuffs, pressurized packages and containers of food intended to be frozen or boiled. Despite their many uses, plastics only account for about 14 % (by volume) of all packaging in municipal solid waste streams. By far the greatest component, at approximately 50 %, is cardboard (paper) – the most widely used consumer packaging material. Light in weight, inexpensive, easily manufactured, printed and sorted, cardboard cartons can be produced in a wide range of shapes and sizes, half of which are used for food packaging. Metallic containers, including tin-plated steel and **aluminium**, account for roughly 16 % of all packaging and are also used primarily for food storage. Being highly resistant to chemical and mechanical damage, tin-plate containers are also used to hold chemical agents such as paints, preservatives and solvents, as well as aerosol products. Lighter and more malleable, aluminium is principally used in packaging for bottle caps and vacuum-sealed easy-open beverage cans. Glass containers, which are easily mass-produced and can be reused, make up approximately 20 % of the

packaging content of municipal solid waste streams. Glass is chemical-resistant, durable and can be kept highly hygienic and is thus ideal for the packaging of solid and liquid food, cosmetics and drugs. While packaging is essential for the promotion and protection of merchandise, it tends to be used excessively by manufacturers, who regard it as a relatively inexpensive way of "adding value" to their products. In Europe today, packaging generally accounts for 25–35 % of the waste (by weight) in municipal solid waste streams. Despite growing concern about its impact on the environment, no real solutions to the problem of reducing packaging waste have been developed. Government-backed initiatives are inevitable, however, and will no doubt focus on programmes of waste prevention that balance recycling, reuse and a reduction of packaging materials. Light-weighting or down-gauging packaging to a safe minimum, as demonstrated by the *Flexible Food Wrap* designed by Pethick & Money for use in fast-food restaurants, is an essential component in waste prevention. Consumers, designers, manufacturers and governments will have to fundamentally and radically rethink the nature of packaging if waste is to be minimized and natural resources preserved.

Pethick & Money,
Flexible Food Wrap
fast-food packaging,
1996 – winner of two
BBC Design Awards,
this packaging offers
a significant reduction
of waste

BIC disposable
lighter, 1990s

Planned obsolescence is a highly contentious issue that lies at the heart of some of the most important debates on consumerism, global sustainability and industrial design. Having first emerged as a major feature of the American economy in the 1950s, planned obsolescence is based on the concept of intentionally limiting the life of products so that consumers are manipulated into consuming more – an approach that continues to form a key part of the strategies of many large manufacturing companies. There are two strongly conflicting views on the morality of planned obsolescence. Advocates claim that it keeps workers (and designers) in employment, is essential to economic growth and is ultimately beneficial to society as a whole. Opponents of planned obsolescence claim that the manipulation of consumers is insidious, that the value for money offered by limited-life products, no matter what the economics, is poor, and that the waste created by their premature replacement is environmentally ruinous. An early and notable opponent of planned obsolescence was Vance Packard (b. 1941), who wrote the seminal book *The Waste Makers* (1960), in which he identified the three principal spheres of obsolescence – function, quality and desirability. Functional obsolescence arises when a new product appears that is perceived to do a better job than its predecessors. The obsolescence of quality, which is directly related to the physical durability of a product, has historically been achieved by manufacturers building in to products key components that have been designed to fail after

BIC disposable
razors, 1990s

a given amount of time. White goods, or domestic appliances, are particularly prone to this type of "built-in obsolescence", with, in most cases, the replacement of the entire unit being more cost-effective than the replacement of the defective component(s). The obsolescence of desirability operates mainly through changes in the appearance of products, fashion and consumer opinion, all of which are driven by **styling** and/or advertising strategies. As early as the 1920s, the chairman of **General Motors**, Alfred Sloan, recognized that aesthetics would play an increasingly important role in the automotive market and instigated a system of annual

stylistic changes so as to minimize the aesthetic durability of cars. While this approach is still common among many automobile manufacturers, those in Germany and Scandinavia have historically added much value to their brands and enjoyed increasing success and brand loyalty by raising the overall durability of their products. Thus annual sales of used **Volvo** cars actually surpass the number of new vehicles it produces each year (around 400,000). The huge and ever increasing secondary market for its vehicles and branded parts is massively profitable for the company. In the case of Volvo, durability equals profitability. While there are clearly good economic arguments against the supposed social benefits of planned obsolescence, the environmental argument is even more compelling, especially given the urgency of the need to take meaningful steps towards achieving global sustainability. Making products more durable reduces the throughput of energy and materials, lowers consumption of finite resources, cuts emissions of pollutants (including greenhouse gases) and produces less residual waste. By doubling the life span of products, their net environmental impact can be halved. Making-less-last-longer is not only good for the environment, it also maximizes value for money and convenience for the consumer. Taken to an extreme, planned obsolescence results in disposable products – the most wasteful and least environmentally justifiable of all consumer product types.

Napkin rings cast from phenolic resin, late 1930s

Synthetic plastics quite literally moulded the material culture of the 20th century. So profound was their effect on mass consumerism that the period could be described as "The Plastics Age". As early as the 15th century, however, natural plastics such as shellac (the resin from a tropical beetle), casein (produced from milk curds) and keratin (a protein found in hair, fur, bone, nails, hoofs and horn) were being used in the manufacture of luxury goods. The development of the first modern plastic is generally credited to the English chemist and inventor, Alexander Parkes (1813–1890). During the 1840s, he discovered that wood-dust or cotton fibre, when dissolved in nitric acid or sulphuric acid and then combined with castor oil and chloroform, produced a dough-like substance that when dried looked remarkably like ivory or horn. This form of cellulose nitrate, scientifically known as pyroxylin, was difficult to work with because of its explosive nature and its inherent brittleness. Parkes subsequently set up a company to manufacture this semi-synthetic plastic, which he named Parkesine, but it went into receivership in 1868. In America, John Wesley Hyatt (1837–1920) continued to search for something better than Parkesine which he could use as a synthetic alternative to ivory to make billiard balls. By mixing camphor under pressure with cellulose nitrate, he was able to solve the problem of brittleness and in 1869 he patented the first practical semi-synthetic plastic under the name of Celluloid. During the 1870s and 1880s Celluloid was used in the manufacture of all kinds of objects, including hair combs, brushes, buttons, hand mirrors, letter openers and dice. It could also be made to imitate ivory, tortoiseshell, mother-of-pearl and amber. Although significantly cheaper than these luxury materials, Celluloid was still relatively expensive and consequently did not suffer from the associations of cheapness that dogged successive synthetic materials. In 1889 George Eastman's firm (later renamed the Eastman **Kodak** Co.) marketed the first commercial transparent, flexible camera film made of Celluloid. From 1904, the Belgian-trained chemist and entrepreneur Leo Baekeland (1863–1944) worked on the development of the first completely synthetic plastic and in 1907 perfected the manufacture of phenol-formaldehyde resin (also known as phenolic resin). Better known under its trade name **Bakelite**, this revolutionary material was produced commercially from 1910 and was marketed as "the material of a thousand uses". Phenolic resin was initially used as a coating material, especially for metals, and as an adhesive, before being employed as a moulding powder. It was remarkably suited to moulding processes and from the 1920s onwards considerably changed the aesthetic of many industrially-manufactured products. In 1928 a method was perfected for "casting" phenolic resins without the need of a filler, such as had previously been required for the production of Bakelite and other synthetic thermoset plastics (i. e. plastics that solidify on heating and cannot be remelted or reformed without decomposing). It was now possible to cast phenolic resins in a wide range of bright colours. These strong, non-flammable and colourful materials were used for numerous articles including napkin rings and jewellery. Urea-formaldehyde thermoset resins were widely used from the late 1920s, including a product marketed under the trade name of Plaskon. With the development of melamine-formaldehyde in the late 1930s, urea-based plastic laminates gave way to Formica laminates. Then, as the demand for better-performing synthetic materials grew, so another major group of plastics began to emerge. These were the thermoplastics – plastics that soften when heated and can be moulded and remoulded repeatedly without any appreciable change in properties. Amongst the earliest of these was polyvinyl chloride (PVC), which was first manufactured as Vinylite by the Carbide & Carbon Chemical Corporation in 1928. Today, PVC is available in two forms, rigid (unplasticized) or flexible (plasticized) – the latter being used extensively for **packaging**. A polymer of PVC marketed under the trade name of Saran can be found in kitchen cling-films, for example. By the end of the Second World War, the range of thermoplastics had expanded to include polyethylene (PE), the most widely used plastic today, polystyrene (PS) and polymethyl methacrylate (PMMA), which is better known under its trade name Perspex. These were followed in the post-

↙ Salt and pepper shakers for BEF Products, England, 1935 – produced using ureum

Chad Valley trainset for Chad Valley, England, 1940s – produced using Bakelite

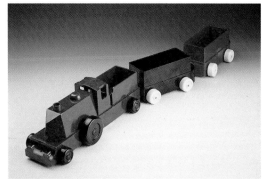

war years by polyurethane (PU), polypropylene (PP), acrylonitrile-butadiene-styrene (ABS) and polyethylene terephthalate (PET). Each of these widely used commodity plastics has its own unique set of properties and is better suited to certain processes and applications than to others. Thus PET, for example, is best used for pressurized beverage containers that are blow-moulded. Plastics can also be reinforced with glass fibres to produce glass-fibre-reinforced plastic (GRP), more commonly known as Fibreglass. Combined with resins such as epoxies or unsaturated polyesters, thermoset glass-fibre reinforced plastic is pound-for-pound stronger than steel and has a wide range of applications, from furniture to car bodies. Today, an extensive array of plastics processing techniques are available to industrial designers and manufacturers, including extrusion (for the manufacture of films, sheets, tubing etc. in which the melted material is pushed through the orifice of a die); compression moulding (in which plastic pellets are heated and compressed into a mould at the same time); injection moulding (in which a molten resin is shot into a mould under considerable pressure, sometimes using nitrogen gas); reaction injection moulding (using a catalyst to speed up the reaction between two polyurethane precursors so that the moulding process requires less pressure); blow moulding (a molten polymer is blown into a mould to create a hollow moulding, e. g. plastic

Erik Magnussen, salad bowls and salad servers for Stelton, 1986 – moulded PMMA (also known as Perspex)

bottles); casting and dipping (inexpensive processes for the production of small objects that require no pressure); rotational moulding (a low-heat, low-pressure process in which a mould is rotated so that the plastic fuses to the interior of the mould to produce hollow objects such as refuse bins); thermoforming (in which a heated sheet of thermoplastic such as polystyrene or PET is pulled by a vacuum into a mould, e. g., for drinking cups); and foaming (in which polystyrene is combined with isopentane to produce a material with gas bubbles that can be moulded or extruded, e. g. for egg cartons and fast-food packaging). Today, plastics account for approximately 14% (by volume) of municipal solid waste, with most of that figure representing packaging. In order to allow more efficient material identification, sorting and ultimately recycling, many plastic objects, especially plastic packaging, now bear the international plastics coding system. Thanks to their easy manipulation, economical production, corrosion resistance and suitability to industrial processes, plastics remain among the most popular and useful materials. The highly innovative treatment of various advanced techno-polymers by some of today's most talented industrial designers has led recently to a significant reappraisal of the aesthetic of plastics in general.

Guido Venturini, *Gino Zucchino* sugar shakers for Alessi, 1993 – moulded PMMA

↖Stefano Giovannoni, *Merdolino* toilet brushes for Alessi, 1993 – moulded techno-polymer

Alvar Aalto, group of plywood and laminated-wood furniture for Artek, on display at Bowman Brothers in London, c. 1938

Charles & Ray Eames, *LCW* (Lounge Chair Wood) for Evan Products and later Herman Miller, 1945 – bent laminated wood frame with compound-moulded plywood seat and back

Plywood and laminated wood have been used in furniture-making since the first half of the 18th century, and possibly even earlier. It was not until the late 1920s, however, that technological innovations in the production of modern wood laminates, and in particular the development of synthetic resin adhesives such as phenol- and urea-formaldehyde, enabled designers such as Alvar Aalto (1898–1976) to begin fully exploiting the technical, formal and aesthetic potential of these materials. So successful was the application of plywood and laminated wood in furniture design that by the mid-1930s they had replaced **tubular metal** as the materials of choice for most avant-garde designers. Plywood is manufactured in panel form by gluing one or more layers of veneer to both sides of a single sheet of veneer or a core of solid or reconstituted wood. Each layer is typically glued with its grain running at right angles to that of the layer above and/or below it, with each layer and grain direction being mirrored on the opposite side of the core. The total number of layers is almost always odd – three, five or more. Wherever there is a requirement to cover a large surface area with a lightweight but strong material, plywood may be used. Plywood has many advantages over solid wood, most importantly its increased dimensional stability and suitability for moulding into curved forms. Some of the largest applications of moulded plywood include aircraft, boat and furniture construction. Moulded plywood is produced by bending and gluing veneer sheets in a combined operation, employing forms in conjunction with either cold press systems, hot press systems, bag or blanket press systems or

Eden Minns, floor receiver for Murphy, c. 1948

radio frequency forming systems. Plywood is most easily moulded into one or more curves on one geometric plane, but can also be moulded into compound or complex curves on two geometric planes simultaneously. Either way, the radius of moulded plywood is normally limited by the thickness of the individual veneers and the construction and thickness of the plywood. The deformation capacity of moulded plywood is almost unlimited – a rule of thumb is that what can be formed with a sheet of paper (i. e. modelled) without deformation can be produced as a moulded plywood element.

In making laminated wood, the veneers are glued so that their grains all run parallel to each other. Curved laminated wood is produced by bending and gluing at the same time. Unlike plywood, however, laminated wood can only be bent on one geometric plane (in the direction of the grain). Laminated wood again possesses several advantages over solid wood: it can be used for large elements of various sizes and shapes that would be impossible in solid wood, and it enables structures to be designed on the basis of required strength – the more veneers there are in either a plywood or laminated wood construction, the less likely critical failure becomes. Laminated wood is primarily used for structural applications, including architectural elements, boat keels and furniture components such as chair arms, legs and frames. Plywood and laminated wood are often used in combination, as famously demonstrated by the revolutionary range of chairs designed in 1945 by **Charles & Ray Eames**. These chairs were among the first examples of production furniture to incorporate compound-moulded seat elements, which provided a high degree of comfort without the need for traditional upholstery. The durability, strength, lightness, versatility, low cost, aesthetic appeal and suitability for industrial production of plywood and laminated wood has ensured that these materials remain an enduringly attractive option to furniture manufacturers and designers today.

Richard Hamilton &
Martin Goody, *House
of the Future* for
Monsanto, 1957

The idea of buildings being designed as mass-producible consumer products has fascinated both architects and industrial designers for many decades. One of the first notable examples of this type of product architecture was developed by **Walter Gropius** and Konrad Wachsmann (1901–1980). Their *Packaged House* system of 1942 rested on the idea of mass-producing modular components for the construction of domestic buildings. Although the concept was patented, it was not a resounding commercial success, with only 200 units being sold before the manufacturing company was liquidated. Between 1944 and 1947 **Richard Buckminster Fuller** designed the **aluminium**-clad *Dymaxion Dwelling Machine* (also known as *Wichita House*). When the prototype of this pre-fabricated building was launched, the company that was specially set up to commercialize it received 38,000 orders. Fuller was not prepared to begin manufacturing the house, however, until its design had been completely perfected. As a result, the project was seriously delayed and eventually cancelled by its financial backers. Undeterred, in 1949 Fuller invented the *Geodesic Dome*, which must be considered the most successful example of product architecture to date. A friend and colleague of Fuller's, **George Nelson**, also began designing a product architecture system in 1951. His *Experimental House* was based on a building system made up of cubes

Matti Suuronen,
Futuro House, early
1970s

(measuring 12" x 12") with translucent plastic domes that could be linked together. Because of its modularity, the *Experimental House* offered a greater degree of flexibility than Fuller's earlier *Dymaxion Dwelling Machine*. Arthur Drexler, the then curator of the Museum of Modern Art, New York, described Nelson's elegant solution as "a product technically superior to its handcrafted competitors". In 1957 Richard Hamilton and Martin Goody designed the Monsanto

House of the Future as an exhibit for Tomorrowland at Disneyland. The four identical plastic elements that made up each wing of this futuristic-looking house predicted the capsule-like product architecture of the Finnish architect Matti Suuronen from the early 1970s. As the journal *Design from Scandinavia* noted, Suuronen's *Futuro House* "took the dream of a plastic house from laboratory to standard production at a single leap ... Ignoring all conventional housing concepts, he has created a huge lozenge which is placed on a fundament of steel piping." Comprising a floor area of 50 m², the *Futuro House* was nevertheless light enough to be slung by helicopter to inaccessible sites. Suuronen's later *Venturo House*, developed for Oy Polykem in Helsinki, was also constructed of highly insulated, glass-fibre-reinforced polyester elements that were similarly easy to transport. The concept of product architecture has more recently been explored by the London-based industrial designer, **Ross Lovegrove**. His *Solar Seed* proposal speculates on a wholly autonomous nomadic structure that uses a minimal amount of material. Like his highly innovative designs for garden lighting, the *Solar Seed* is also intended to be solar-powered and produced with state-of-the-art manufacturing techniques. While the industrial production of complete architectural structures has remained elusive, mobile forms of living space such as the famous **Airstream** trailer and the **Volkswagen** camper have become design icons of the 20th century. Given the widespread and increasing need for affordable, flexible and transportable housing, the future success of product architecture will no doubt depend less on technical issues of construction and manufacture than on problems associated with local building regulations, siting and land ownership.

Ross Lovegrove,
Solar Seed, 1999 –
concept for a wholly
autonomous
nomadic structure

Look right through

Early advertisement
for Pyrex ovenware

PYREX

DEVELOPED IN 1912
BY THE CORNING GLASS WORKS

Having established its research laboratories in 1908, in 1912 the **Corning** Glass Works developed a revolutionary new borosilicate glass. Given the name of Pyrex, it was the first glass to be capable of withstanding sudden exposure to either heat or cold. Pyrex was also resistant to fire, chemicals and electricity, making it suitable for a multitude of applications – from laboratory apparatus and ovenware to thermometers and piping. These inherent advantages stem from the fact that borosilicate glass (Pyrex) expands approximately 66 % less than normal silicate glass and is therefore less prone to breaking from sudden expansion. Corning proceeded to launch its first Pyrex-branded cookware and laboratory glassware in 1915. Pyrex ovenware was hugely popular, as users could "Look right through" to see how a meal was cooking. It was also relatively easy to clean and did not "crack, chip or craze" like traditional **ceramics** and glass. Yet another advantage was that Pyrex casserole dishes, ramekins etc. could be used from oven-to-table, thus promising less washing up in an era when people did not have the luxury of dishwashers. The production of Pyrex was later licensed to companies in England, France, Germany and Japan. During

Mrs Catherine
Huber with Pyrex
casserole dish,
c. 1915

the 1920s, Pyrex wares were mass-produced using a completely automated production system. The 1930s saw the introduction of Pyrex *Rangetop* cookware and Pyrex opal ware. Pyrex architectural glass panels were also used for several important buildings, including the Rockerfeller Center (1929–1940). During the Second World War, Pyrex was employed industrially for radar tubes, glass piping and radio insulator, and domestic Pyrex subsequently remained extremely popular with the new generation of homemakers after the war. Commercial-use *Pyrex Double Tough* tableware was launched in 1953, and other ranges were introduced in the 1970s, including microwave browning dishes. The longevity of Pyrex ovenware and laboratory equipment is rooted not simply in the extraordinary qualities of its material, but also in the purely utilitarian character of its design, which makes it virtually impervious to fashion.

Early Pyrex
advertisement

Look right through

You can see the food bake on the bottom as well as the top in a *Pyrex* dish, without taking it from the oven. And the food bakes quickly and evenly, has a better flavor, and does not burn.

All this saves time, labor, fuel. It makes the food more appetizing, and the table more inviting—for you serve in the same dish.

Pyrex will not crack, chip nor craze. The hottest oven doesn't affect it. *Pyrex* is everlastingly sanitary, durable, easy to wash, a constant source of satisfaction in the well-appointed home.

PYREX Transparent
OVEN WARE

Trade mark reg.

Has the name on every piece

Many shapes and sizes from ramekins at 15c to large casseroles at $2. Sold by housewares dealers everywhere. Ask them for booklet.

CORNING GLASS WORKS, 111 Tioga Ave.
CORNING, N. Y., U. S. A. Established 1868

Pyrex advertisement,
1963

PYREX
Regd. Trade Mark.

for good looking cooking and serving

Figaro car for
Nissan, 1991 –
produced as a
limited edition
of 20,000

Retro design is a term that was first used in the mid-1970s to describe a tendency in popular design to embrace previous historical styles. Throughout the 1960s and early 1970s there was a huge revival of interest in Victoriana and Art Nouveau, with – for example – Victorian fair ground typography being used extensively for pop posters and album covers. Retro design really came of age in the early 1980s, however, when Post-Modernism became an international style. Many designers aligned with this movement looked to the kitsch products of the 1950s for inspiration, resulting in the emergence of a plethora of retro-designed products, including radios in sickly pastel colours and asymmetrical furniture on spindly legs. The allure of 1950s style continued well into the 1990s, with the *Figaro* car launched as a limited edition by **Nissan** in 1991. This diminutive vehicle is an almost cartoon-like representation of 1950s design, while the *Royal Star* motorcycle designed by **GK Design** for **Yamaha** is a more direct interpretation of the period style. More recently, **Jaguar**, **BMW** and **Chrysler** have also produced blatantly retro-designed vehicles. Clearly a big business, retro design today involves the combination of historicizing **styling** with up-to-the-minute technology so as to produce hybrid products that function well while projecting a strong sense of character.

GK Design Group,
Royal Star motor-
cycle for Yamaha,
1996

Hot-water bottle
made by Haffenden
Moulding Company,
1996

The elasticity, resilience and strength of rubber make it
ideally suited to a wide range of applications, from vehicle components to electrical insulation. Natural rubber,
a polymer of isoprene, is obtained from plants such as
the *Hevea braziliensis* tree, which produces a latex containing around 35%
rubber that is tapped from grooves specially cut into its bark. The milky latex
is then strained of impurities and diluted with water before being coagulated
into its solid form. Although natural rubber was first scientifically described
by François Fresneau and Charles-Marie de la Condamine in 1735, its use
was not fully exploited until the American inventor Charles Goodyear (1800–
1860) developed the vulcanization process in 1839. This method of treating
rubber with sulphur under heat and pressure greatly improved the material's
strength and elasticity. Vulcanized rubber found one of its first applications
in the manufacture of car and bicycle tyres. Gutta-percha, which has the
same elemental composition as natural rubber but a different molecular

Gutta-percha *Ocobo*
golf balls made by
James B. Halley,
c. 1920

Calculator produced by Marksmark Products, c. 1997 – combining advanced synthetic elastomer with plastic to give a greater degree of tactility

structure, was also widely used in the 19th century. Considered a natural plastic, gutta-percha is a rubber-like substance that can be moulded easily when heated but which becomes hard and leathery at room temperature. Because of its good heat-resistance, natural rubber is still used today for high-performance tyres such as those used on racing cars and aircraft. During the late 19th and early 20th century, several attempts were made at producing synthetic rubber from isoprene. It was not until the First World War, however, that scientists in Germany successfully developed the first synthetic "methyl rubber" by polymerising butadiene. By the 1920s and 1930s synthetic rubber was being manufactured by several polymerisation methods and during the Second World War large amounts were produced from butadiene polymers. Like natural rubber, synthetic rubber can be vulcanized and reinforced with fillers and is especially suited to casting and compression-moulding processes. Over half the rubber currently produced is used for the manufacture of vehicle tyres, while the rest is used for the manufacture of mechanical components and consumer products, ranging from hot-water bottles and garden hoses to shoes and toys. Today, advanced synthetic elastomers are frequently used in conjunction with **plastics** for co-injecting moulding in order to give the casings of electronic products an enhanced tactile quality.

Design Research
Unit, signage for the
Festival of Britain,
1951

Like **packaging**, signage can be considered a discipline within the field of "industrial" graphics. Its design imperatives are thereby visual clarity and a logical and aesthetic coherence. Signage frequently forms the subset of a larger **corporate identity** or municipal design programme, such as that developed by Edward Johnston in 1916 for **London Transport**. In the 1930s, Continental Europe responded to the need for clearer signage on its increasingly busy roads with the formulation of protocols governing road signage design. Later, in the 1960s, the burgeoning of the motorways led the British government to commission the *Worboys Report* on road signage. The Ministry of Transport subsequently commissioned Jock Kinneir (1914–1974) and Margaret Calvet (b. 1935) to design a new road signage system for use in Great Britain. It is a tribute to the designers' skill that the resulting Motorway Signage System (1964) is still in use today. Many multi-disciplinary design consultancies have designed fully-integrated signage systems for public events, public buildings and local government, as in the case of the Festival of Britain (**Design Research Unit**, 1951), the Victoria and Albert Museum (**Pentagram**, 1988) and the City of Rome (**Ninaber, Peters & Krouwel**, 2000).

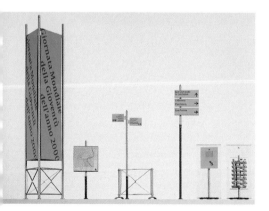

Ninaber, Peters &
Krouwel (NPK),
signage for the City
of Rome, 2000

During the mid-1980s the term "soft-tech" was coined to describe products that used rounded or "softened" sculptural forms. One of the first manifestations of soft-tech was in consumer electronic equipment, most notably in products by **Sharp** and **Yamaha**. Post-modern designs like these opposed the pervasive rationalism

Frazer Designers,
Cobra microscope
for Vision Engineer-
ing, 1997

and "good forms" promoted by manufacturers such as **Bang & Olufsen** and were inspired instead by period styles, especially 1950s' biomorphism. This type of American retro **styling** became a widespread phenomenon in the mid to late 1980s. It gave way in the early 1990s, however, to a more considered, organic and holistic approach to design. Described by **Ross Lovegrove** as "a new naturalism", soft design (as it became known) was emotional, gestural and human-centric. Remarkably, the car industry was amongst the first to embrace the new approach, as demonstrated by the

Fiat *Multipla*, 1999

Renault *Twingo* (1992) and the **Nissan** *Micra* (1992). The emergence of

ALESSI

soft design can be attributed to a number of factors: the outright rejection, by the late 1980s, of **Bauhaus**-style functionalism and the re-appraisal of organic design in general; the improved application of ergonomic data and the realization of more complex curved shapes made possible by advanced **CAD/CAM** software; and the availability of new materials, especially exotic techno-polymers, offering products the potential for greater tactility. By the late 1990s soft forms had made significant in-roads not only into the car industry, as the **Fiat** *Multipla* (1999) shows, but also into mainstream design practice, and products ranging from microscopes and calculators to furniture and even MRI body scanners took on seductive sculptural forms. Soft design is often used stylistically and should not be confused with organic **essentialism**, which although bearing striking visual similarities is purely design-led.

George Sowden,
Dauphine desktop
calculator for Alessi,
1997

Karl Trabert, indus-
trially-produced
standardized lamp,
Frankfurt School,
1920s

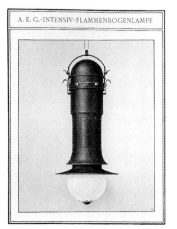

Peter Behrens, light
fixture for AEG, 1908

Marcel Breuer,
adjustable armchair
for Metz & Co.,
1931–1932

STANDARDIZATION

Standardization is a crucial aspect of industrial mass
production. The use of standardized components, which
can be fitted together with little or no adjustment and
interchanged from one product to another, increases efficiency and output. Early advocates of standardization
included members of the **Deutscher Werkbund**, such
as Hermann Muthesius (1861–1927), who saw it as a
powerful tool for the democratization of design. One of
the first companies to implement a coherent system of standardization was
AEG, whose integrated product line, designed by **Peter Behrens,** reflected a
deep understanding of Modern manufacturing techniques. The importance
of standardization was again stressed at the Dessau **Bauhaus**, and associated designers such as **Marcel Breuer**, **Gerhard Marcks** and **Wilhelm Wagenfeld** produced standardized designs intended for large-scale industrial production. In France, Le Corbusier (1887–1965) designed a standardized housing unit (1925) and a range of systemized furniture (1928) which included

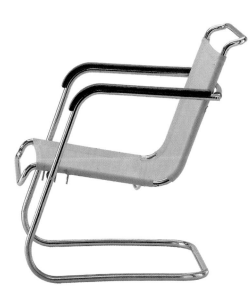

standardized modular storage units.
The industrial designers of the post-
war era fully embraced standardization, which offered the optimum
means of manufacture and allowed
the design of cost-effective product
systems. **Charles and Ray Eames'**
plastic shell group of chairs (1948–
1950) and **Robin Day**'s *Polyprop*
series (1962–1963), for instance,
both employed single standardized
seat shells that could be attached to
a variety of bases to create different
options. Today, this standardized
approach is common practice in
all areas of design, from computer
systems to **corporate identity** programmes.

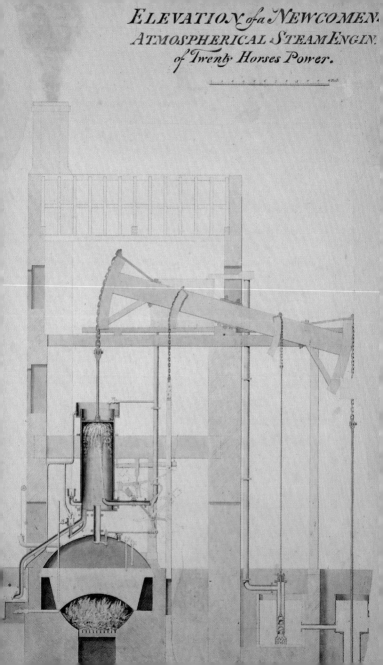

ELEVATION of a NEWCOMEN.
ATMOSPHERICAL STEAM ENGINE
of Twenty Horses Power.

Aveling Porter steam roller, early 20th century

←Elevation of the Newcomen 20 hp atmospheric engine, c. 1826 – designed by Thomas Newcomen, who invented the first practical steam engine

The principles of steam power were first advanced by the French-born British physicist Denis Papin (1647–c. 1712), who while developing the steam digester (pressure cooker) observed that steam actually had the power to raise the lid of his cooking pot. Papin's ideas inspired the English military engineer Thomas Savery (c. 1650–1715), who subsequently built and patented the first primitive steam engine in 1698. His machine comprised a closed water-filled vessel into which steam was introduced at high pressure, forcing the water to a higher level; when the water was expelled, a sprinkler condensed the steam, thus producing a vacuum capable of raising water through a valve positioned below. Savery's machine was designed specifically for the raising of water and was used as a pump in mines and for supplying water to large buildings. In 1705 the German physicist and mathematician Gottfried Wilhelm von Leibniz (1646–1716) sent Papin a drawing of Savary's steam engine, which in turn prompted Papin to design his own improved version (never actually built), which incorporated an innovative cylinder and piston mechanism. In 1712 Thomas Newcomen (1663–1729), assisted by Savery, adapted Papin's concept and subsequently erected the first recorded "fire engine" (as steam engines were then called) at the Dudley Castle mine in Staffordshire. Newcomen's improved atmospheric-pressure piston engine was so effective at pumping

James Watt, double-action rotary steam engine, 1782

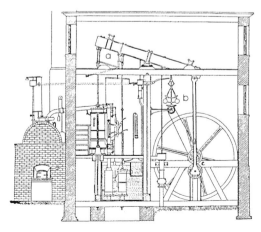

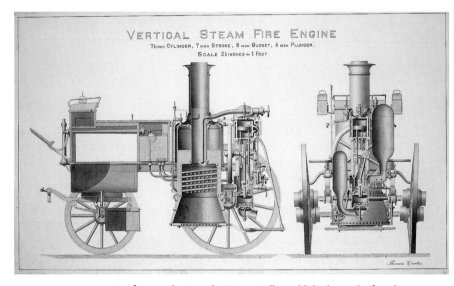

Shand Mason
vertical steam fire
engine, 1863

water from coal mines that it reputedly could do the work of 110 horses or over 2,500 men. Newcomen's engine was nonetheless somewhat inefficient as it condensed the steam in its cylinder. In 1764, while repairing a Newcomen engine, **James Watt** correctly identified that the design wasted power through its loss of latent heat. Realizing that this problem could be solved with the addition of a separate condenser, Watt began developing his own engine and in 1769 patented his revolutionary "New Invented Method of Lessening the Consumption of Steam and Fuel in Fire Engines". In 1775 Watt built his first successful fire engine, which was used to pump water from mines. He went on to design the rotative "Sun-and-Planet" engine in 1781 and the first double-action rotary steam engine in 1782. Three years later, Watt and his partner Matthew Boulton (1728–1809) installed a rotative engine in a cotton-spinning works at Papplewick, Nottinghamshire. Significantly, this was the first time that motive power had been harnessed to drive machinery, in a step heralding the revolutionary transition from water power to steam power. By the early 1800s, steam power was finding one of its most important and far-reaching applications in the powering of locomotives – **George Stephenson**'s *Rocket* of 1829 being among the first and most famous. Continuous design improvements produced more and more powerful engines, such as the Shand Mason vertical steam fire engine of 1863. By the end of the 19th century, the steam engine had become the main power source for both industry and transport and was used well into the 20th century, until eventually being replaced by cleaner and more efficient **electric power**.

Norman Bel Geddes, model of a stream-lined bus, 1939

Streamlining involves the contouring of objects into rounded, smoothly finished and often teardrop-shaped aerodynamic forms so as to reduce their drag or resistance to motion through air. Streamlining was first used in the early 20th century to improve the performance of aircraft, locomotives and automobiles when moving at high speeds. By the 1930s, however, industrial designers were using streamlining less for functional reasons than to make household products look sleeker and thereby more appealing to the consumer. In America, the Wall Street Crash of 1929 and the ensuing Great Depression, together with the implementation of the price-fixing National Recovery Act of 1932, meant that manufacturers were operating within a fiercely competitive marketplace. Rather than investing in the development of entirely new products, many manufacturers preferred to re-style or "streamline" their existing products so as to make them appear new. Streamlining also helped manufacturers differentiate their products from those of their competitors, while annual restyling programmes – such as that implemented by **Harley**

Chrysler *Airflow* (1934) with Union Pacific *City of Selina* locomotive (1934)

Earl at **General Motors** –became a deliberate means of accelerating the aesthetic obsolescence of products and thereby increasing sales. Interestingly, many of the American designers who became renowned for their streamlined designs, such as **Raymond Loewy**, **Norman Bel Geddes**, **Henry Dreyfuss** and **Walter Dorwin Teague**, had previously worked as fashion illustrators, stage designers or commercial artists. Using clay models, such designers created sleek, modern-looking forms for a whole range of consumer goods, including refrigerators, vacuum cleaners, radios, cameras and telephones. Many of these products featured casings of **Bakelite**, a thermoset plastic eminently suited to the moulding of streamlined forms. In 1934 Loewy's *Coldspot* streamlined refrigerator for Sears became the first domestic appliance to be marketed on its looks rather than on its performance. The use of streamlining rapidly became widespread, and its practitioners highly celebrated. As **Harold van Doren** observed in 1940: "Streamlining

has taken the world by storm ... The manufacturer who wants his laundry tubs, his typewriters, or his furnaces streamlined is in reality asking you to modernise them, to find the means for substituting curvilinear forms for rectilinear forms." In 1949 Raymond Loewy became the first designer to be featured on the front cover of *Time* magazine, his picture accompanied by the telling copy-line, "He streamlines the sales curve". By "adding value" to products at relatively little cost and stimulating sales, streamlining helped American manufacturing industries regain strength and profitability.

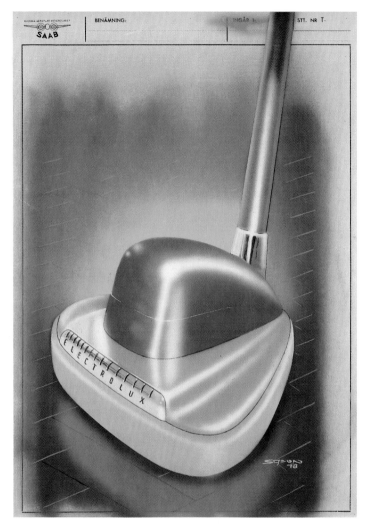

Sixten Sason, *B9*
floor polisher for
Electrolux, 1948

Familial radio, mid-1950s

While design and styling are completely distinct disciplines, styling is often a complementary element of a design solution. Styling is concerned with surface treatment and appearance – the expressive qualities of a product. In contrast, design is primarily concerned with problem-solving – it tends to be holistic in its scope and generally seeks simplification and essentiality in products. Historically, styling has been used either to disguise the inherent mechanical aspect of a product, or to highlight it through the application of exaggerated symbolic forms. **Raymond Loewy** regarded stylistic "sheathing" as a means of developing "the self-expression of the machine". Very often, styling is used by manufacturers as a means of "adding value" to products because it can dramatically enhance consumer appeal and increase product differentiation. As Raymond Loewy insightfully noted: "Between two products equal in price, function, and quality, the better looking will outsell the other."

The prevalence of design over styling – or vice versa – is something which has fluctuated over the course of the 20th century in line with the economic cycles of Western economies. Thus design (rationalism) tends to come to

KNR radio, mid-1950s

Raymond Loewy,
Coca-Cola dispensers,
1940s

the fore during economic downturns, while styling (anti-rationalism) is apt
to flourish in periods of economic prosperity. Styling found early expression
in the 1920s, with the flourishing of Art Deco, and in the late 1930s and
1940s, when **streamlining** became a widespread phenomenon in American
industrial design. The mid to late 1950s saw the emergence of biomorphic
styling in opposition to the "good forms" perpetuated by the international
design establishment through its well-established canons of "good taste",
while the Pop-influenced products of the 1960s focused on short-lived stylistic gimmicks rather than on the long-lasting design solutions. With the rise
of Post-Modernism in the applied arts in the 1980s, the transmission of
meanings and values through aesthetics (i. e. surface treatment) became
more important to the avant-garde than technical function.

As well as being employed to make a product more attractive or symbolically
meaningful, styling has also been used as a vehicle of **planned obsolescence**.
Having first emerged in the American automobile industry in the 1920s, stylistic planned obsolescence significantly accelerated product lifecycles. Annual re-styling programmes ensured that what was today the "latest thing"
would be completely out of date within just a couple of years. This trend culminated most spectacularly in the decorative **chromium** flourishes of 1950s
American automobile styling – and most worryingly in the actual compro-
mising of vehicle safety for the sake of stylist devices. Although car manufacturers have toned down annual appearance changes in more recent years,
styling continues to play an important role in the automotive industry, as
evidenced by the prevalence of **retro design**.

Today, styling is integrated into the whole of the design process and is applied from the beginning of a product's development rather than as an afterthought. In defining the differences between design and styling, the famous Italian industrial and furniture designer Vico Magistretti (b. 1920) stated: "Design does not need drawing, but styling does. What I mean by this is that an object of design could be described ... by spoken or written words, because what materializes through the process is a precise function, and, in particular, a special use of materials which, as a matter of principle, leaves all aesthetic questions out of consideration because the object is to achieve a precise practical aim. That does not of course mean that a precise image cannot be produced that will reflect and express 'aesthetic' qualities proper to the new methodology used in the conception of the object. Styling, on the other hand, has to be expressed by the most exact drawings, not because it disregards function but simply because it wraps that function in a cloak of essentially expressed qualities that are called 'style' and that are decisive in making the quality of the object recognizable."

Preston Tucker and Harry Miller, car designs for Tucker of Chicago, 1946–1948

Early time & motion study

Taylorism is a term used to describe an approach to mass production that is based on an industrial management system pioneered by the mechanical engineer Frederick Winslow Taylor (1856–1915). In 1881 Taylor developed and implemented the concept of time studies, having realized that productivity could be enhanced if tasks performed by workers were broken down into constituent parts and then scientifically analyzed so as to eliminate any waste of time or motion. His book, *The Principles of Scientific Management* (1911), set forth his common-sense principles not only for the organization of specific tasks but also for the overall running of factories, and was highly influential upon Henry **Ford**, amongst others. Taylor believed that his approach would bring "the elimination of almost all causes for dispute and disagreement" between employer and employee. While Taylorism was initially regarded as dehumanizing, and led to the greater implementation of automation, it resulted in the rise of a new generation of highly skilled and empowered machine operators.

Still from Charlie Chaplin's *Modern Times*, a film reflecting on the alienating effect of the production line

Cover of Mauser
catalogue, 1939

Tubular metal was first manufactured in Germany, where Max and Reinhard Mannesmann patented processes used in its mass production in 1885. Five years later Reinhard Mannesmann (1856–1922) founded the Mannsmannröhren-Werke in Düsseldorf, which became the leading producer of tubular metal. The Mannesmann process involved the spinning of a solid rod of hot steel between two inclined rollers that rotated in the same direction so as to pull the rod over a mandrel bar (a spindle-like element) to produce a tubular section. Unlike earlier piping, Mannesmann tubular metal was seamless and consequently possessed greater strength as well as better aesthetic qualities. Tubular metal found one of its first successful commercial applications in the construction of bicycles, including models manufactured by the German company, Adler. At this stage, the tubing used for bicycles had an outside diameter ranging from 95 mm to 380 mm. During the early years of the 20th century, tubular

Marcel Breuer,
Model No. B3 Wassily
chair, for Standard-
Möbel & Gebrüder
Thonet, 1925–1927

Arch. Marcel Breuer

metal began to be used in the construction of other types of vehicles, as its strength, lightness and resilience made it a good substitute for wood. Furthermore, angled sections that were traditionally riveted could now be replaced with welded sections of tubing. The well-known Dutch aircraft designer and manufacturer Anthony Fokker (1890–1939) used welded tubular metal in the construction of his first plane, the *Spin Mark I* (1910). Subsequent models of his aeroplanes produced in Germany during the First World War also incorporated tubular metal in their construction. Portable welding equipment developed at that time made tubular metal designs even more adaptable. In the early 1920s both Maschinenfabrik Sack GmbH and Josef Gassen were granted patents relating to manufacturing processes that produced improved tubular metal with thinner walls. In 1925 **Marcel Breuer** became the first designer to use tubular metal in the construction of furniture. His famous *Model No. B3 Wassily* chair (1925–1927) powerfully exploited the machine aesthetic of tubular metal as well as its inherent material qualities. A year later, the Dutch architect Mart Stam (1899–1986) con-

Marcel Breuer, *B11*
armchair, 1926–1927
(1st version) and *B11*
armchair, 1927 (2nd
version)

structed a cantilevered prototype chair made of rigid gas pipes welded together, which inspired later tubular-metal designs, most notably by Marcel
Breuer and Ludwig Mies van der Rohe (1886–1969), that were nickel-plated
or **chromium**-plated. During the late 1920s and 1930 several manufacturers
became renowned for their Modernist tubular-metal furniture, including
Gebrüder **Thonet**, Standard-Möbel and PEL. Tubular metal also became a
material of choice for Moderne designers of the 1930s, who exploited its
gleaming aesthetic for stylistic purposes. Until the widespread availability
of **plastics** in the 1950s and 60s, tubular metal was used extensively for
contract furnishings. Today it continues to be used for a plethora of applications, ranging from motorcycle frames to golf-club shafts.

Cover of the first
Utility Furniture
catalogue, 1943

→James Leonard,
stacking aluminium
and plywood school
chairs, 1948

Utilitarian design is based on the concept that the primary criterion of virtue is utility. For centuries many different types of artefacts – from agricultural tools to cookware – have been designed purely for use rather than beauty. But in fulfilling technical practical requirements as logically, efficiently and inexpensively as possible, it has long been recognized that utilitarian designs, whether handcrafted or machine produced, project a distinct aesthetic or kind of beauty based on functional purity. The honesty and fitness for purpose that is characteristic of utilitarian design became the basis of the Modern Movement's dictum, "form follows function". During the 1920s and 1930s, Modernists such as the Dutch architect Jacobus Johannes Pieter Oud (1890–1963) produced utilitarian furniture purged of all ornament. Designs like these were strongly inspired by the socialist ethos of the Modern Movement. It was argued that the more rational a design, the cheaper its manufactured cost would be and hence the more accessible it would be to the working classes. This approach to the design of consumer products, however, did not take into account the conservative taste that permeated that sector of society at which the designs were primarily aimed. The most

Aynsley China coffee
cup and saucer, 1956

notable large-scale programme of utilitarian design was implemented in Britain between 1941 and 1951, when government-approved Utility furniture and textiles were made available to the British public in an effort to boost the domestic manufacturing economy at a time when many materials were still rationed. In recent years, avant-garde designers such as **Jasper Morrison** have begun producing essentialist product designs which project an aesthetic that is again very much utilitarian in nature.

CASE STUDIES

An early Raleigh safety bicycle, 1887

→ Rudge "ordinary" bicycle, 1884 – this bicycle was built for racing and was therefore very light. Its basic format was based on James Starley's early "penny-farthing" design.

John Kemp Starley, *Rover* safety bicycle, first designed in 1885 – Starley's design provided the blueprint for subsequent commercially-produced bicycles.

The bicycle is almost certainly the most efficient means yet devised of converting human energy into propulsion. The first known patent for a rider-propelled machine was granted to the Frenchman Jean Theson in 1645. This design, like other similar machines from the 17th and 18th centuries, was configured with four wheels. The first two-wheeled rider-propelled machine – the bicycle – was invented by Baron Karl de Drais de Sauerbrun and was exhibited in Paris in 1818. This steerable though relatively cumbersome wooden bicycle was propelled by the seated rider's feet pushing against the ground. It was not until 1839 that the first self-propelled bicycle was invented by the Scottish blacksmith Kirkpatrick Macmillan. His design, known as the *Hobbyhorse*, required the rider to move swinging cranks back and forth with his feet so as to drive the rear wheel via a pair of moving rods. In 1862 Pierre and Ernest Michaux improved on Macmillan's design with their *Vélocipède*, which had two cranks with foot pedals attached to the front wheel. This design was extremely popular even though its construction of wood and iron made it heavy and a veritable "boneshaker" when

Alex Moulton,
Moulton bike,
mid-1960s

ridden. In the 1870s, the British inventor James Starley (1830–1881) managed to reduce the bulk and weight of bicycles with various innovative designs including the patented *Ariel* (1871), which had a large front wheel and a small back wheel and was nicknamed the "penny-farthing" after the smallest and largest copper coins used in England at the time. Starley made various improvements to his penny-farthing models over the years, including the introduction of a gear mechanism that revolved the drive wheel twice for every pedal revolution. He also managed to significantly reduce weight by using wire rather than iron for the spokes of the wheels. Penny-farthings were also made by other manufacturers and generally weighed around 50 lbs. Racing models, however, weighed half this much. In 1874 H. J. Lawson designed the first chain-driven bicycle, which featured two medium-sized wheels of equal diameter and an endless chain running from a main sprocket to a second sprocket on the rear wheel. This revolutionary design provided greater stability and better braking control than the large front-wheeled penny-farthing and came to be known as a "safety" bicycle. In 1885 James Starley's nephew, John Kemp Starley, designed and built the landmark *Rover* safety bicycle, which became the blueprint for many subsequent models and still provides the basic constructional format for the majority of bicycles today. Three years later, riding comfort was greatly improved by the development, by John Boyd Dunlop (1840–1921), of pneumatic **rubber** tyres, which led to the predominance of safety bicycles over high front-wheeled "ordinary" models with solid rubber tyres. By the early 1890s bicycle design had stabilized into the

Seymour Powell,
prototype *Nexus*
motorized bicycle,
1992 – internal
design study

modern **tubular-metal** diamond-shaped frame with roller-chain drive and pneumatic tyres, and the bicycle industry started mass-producing low-cost practical models. The next major step forward in bicycle design came with the introduction of gears. In the early 1900s, H. Sturmey and J. Archer patented a number of gear mechanisms based on the epicyclic principle. Incorporated in the rear hub, these gears altered the speed of the drive sprocket relative to the rim of the

wheel. This innovation led to the development of *dérailleur* systems, which moved the drive chain from one sprocket to others of varying sizes. From the early 1900s until the 1960s, the diamond frame (for ladies) and cross frame (for men) dominated bicycle design. In 1962, however, the British designer **Alex Moulton** developed an entirely new type of bicycle, characterized by small wheels, high-pressure tyres and a full suspension system. The Moulton bicycle was easier to control than earlier models because it had a lower centre of gravity. It was also much easier to transport and to store. Since the mid-1970s, bicycles have fallen into four main categories – utility, touring, racing and mountain bicycles. Depending on their function and cost, they employ a variety of materials in their frame construction, such as steel alloys, chrome-molybdenum and **aluminium**. While weight reduction has always been a primary concern in bicycle design, it was not until the 1990s that ultra-lightweight yet strong advanced composite materials, such as kevlar and **carbon fibre**, began to be used in the construction of competition and other highly specialized bicycles. One of the first and most interesting applications of carbon fibre in bicycle design was seen in the revolutionary *Mono* superbike (1992) designed by **Mike Burrows**. Throughout the 1990s, the drive for new and better-performing bicycles resulted in other highly innovative experimental designs, such as **Seymour Powell's** motorized *Nexus* (1992), a lightweight hybrid vehicle that combined bicycle and motorcycle elements, and **Jean-Pierre Vitrac's** extraordinary plastic model (1995) for Diam that redefined the aesthetic parameters of bicycle design. As an inexpensive, non-polluting and healthy mode of transport, the future of cycling and of bicycle design looks very bright indeed.

Jean-Pierre Vitrac, prototype plastic bicycle for Diam, 1995

19th-century iron
made of cast iron

Crompton's electric
iron, 1895 – this
model had a wooden
plug casing

Siemens' *Excel*
electric iron, 1926

CASE STUDY 2

THE IRON

While the function of the iron has remained utterly constant throughout its history, over the last 100 years its design has been subject to many radical improvements. During the 18th and 19th centuries, heavy cast-iron designs were manufactured in graduated sizes, with the smallest being used for delicate lace trimmings. Relatively cumbersome, these irons were heated on coal or wood-burning stoves and were the bane of domestic servants' lives, as their temperature was difficult to control and they needed constant re-heating. The task of ironing was transformed, however, by the introduction of electric irons in the 1890s. The earliest of these irons were expensive and were generally only used in commercial laundries. Electric irons for the home were nevertheless pioneered by the German company **AEG**, whose catalogue of 1896 featured eight – still fairly expensive – models with turned, carved and cantilevered handles. In 1912 the American Heater Company of Detroit launched its *American Beauty* iron, which heralded the advent of lower-cost electric irons for domestic use. This popular design was later succeeded by more advanced designs such as **Siemens'** *Excel* iron of 1926. In the 1930s, HMV produced an electric iron with a ceramic casing to reduce heat loss. This influential design had a heavily chrome-plated sole plate that conducted heat well and reduced friction so that the iron could glide more smoothly over garments. Around the same time, Morphy Richards produced a similar ceramic-cased electric iron that was exhibited at the Ideal Home exhibition in London. During the 1950s, the weight of electric irons was significantly reduced by the introduction of plastic handles and **aluminium** sole plates. The "featherlight" *Litomatic* iron manufactured by Revo in 1953, and an iron introduced by **General**

HMV's ceramic electric iron, c. 1936 – this heat-control iron incorporated a powerful heating element and featured a heavily-chromed sole plate designed to reduce heat loss. Its integrated handle and body were made of glazed porcelain.

Electric in 1957, both had ergonomically-designed plastic handles that enhanced their formal homogeneity. Over the succeeding years, electric irons became increasingly sophisticated with the addition of reliable thermostat controls and mechanisms for steam ejection and water spraying. The **styling** of iron casings also became increasingly advanced, with Rowenta's *Surline* (1994) and **Kenneth Grange**'s *ST50* travel steam iron for Kenwood (1995) featuring streamlined forms and colourful translucent plastic elements that allowed the water level to be viewed easily. Thanks to the extraordinary advances in materials and technology over the last 100 years, the iron has been completely transformed from a heavy cast-iron object into a lightweight and sculptural electrical appliance.

Advertisement for
Litomatic electric
iron, 1953

Silver Streak iron,
1942–1946

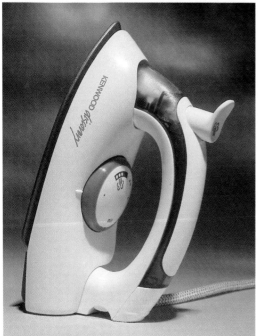

Kenneth Grange,
ST50 travel steam
iron for Kenwood,
1995

Robert Hooke's compound microscope, c. 1675

→Indian Medical Service's portable microscope, c. 1895

John Priestley's microscope, made by Benjamin Martin of New Invented Visual Glasses, Fleet Street, London, 1767

The history of optical instruments has recently become a hotly debated issue, with several historians claiming that the use of relatively sophisticated optics can be traced back to antiquity. There is certainly irrefutable evidence that so-called "simple" microscopes, comprising single lenses, were in use from as early as the mid-15th century. The compound microscope, which comprises two lenses, was invented between 1590 and 1609 in the Netherlands, but tended to suffer more than simple microscopes from problems of chromatic aberration. In the mid-1670s, the Dutch naturalist Antonie van Leeuwenhoek (1632–1723) developed high-quality optical lenses for single-lens microscopes that were powerful enough to enable the viewing of bacteria and protozoa measuring only two to three microns in diameter. In 1665 the English physicist and first curator of the Royal Society, Robert Hooke (1635–1703), published *Micrographia*, which was the first important text on microscopy. In this seminal work he illustrated a compound microscope which he used to examine a flea, a piece of cork, a snowflake and other specimens. This early compound design incorporated a small oil lamp and a liquid-filled globe to focus light onto the specimen. By the late 18th century, precision engineering had advanced the design of microscopes considerably. The microscope designed by Joseph Priestley (1733–1804) in 1767, for example, was altogether more "scientific" in appearance than its 17th-century antecedents and was much easier to focus. Its mirrored illuminating system, which used natural light, was also considerably more practical. In 1830 the English microscopist Joseph Jackson Lister (1786–1869) began grinding his own high-quality lenses, and discovered a way of combining them that eliminated various chromatic and spherical aberrations. Lister's subsequent law of *aplanati foci* became the basis of modern scientific microscope construction. By the late 19th century, scientists could choose from a wide range

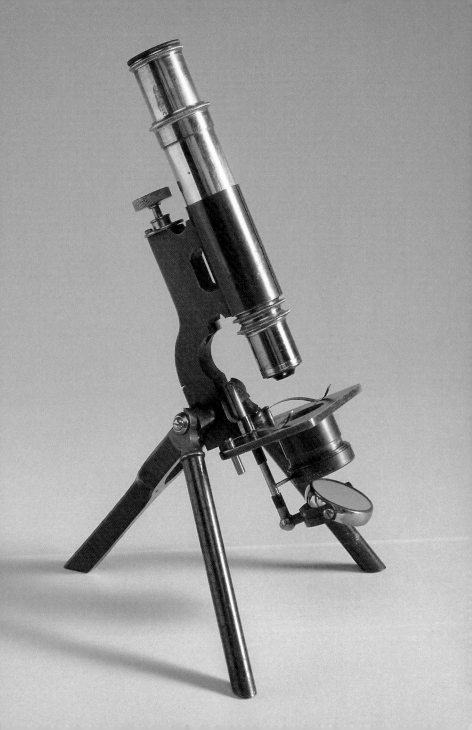

of microscopes able to achieve excellent resolutions. Around 1895, the British bacteriologist Ronald Ross (1857–1932) proposed the construction of a simple portable microscope that could be used by officers in the Indian Medical Service. The resulting lightweight and practical design, manufactured by C. Baker of High Holborn, London, was specifically adapted for the diagnosis of malaria. No further significant advances occurred in microscopy, however, until 1924, when the French physicist Louis-Victor de Broglie (1892–1987) discovered the wave-like nature of electrons and showed that they had considerably shorter wavelengths than light. This groundbreaking research into electron beams led directly to the development of the electron microscope in the 1930s. With electron microscopes, the specimen is illuminated using an electron beam that is focused with either an electrostatic or an electromagnetic field. Electron microscopes are vastly more powerful than conventional models and today provide magnifications in excess of x 250,000. The early 1980s saw the development of electron scanning tunnelling microscopes, which measure the variations of electric current between the surface of the specimen and the microscope's probe. The first commercially-produced scanning tunnelling microscope was constructed by W. A. Technology of Cambridge in 1986. Highly sophisticated instruments such as these are capable of producing such high magnifications that they are able to detect individual atoms. While electron microscopes provide ever greater degrees of magnification, they are expensive, and specimens are destroyed by the processes required to view them. Conventional electric microscopes, therefore, remain much more widely used. As it is not unusual for electric microscopes to be operated by researchers and clinicians for up to ten hours a day, ergonomic criteria are vital considerations in their design. **Frazer Designers**' moulded ABS *Cobra* microscope (1997) for Vision Engineering was one of the first microscopes designed from an ergonomic perspective, and its sculptural soft form signalled a new direction in the design of scientific equipment.

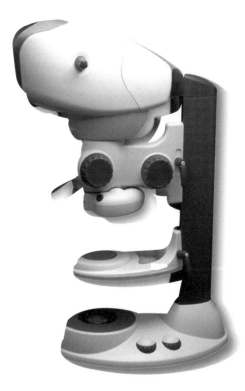

Frazer Designers,
Cobra microscope
for Vision Engineering, 1997

ICE SAFES.

THE NEW DUPLEX REFRIGERATOR,
Registered; for Wenham Lake or Rough Ice. Prize
Medal Refrigerators, fitted with Water Tanks and Filters.
The New American Double-Wall Ice Water Pitcher, suit-
able for Wine Cup, &c. The American Butter Dish, with
revolving lid, for use with ice. Wenham Lake Ice delivered
in town for less than 1d. per lb., or forwarded into the
country in packages of 2s. 6d., 4s., 8s., and upwards, by
"Goods Train," without perceptible waste. Illustrated
Price Lists free on application.

WENHAM LAKE ICE COMPANY
125, STRAND, LONDON (Corner of Savoy-street).

Ice-safe made by
Wenham Lake Ice
Company, London,
1875

Self-feeding ice-safe
made by the Piston
Freezing Machine
& Ice Company,
London, 1875

THE SELF-FEEDING ICE SAFE.

The PISTON FREEZING MACHINE and
ICE COMPANY, 314 and 315c, Oxford-street, direct attention to

MR. ASH'S NEW PATENT
SELF-FEEDING ICE SAFE.

It produces perfectly dry, cold, ventilated air; no increase
of temperature effected with, say 50lb. of ice, is fully maintained in every part of the Safe, even though the ice may
have diminished to 10 or 12 lb. In operation daily, and particulars detailed at the

OFFICES OF THE COMPANY,
314 and 315c, OXFORD-STREET (near
Harewood Gates).

Throughout the history of civilization, various methods have been used for the preservation of food, ranging from salting and drying to freezing. In classical antiquity, the Greeks and Romans transported ice from mountain tops and stored it in cellars and pits that were insulated with straw, wood and earth, so that food could be preserved during the hot summer months. In later centuries, many large European country estates included icehouses that worked on much the same principle, and which were often ingeniously designed – with features such as angled entrances that did not allow sunlight to penetrate, and timber-lined inner chambers – to achieve maximum insulation. The 19th century saw the development of various types of domestic ice-safes, which were essentially insulated boxes cooled with ice blocks. Since these melted away over time, they needed frequent replenishing. Also known as ice chests or caves, these proto-refrigerators used charcoal for insulation and preserved food reasonably well. Although numerous models were shown at the **Great Exhibition** of 1851, ice-safes remained much more popular in America than in Europe. Meanwhile, the first steps were being taken towards artificial refrigeration. The origins of artificial refrigeration lie in the remarkable phenomenon of evaporative cooling, as traditionally used for the production of ice in hot countries such as India and Egypt. This process relies on the rapid expansion of gases when heated and their drawing of energy from surrounding molecules to produce a cooling effect – ice can be made with this method even in desert conditions. As early as 1748, the Scottish physician William Cullen (1710–1790) had demonstrated artificial refrigeration by boiling ethyl ether in a partial vacuum, but his discovery was not given any practical application.

It was not until 1805 that the American inventor Oliver Evans (1755–1819) proposed the first refrigerator design to employ a vapour-cooling system. Although Evan's design was never built, a similar cold-air machine was designed and constructed in 1844 by the American physician, John Gorrie (1803–1855). Gorrie's refrigeration system used compressed heated gas that was cooled through radiating coils and then expanded to produce a cooling effect. Although Gorrie patented his mechanical refrigeration process in 1851, he was unable to find the necessary funding to manufacture the machines. Five years later, the American entrepreneur Alexander Twinning began producing commercial-use vapour-compression refrigerators. In 1859 the Frenchman Ferdinand Carré introduced ammonia, as an alternative coolant to air, since its lower liquefaction temperature provided a more efficient system of refrigeration. It was not until 1913, however, that the first domestic-use refrigerator, the *Domelre* (Domestic Electric Refrigerator), designed and manufactured by the refrigeration engineer Fred Wolf Jr., went on sale in Chicago. Kelvinator introduced its own model three years later, followed by Frigidaire in 1917. These early ammonia refrigerators were extremely bulky, as their compression system accounted for about half their volume. During the 1920s, better synthetic coolants were developed, most notably an odourless gas marketed under the name of Freon. Although refrigerators became widely accepted in America over the course of the 1920s, in other countries (such as Britain) they were deemed an unnecessary luxury. It was not until after the Second World War that refrigerator sales in Europe really started to take off, even though such well-known companies as **Electrolux** and **Bosch** were pioneering refrigerator designs from the late 1920s onwards. The Electrolux model from this period was marketed as "the only refrigerator in the world to operate continuously by electricity, gas or paraffin", and it spuriously claimed to keep "food fresh for an indefinite period". The Bosch models, which included a cylindrical design with a circular door, were more compact and therefore better suited for smaller homes. In America, industrial design consultants began using streamlined

➤Ate fridge, 1950s – showing typical Fifties styling with chromed elements (the fridge body would originally have been painted white)

Bosch refrigerators, 1933 (left) & 1936 (right)

styling to make their refrigerators more visually appealing, as illustrated by **Raymond Loewy**'s well-known *Coldspot* refrigerator (1934) for **Sears, Roebuck & Co.** – the first domestic appliance to be marketed on the strength of its looks. By the mid-1950s the majority of homemakers in America had a large refrigerator in their kitchen, and the rounded opulent forms of these appliances came to epitomize the unprecedented abundance of the period. During the 1960s and 1970s refrigerators became sleeker in form and were often integrated into the overall design of a kitchen. It was not until the 1980s, however, that the convention of the white box was turned on its head with the introduction of post-modern appliances, such as Roberto Pezzetta's *Black* fridge (1987) from the *Wizard's Collection* for **Zanussi**. Pezzetta subsequently produced a similarly groundbreaking design in his sculptural and boldly formed *Oz* refrigerator (1994). Although its basic function remains constant, the refrigerator is increasingly subject to the latest aesthetic trends. **Siemens**' *KS39V97* refrigerator, for instance, displays the current fashion for crisp lines and stainless-steel surfaces. Today,

Roberto Pezzetta, *Black* fridge from the *Wizard's Collection* for Zanussi, 1987 – selected for a Compasso d'Oro award in 1987 and received a Gold Medal BIO 12 in Lubljana in 1988

the use of CFCs (chlorofluorocarbons) as refrigeration coolants has been largely outlawed due to their depleting effect on the ozone layer, and more environmentally-friendly coolants have taken their place. Within the space of 100 years, the refrigerator has been transformed from a luxury item into an absolute household necessity in most societies.

Siemens *KS39V97* refrigerator, late 1990s

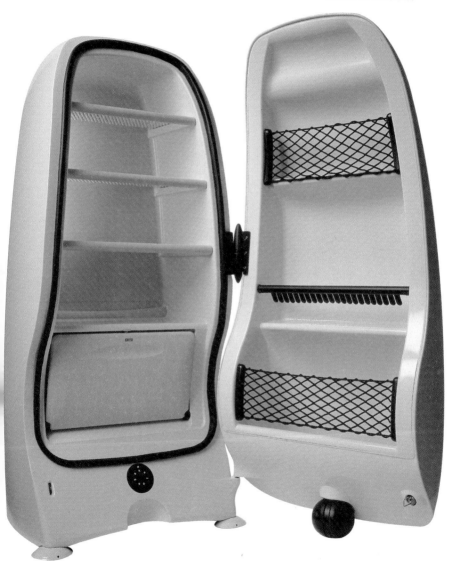

Roberto Pezzetta & Zanussi
Industrial Design Center, *Oz*
prototype refrigerator for Zanussi,
1994 – won Design Prestige '97
award in Brno, Czech Republic
and Goed Industrieel Onterp '99
award in the Netherlands

National Telephone Company, *Model No. 1* candlestick telephone, 1914 (a version of the Bell Telephone Company's model)

Since its invention in the mid-1870s the telephone has become an increasingly important and vital communications tool. While continual technological advances have driven its evolution, its primary requirements have remained constant – the telephone must provide a terminal through which speech and hearing can extend over great distances. The challenge of telephone design thus remains the improvement and expression of the telephone's three basic functions: the transmission of speech, the reproduction of speech, and the management of signal. Aesthetically, the telephone is unusual compared to most other industrially-produced consumer products in that, until relatively recently, it has escaped the vagaries of taste and the manipulations of the marketplace. **Styling** has not had much of an impact on its evolution, mainly because the production and distribution of telephones has historically been controlled by monopolies, such as the **Bell Telephone Company** in the United States. With virtually no competition to contend with, there was very little need for the continuous re-styling that has so characterized the automotive industry, for example.

The origins of the telephone are linked to the history of the telegraph and can be traced to 1667, when the English physicist and microscopist Robert Hooke (1635–1703) described how sounds could be transmitted through a tightly stretched wire. It was not until 1794, however, that the French engineer Claude Chappe (1763–1805) constructed the first working telegraph machine, the tachygraphe. This early communications device was a semaphore that transmitted visual signals, and was therefore dependent on weather conditions with good visibility. The invention of the first electric battery by Alessandro Volta (1745–1827) in 1800, and the later discovery of the link between electricity and magnetism by the Danish physicist Hans Christian Øersted (1777–1851), paved the way for the development of the electric telegraph. In 1839

Bosch table-telephone, 1887

Bell Telephone Laboratories, *Model No. 162* telephone, c. 1932

GPO *Model No. 332* telephone with drawer, c. 1937 – manufactured in Britain under license from L. M. Ericsson

Henry Dreyfuss,
Model No. 500 for
Bell Telephone
Laboratories, 1949

Henry Dreyfuss,
Trimline telephone
for Bell Telephone
Laboratories, 1968

Britain became home to the world's first commercial telegraph line, installed between Paddington and West Drayton for the Great Western Railway. Significantly, this also marked the first-ever commercial application of **electric power**. Five years later, Samuel Finley Breeze Morse (1791–1872) used his dot-dash coding system to send his famous message – "What God hath wrought" – over the first American telegraph line, which ran between Washington and Baltimore. The first transatlantic telegraph cable was laid in 1858, and three years later the German electrician Johann Philipp Reis (1834–1874) developed a primitive phone-like apparatus that could transmit musical and other sounds by electrical means. It was not until 1875, however, that the American audiologist **Alexander Graham Bell** constructed his first experimental telephone, which was patented in 1876. That same year, Bell transmitted the first recognizable vocal sounds using his cone-like cylinder apparatus that bore little resemblance to today's telephones. Reputedly, these first words – "Mr Watson, come here, I want you" – were uttered by Bell to his assistant after he had spilled sulphuric acid on his clothes, and were transmitted over a wire measuring 100 feet. Bell promptly established the Bell Telephone Company in New York to commercialize his telephone patent. In 1877 **Thomas Alva Edison** patented a carbon telephone transmitter using magnetic current, which was superior to Bell's system. The following year, the US Coast Survey officer Francis Blake developed a transmitter that equalled Edison's in performance and reliability, and which was subsequently used in Bell systems. At this early stage telephones comprised two distinct elements – the transmitter and the receiver – and were for the most part wall-mounted with a fixed mouthpiece and a hand-held receiver. In 1884 the Swedish company **L. M. Ericsson** combined the transmitting and receiving elements to produce the first-ever telephone handset. Three years later, **Bosch** produced a table-telephone employing a similar "French-style" layout. This design format was not adopted in America and Great Britain, however, where the Bell Telephone Company introduced "candlestick" desk telephones in 1914. The early candlestick model was initially made of enamelled brass and was redesigned in 1919 so as to incorporate a dial. The candlestick-style telephone remained the standard design in America and Great Britain until 1927, when it was replaced by the Bell "French phone" (1928), which (like the earlier Ericsson and Bosch models) combined the receiver and transmitter into a single hand-held

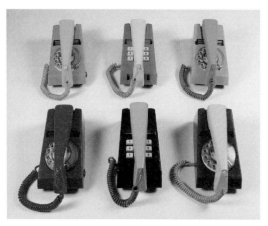

Lord Snowdon,
Phoenix range of
Trimphones for
British Telecom,
c. 1982

Eiger telephone for
British Telecom,
c. 1982

Mickey Mouse push-
button telephone
for British Telecom,
c. 1983

Vodaphone transportable mobile telephone, 1985

unit. With its four-pronged cradle, slope-faced base and angled mouthpiece, this telephone established the blueprint for many subsequent telephone designs.

In 1929 Bell held a competition for the "phone of the future", which led to a longstanding relationship with the design consultancy **Henry Dreyfuss** Associates. In 1937 Dreyfuss and his office developed a radically new telephone for Bell Telephone Laboratories, the *Model 300*. This compact **Bakelite** telephone was based on the sculptural *DHB 1001* telephone launched five years earlier by L. M. Ericsson and designed by **Jean Heiberg** and Johan Christian Bjerknes. Dreyfuss later designed the *Model 500* telephone, which was launched in 1949. This design was visually more unified than previous models and its curvaceous form reflected the contemporary taste for organic shapes. The *Ericofon* introduced by L. M. Ericsson in 1954 revealed a similar interest in sculptural organic forms, its highly innovative design integrating the earpiece, mouthpiece and dial into a single form. In 1959 Bell launched the compact *Princess* telephone as a "boudoir phone" specifically intended for the female market. Its innovative layout enabled the handset to be balanced across the telephone dial, and inspired Dreyfuss' more unified *Trimline* telephone, which was introduced in 1964. The *Trimline* also incorporated the dialling or push-button apparatus in the handset, so that when not in use the telephone took on a sleek and sculptural lozenge form.

Throughout the 1970s and 1980s telephones proceeded to assume a host of innovative and colourful forms, such as the *Phoenix* range (c. 1982) designed by Lord Snowdon and based on the earlier *Trimphone* design first introduced by the GPO in 1965, the *Mickey Mouse* push-button telephone (c. 1983) and the sculptural one-piece *Eiger* phone (c. 1982). Combination telephone/facsimile/answer machines began making an appearance from

Panasonic *I-series*
ETACS mobile phone,
manufactured by
Matsushita Commu-
nication, 1993

Nokia *6210* mobile
phone with in-built
WAP browser, 2000

the early 1980s, along with cordless domestic telephones. The mid-1980s
saw the introduction of transportable mobile phones, marketed – as in the
case of Vodaphone's early model, for example – both as a mobile unit for
installation in cars and a personal portable unit. Though heavy and cumbersome, these early mobiles were aimed primarily at professionals whose jobs
involved travelling or working in the field. By the early 1990s, advances in
mobile phone technology allowed the introduction of more compact models, such as the Panasonic *I-series ETACS* telephone (1993) manufactured
by **Matsushita** Communication. The increasing **miniaturization** of handsets,
combined with better and wider communications networks, led to an explosion in demand for mobile phones in the mid to late 1990s. More recently,
WAP (Wireless Application Protocol) mobile telephones have been launched which allow Internet access. Today the communications industry is on
the threshold of a whole new era, and the telephone appears to be the ideal
medium through which the cutting-edge digital technologies and "life services" of tomorrow will be delivered to users.

RCA *Kinescope* television receiver, 1936

→RCA *TT-5* television, 1939 – the world's first commercial television receiver

↘RCA *Kinescope* television receiver, 1936

HMV *Model No. 905* television and radio receiver, 1938

No other medium has brought about a greater globalization of culture or so dramatically changed the way in which people perceive the world around them than television. The concept of television was first explored in the late 19th century, both in theoretical discussions and in practical experiments. One of the most important principles of this new technology, which was subsequently adopted in all forms of television, was proposed in 1880 by both W. E. Sawyer in the United States and Maurice Leblanc in France and involved the rapid scanning of each element of a picture in succession, line by line and frame by frame. This was followed by the crucial development of a simple and effective method of mechanical image scanning that was patented in 1884 by the German engineer, Paul Gottlieb Nipkow (1860–1940). Between 1900 and 1920, several events took place that were also critical to the evolution of television, including the development of early picture tubes, the devising of methods that increased the amplification of electronic signals, and the formulation of theoretical principles for the electronic scanning of images.

Bush *TV22* television, 1950

In 1922 the Scottish engineer **John Logie Baird** began developing television equipment, which three years later transmitted the first recognizable images of human faces. In 1926, at the Royal Institution in London, Baird demonstrated the first true television system by electrically transmitting moving images. In the late 1920s **General Electric** became a pioneer of television broadcasting using technology developed at its research laboratory by Ernest Alexanderson (1878–1975). Baird similarly commenced the development of a television service for the German Post Office in 1929, while **Marconi** Electric & Musical Industries also worked on a competitive system. In 1936 the British Broadcasting Company (BBC) began broadcasting the world's first high-definition regular television service, and a year later fully adopted Marconi's EMI system in preference to Baird's. In America, **RCA**'s research laboratory, headed by the electronic engineer Vladimir Zworykin (1889–1982), demonstrated an all-electronic television in 1932. This design incorporated a cathode-ray tube and a camera tube known as an iconoscope, which Zworykin had patented in 1923 (Zworykin also developed a colour television system that was patented in 1928). In December 1936 the RCA laboratory unveiled one of the world's first practical televisions, the RCA *Kinescope*, which had a mirror mounted in its lid so that the inverted image could be viewed the right way up. In April 1939 RCA introduced the world's first commercial television receiver. Shown at the **New York World's Fair**, this landmark television was produced in four different versions – three console designs and one table-top design (which had a 5-inch screen and was known as the *RCA TT-5*) – all of which were housed in hand-crafted walnut cabinets. Although television production was interrupted by the Second World War, much dedicated

Bush *TV24* television, early 1950s

research was undertaken by the military into the manufacture of television tubes. Manufacturers after the war benefited from this research, and by the early 1950s a practical colour television system had been invented that separated black and white signals from colour signals. (It would nevertheless be many years before colour televisions became the norm.) The gradual **miniaturization** of technology enabled casings to become smaller and less obtrusive, while the size of television screens was proportionally increased. The well-known plastic-housed *TV22* television (1950s) manufactured by the British company, Bush, epitomised the "new look"

Perdio *Portarama MkII* television, 1962

television, although it was not until the mid to late 1950s that televisions began to be purchased in significant quantities in Europe. In Britain, for example, many people bought television sets for the first time specifically for the broadcast of Queen Elizabeth II's coronation in 1953. During the late 1950s, the American manufacturer **Philco**, inspired by the Russian *Sputnik* satellite, used futuristic **styling** for its television sets. Called Philco *Predicta*, these space-age designs were some of the first televisions to break with the convention of the furniture-like box cabinet. In 1960 the Japanese company **Sony** launched the world's first transistor television, the *TV8–301*, which was subsequently followed by other portable designs, such as the 8-inch *Portarama Mk II* (1962) produced by Perdio. In 1968 Sony introduced the first of its revolutionary *Trinitron* colour televisions, which were marketed on the strength of their portability – one Sony advertisement even featured a 12-inch television that weighed just 19 lbs resting on a rather large belly with the line, "Ah, at last the Tummy Trinitron". Another portable design that was extremely popular during the early 1970s was the JVC *Nivico*

Philco *Predicta*, 1958
– marketed as the
"world's first swivel
screen television"

↗Philco advertise-
ment showing
Predicta Decorator
television, early
1960s

JVC *Nivico 3240 GM*
television produced
by the Victor Com-
pany of Japan, 1970

Sony *Trinitron*
television, 1970

Sony *Trinitron*
television, early
1990s

3240 GM television (1970) produced by the Victor Company of Japan. This quintessentially Pop design (also known as the *Videosphere*) was available in several colours and rendered the television a space-age fashion accessory. Several other companies also produced spherical televisions, including a large white model designed by Arthur Bracegirdle for Keracolor in 1969, which reflected the optimistic playfulness of the period and the continuing fascination with space technology. The 1980s and early 1990s saw the adoption of a more sombre form of styling that was exemplified by Sony's sleek large-screened *Trinitron* models. One of the most interesting television designs of the 1990s was **Philippe Starck**'s *Jim Nature* portable television (1994) for Saba, which used a moulded high-density chipboard casing as an environmentally-friendly alternative to **plastics**. The majority of modern televisions still employ a high-tech style, however, as reflected by **Bang & Olufsen**'s wide-screen *BeoCenter AV5* model (1997), which also has an integrated CD player and radio. With the advent of digital and flat-screen technology, image definition has been dramatically enhanced, while at the same time the functional potential of the television continues to grow. Although the televisions of tomorrow will function as portals to

Philippe Starck, *Jim Nature* television for Saba, 1994

other digital technologies, they will continue providing access to both
entertainment and knowledge for literally billions of people around
the world.

Typewriter, 1836

← Sholes & Glidden typewriter,
c. 1873 – produced as the
Remington *No. 1*

Blickensderfer (Blick) machine,
c. 1893

Hammond typewriter, c. 1895

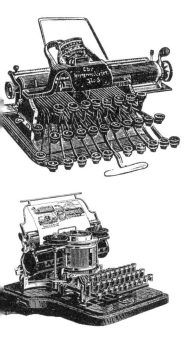

The invention of the typewriter completely revolutionized office work and the office environment during
the latter half of the 19th century. As early as the 1710s,
the inventor James Rawson had constructed a keyboard
instrument which "amazed all beholders" and which
incorporated engraved brass letters, an inked ribbon
and a spring-driven mechanism. This experimental design was followed in the early 19th century by the invention of a number of typing machines, including one of
1836 that used a rotating arm to select the chosen character, which was then depressed to print the letter. Such
primitive machines were extremely laborious to use and
certainly could not compete with handwriting in terms
of speed. The success of the typewriter depended on a
quicker and more efficient method of writing, the reproduction of characters that were comparable with traditional printers' types, and a mechanism that enabled
the paper to be moved along as each character was
written.

In 1864 the American inventor Christopher Latham
Sholes (1819–1890) patented a page-numbering machine that he had co-designed with his friend, Samuel
Soulés. At the suggestion of fellow-inventor Carlos
Glidden, who had brought to his attention a written
account of a letter-printing machine designed by John
Pratt of London, Sholes turned his device into the
first-ever practical typewriter. He subsequently focused
his efforts on the improvement of this machine and
in 1868 patented a second model with Soulés and
Glidden. In the years that followed, however, he was
unable to raise sufficient capital to fully commercialize
his design. While Sholes' "typewriter" – it was he who
first coined the term – was limited in that it could only
print capital letters, it featured the "QWERTY" keyboard

Camillo Olivetti, M1
typewriter for Olivetti,
1910–1911

layout which is still in use today. In 1873 Sholes sold his patent rights to **E. Remington** & Sons for the sum of $12,000. The *Remington No. 1* typewriter, which was based on Sholes and Glidden's design, was introduced that same year, becoming the world's first commercial typewriter. 1878 saw the launch of the *Remington No. 2*, which was the first typewriter to have a shift-key mechanism that enabled the printing of both capital and lowercase characters.

The demand for typewriters increased dramatically during the 1880s, with an article in the *Phonetic Journal* declaring in 1887 that, in America, "typewriting has become quite fashionable even among the upper classes". Typing came to be seen as a suitable career for young ladies, and teachers of touch-typing strongly urged for the retention of Sholes' now familiar "QWERTY" keyboard over the "ideal" character layout devised by J. B. Hammond, despite demonstrations that the latter was easier to use. The 1890s saw the introduction of several "improved" machines featuring different inking and type mechanisms, including the Blickensderfer machine

→Olivetti poster
advertising the *MP1*
typewriter designed
by Aldo Magnelli in
1932

Olivetti poster
advertising the
Studio 42 typewriter
designed by Xanti
Schawinsky, Luigi
Figini and Gino
Pollini in 1935

↗Marcello Nizzoli,
Lettera 22 typewriter
for Olivetti, 1950

Marcello Nizzoli,
Diaspron typewriter
for Olivetti, 1959

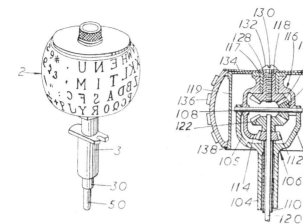

British patent drawing for IBM "golfball" typewriter, 1960 – patent first filed in America in 1955

↓Eliot Fette Noyes, *Selectric I* "golfball" typewriter for IBM, 1961

Ettore Sottsass &
Perry King, *Valentine*
typewriter for Olivetti,
1969

Ettore Sottsass &
Perry King, *Valentine*
typewriter for Olivetti,
1969

↖Ettore Sottsass &
Hans von Klier, *Praxis
48* typewriter for
Olivetti, 1964

←Ettore Sottsass &
Hans von Klier, *Editor
4* typewriter for
Olivetti, 1969

(c. 1893), which employed a type drum and an ink pad, and the Hammond typewriter (c. 1895), which had a ring-like type element positioned between the spools of inked ribbon. In 1908 Camillo **Olivetti** established the first Italian typewriter factory and three years later introduced the robust and elegant *MI*, which was celebrated for its speed and operative smoothness. Other similar machines began to be marketed widely during that era. Although **Thomas Alva Edison** had been granted a patent for the design of an electric typewriter as early as 1871, it was not until 1920 that James Smathers developed the first practical electric model. Five years later, Remington launched its first all-electric typewriter, but it proved an unreliable design and only served to ensure the continued popularity of manual models over the succeeding decades; models such as the stylish and portable *Lettera 22* (1950) designed by **Marcello Nizzoli** for Olivetti. After the Second World War several better-performing electric models made an appearance, but it was not until 1961 that the electric typewriter market was utterly transformed with the introduction of the hugely successful **IBM** *Selectric I*. This revolutionary typewriter, which was designed by **Eliot Fette Noyes**, incorporated a spherical "golfball" typing head and a static carriage that meant that the typewriter took up less space than conventional models and vibrated less, thereby reducing noise.

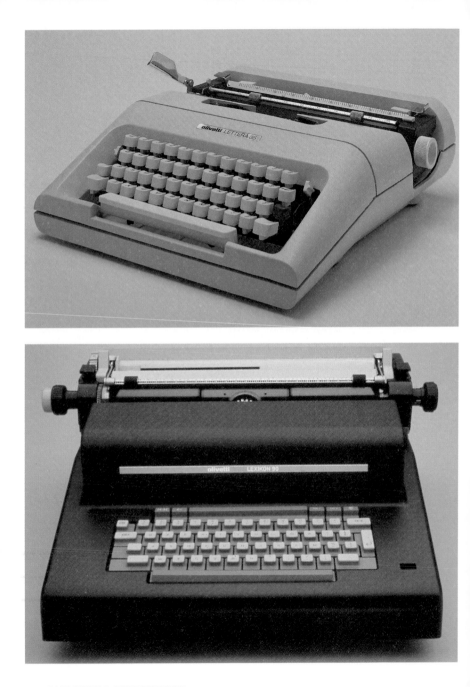

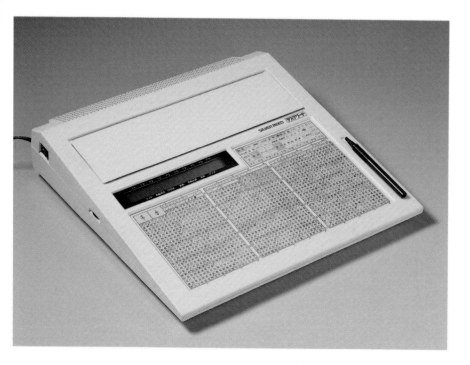

Silver Reed Japanese-
language electric
typewriter, 1985

↖ Mario Bellini,
Lettera 35 typewriter
for Olivetti, 1974

← Ettore Sottsass &
A. Leclerc, *Lexicon 90*
typewriter, 1975

In contrast to the office-style aesthetic projected by typewriters such as
the *Selectric I*, Marcello Nizzoli's *Diaspron* (1959) and **Ettore Sottsass** and
Hans von Klier's *Praxis 48* (1964), the famous bright red *Valentine* portable
typewriter (1969) designed by Sottsass and Perry King became an iconic
Pop accessory. This relatively diminutive machine not only reflected the
1960s spirit of fun and freedom but also testified to the increasing use of
plastics in typewriter construction. The typewriters of the 1970s and early
1980s became more visually unified, and electric models – such as the
Japanese-language typewriter produced by Silver Reed in 1985 – grew increasingly popular. The advent of the personal computer in the mid-1980s,
however, saw a massive shift away from typing towards word-processing,
and from the early 1990s onwards lap-top computers such as Olivetti's
Philos (1993) and **Apple Computer**'s candy-coloured *iBook* (1999) offered
better and more powerful alternatives to the portable typewriter. The digital
age has rendered the traditional typewriter all but obsolete, and whether its
keyboard system will be completely replaced in the future by voice recognition technology only time will tell. The typewriter has nevertheless played a
significant role over the last 100 years in the structural organization of the

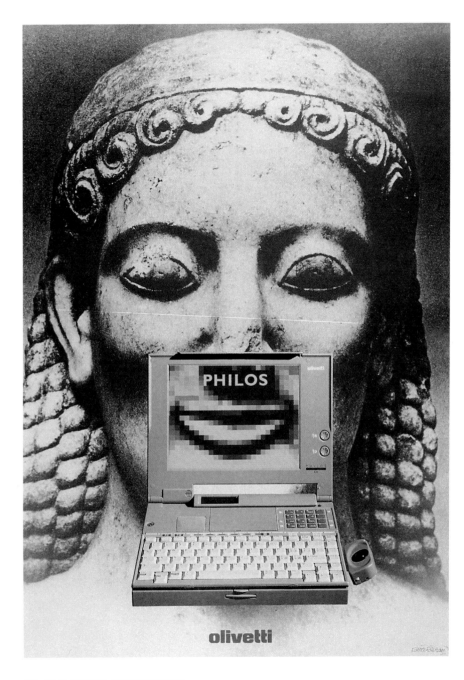

Jonathan Ive, *iBook*
for Apple Computer,
1999

←Olivetti poster
advertising the *Philos*
laptop computer de-
signed by Ettore
Sottsass in 1993

office as well as in the emancipation of women, for whom it offered an important means of self-sufficiency.

Hubert Cecil Booth's *Puffing Billy* cleaning machine (invented 1901) used on a mobile unit by the Vacuum Cleaner Company of Victoria Street, London

CASE STUDY 8

THE VACUUM CLEANER

Prior to the invention of the vacuum cleaner the task of carpet cleaning was an extremely frustrating exercise, for no matter how much brushing and sweeping was undertaken, the dust just tended to settle somewhere else in the house. The only answer to this problem was for the carpet to be taken outdoors and shaken and beaten vigorously, although this was not always a practical solution – especially if the carpet was of a large size. In the 1840s, the British mechanical engineer Joseph Whitworth (1803–1887) invented a street-sweeping machine with brushes set on a revolving drum. This subsequently inspired the development of the first domestic carpet sweeper by the American inventor Melville Reuben Bissell (1843–1889), patented in 1876. Bissel's *Grand Rapids* carpet sweeper, which featured a central rotary brush, enjoyed significant commercial success, but was eventually overtaken – as were other machines like it – by better performing and more efficient suction-based cleaners.

The first successful vacuum cleaner was designed in 1901 (and eventually named) by the British engineer Hubert Cecil Booth (1871–1955), and came to be known as the *Puffing Billy*. This revolutionary machine had a petrol-driven 5-horsepower pump, but was too large to be taken into most domestic interiors. It was therefore mounted on a vehicle and parked in the street

Hoover advertise-ment, 1909

Try this Electric Suction Sweeper 10 Days Free

THIS LITTLE MACHINE will take up all the dust and dirt from carpets, furniture and portières as perfectly as many of the more expensive vacuum cleaners.

We send it on 10 days' free trial.

At a cost of **less than one cent,** you can thoroughly clean any room. Simply attach the wire to an electric light socket, turn on the current and run it over the carpet. A rapidly revolving brush loosens the dust which is sucked back into the dirt bag.

There are attachments for cleaning curtains, portières and pictures, without removing them from the walls. Nothing need be disturbed. Anyone can operate it.

This machine is substantially made —will last a lifetime. Repairs and adjustments are never necessary.

Try this machine for 10 days. It will cost you nothing. We pay all

express charges. If, after you have used the Electric Suction Sweeper for 10 days, and are not satisfied that it is worth the price we ask, return it to us at once at our expense.

Orders for machines to be sent on trial will be filled in the order received. Do not delay. Write today for full information about the free trial plan and booklet, "Modern Sweeping by Electricity."

ELECTRIC SUCTION SWEEPER CO., Dept. 11, New Berlin, Ohio

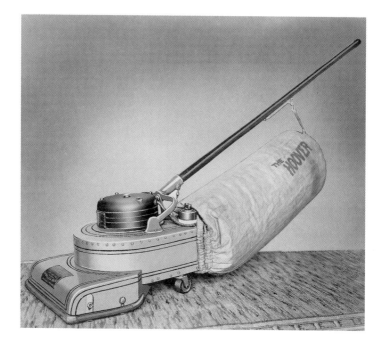

while a suction hose was extended into the building to be cleaned. Booth's machine was so noisy when it was operated that he was sued for frightening horses, but when he lodged an appeal, the Lord Chief Justice upheld his right to operate the vacuum cleaner in the streets. A similar machine was sold in America, and the early 1900s saw the launch of other, smaller domestic-use models whose vacuum effect was produced by means of bellows-like devices, as in the hand-operated *Baby Daisy* of c. 1908 and the *Star* vacuum cleaner of 1911.

The first practical electrically-powered vacuum cleaner was developed in 1907 in the United States by James Murray Spangler (1848–1915). This primitive wood and tin design incorporated a broom handle and a pillow-case-bag together with a rotating brush and an electric motor coupled to a fan. Spangler, who was a janitor, did not have sufficient money to commercialize his invention, but succeeded in persuading his friend, William "Boss" **Hoover**, to fund its development. In 1908 the Hoover Suction-Sweeper Company was founded and began producing a redesigned version of Spangler's cleaner featuring an **aluminium** casing. From 1909 a new, lighter motor was used for Hoover vacuum cleaners that reduced their weight to some 5–6 lbs and so heralded the age of "Modern Sweep-

Electrolux cylinder
vacuum cleaner,
1926 – "so simple
that even a child
can use it"

ing by Electricity". Within a few years, other manufacturers such as Goblin, Magnet and Universal were producing similar devices, but Hoover continued to dominate the market it had almost single-handedly established. In 1915 **Electrolux** introduced the first of its revolutionary cylinder vacuum cleaners, which came with a range of tools that could be attached to a flexible hose. In 1926 Hoover added a rotating "Agitator" bar with spiralling bristles to its machines, which produced a beating action that assisted in the loosening of dirt from carpet fibres. That same year, Electrolux introduced a cylinder vacuum cleaner which was marketed on its ease of use and the spurious claim that "it disinfects as it cleans".

The popularity of the vacuum cleaner rested on the principle of hygiene and the belief that the elimination of dust meant the elimination of germs and thereby the elimination of disease. As a commentator in the *Electrician* magazine noted in 1927: "The universal ideal behind the vacuum cleaner ... is much larger, much more far reaching than asking the housewife to exchange a hand broom for a power sweeper. It is asking her to accept a higher, safer, more desirable health standard for her entire family." During the 1930s and 1940s, vacuum cleaners were given streamlined forms to reflect this concept of "cleanliness", as demonstrated by the Hoover *Model 150* designed by **Henry Dreyfuss** in 1936 and the Electrolux *Model 248* de-

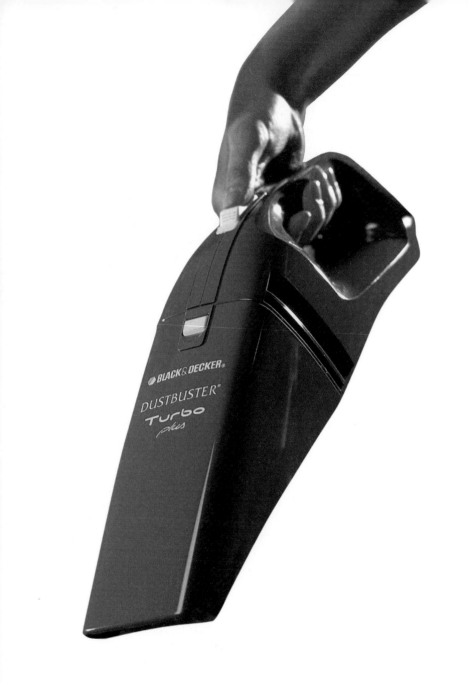

Electrolux *Oxygen* vacuum cleaner, 1999 – demonstrating the increasingly sculptural styling being used for such appliances

signed by **Sixten Sason** in 1943. Although vacuum cleaners became lighter in weight with the adoption of plastic casings, the following decades brought few real technical advances. The 1960s and 1970s saw the introduction of various hand-held models, such as the *Hoover Dustette* and **Black & Decker**'s highly successful *Dustbuster* (1979), which was the first hand-held cordless model.

During the 1980s and 1990s **styling** rather than technological innovation played an increasingly important role in the design of vacuum cleaners, particularly as it offered manufacturers a relatively inexpensive means of differentiating their products from those of their competitors. The notable exception to this trend was **James Dyson**'s revolutionary bagless cleaner, which was based on his innovative "dual cyclone" system which used centrifugal force to provide a high level of continuous suction. Launched in 1993, Dyson's high-performance *DC01* and subsequent *DC02* cylinder model (1995) took the vacuum-cleaner industry – which for decades had progressed little beyond Spangler's original fan-and-bag concept – by storm. If the handsome profits generated by replacement bag sales perhaps delayed the evolution of the vacuum cleaner in the past, the consumer can only benefit from the latest technological developments, particularly in view of modern

←Black & Decker *Dustbuster*, 1990s – first version designed in 1979

→ Dyson *DC03* dual-
cyclonic vacuum
cleaner, 1998 – em-
ploys a "hepa" filter
(high efficiency partic-
ulate air filter) to give
a higher level of filtra-
tion and is therefore
recommended for
people with allergies
and asthma.

Sanyo *Model SC845*
vacuum cleaner, 1999
– showing the use of
coatings to produce a
metallic-like finish for
plastic housings

awareness about the importance of hygiene and the fact that house dust,
which commonly comprises bacteria, animal and human hair, dust mites,
pollen and dead skin tissue, can be extremely injurious to health, particularly
for asthmatics and children.

The laundering of clothes was one of the most tiresome household chores prior to the invention of the washing machine. On "wash day", a great deal of physical exertion was required to get clothes really clean – typically by agitating them with a "dolly" (a pounder with short stool-like legs), scrubbing them on a wash board (introduced c. 1860) and wringing them with a mangle. It is not surprising that, by the 19th century, many inventors were looking for a means of harnessing either water power or **steam power** in order to alleviate the drudgery of washing clothes, especially in commercial laundries. In America, patents for human-powered washing machines employing swing-like "agitating" elements were granted to K. Hinckley in 1831 and E. D. Wilson in 1846. While these designs remained somewhat primitive, a number of more practical designs were shown at the **Great Exhibition** of 1851, including a steam-driven tub machine designed by a Mr MacAlpine. Between 1851 and 1871, around 2,000 patents were granted in Britain and America for a variety of washing appliances, which had to be filled with heated water and used either rotating drums, gyrators or dollies to agitate the clothes. The majority of these machines needed to be hand-cranked and clothes tended to get tangled around the rotating dolly elements. A machine devised by S. S. Shipley in 1855 incorporated an outer tub which used water heated by gas burners, while another invented in 1860 used an oscillating movement so as to prevent the tangling of clothes. Another machine manufactured in Manchester by Chatterton & Bennett, which was retailed as the *Float Washer* in the Montgomery Ward mail-order catalogue, used an action to replicate "exactly the knuckles of a vigorous washerwoman". A more gentle action was developed by Mr Bradford of Slatram for his *Vowel* machines designed around 1880, which utilized a drum element that was cranked by hand. It was not until the first decade of the 1900s that electric motors were incorporated into the design of

Bradford's *Vowel* washing machine, c. 1880 – with a handcranked drum

→Wooden domestic washing machine, c. 1920 – with a dolly-style agitator

Washer Company advertisement, c. 1900

Just Six Minutes to Wash a Tubful!

This is the grandest Washer the world has ever known. So easy to run that it's almost fun to work it. Makes clothes spotlessly clean in double-quick time. Six minutes finishes a tubful.

Any Woman Can Have a **1900 Gravity Washer on 30 Days' Free Trial**

Don't send money. If you are responsible, you can try it first. Let us pay the freight. See the wonders it performs. Thousands being used. Every user delighted. They write us bushels of letters telling how it saves work and worry. Sold on little payments. Write for fascinating Free Book today. All correspondence should be addressed to **1900 Washer Co., 1702 Court St., Binghamton, N. Y.** If you live in Canada, address Canadian 1900 Washer Co., 357 Yonge St., Toronto, Canada.

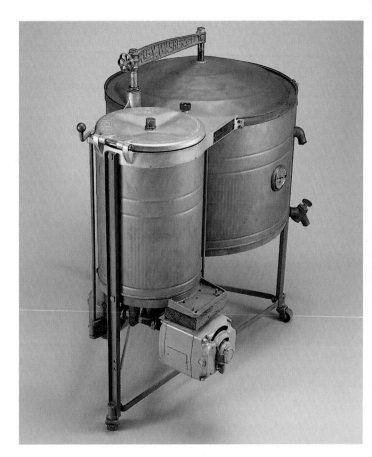

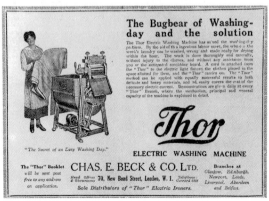

The Bugbear of Washing-
day and the solution

The Thor Electric Washing Machine has solved the washing-day problem. By the aid of this ingenious labour saver, the whole of the week's laundry can be washed, wrung and made ready for drying within the hour. The work is done thoroughly and carefully, without injury to the clothes, and without any assistance from you or the antiquated scrubbing board. A cord is attached from the "Thor" to the electric light fixture, the clothes placed in the space allotted for them, and the "Thor" carries on. The "Thor" method can be applied with equally successful results to both delicate and heavy materials, and 1d. easily covers the cost of the necessary electric current. Demonstrations are given daily at every "Thor" Branch, where the mechanism, principal and general capacity of the machine is explained in detail.

Thor

ELECTRIC WASHING MACHINE

"The Secret of an Easy Washing Day."

The "Thor" Booklet
will be sent post
free to any address
on application.

CHAS. E. BECK & CO. LTD.

Head Offices
& Showrooms 70, New Bond Street, London, W.1. Telephone Gerrard 6488

Sole Distributors of "Thor" Electric Ironers.

Branches at
Glasgow, Edinburgh,
Newport, Leeds,
Liverpool, Aberdeen
and Belfast.

washing machines, one example being that designed by the American inventor A. J. Fisher in 1909. Manual systems nevertheless predominated until well into the 1920s, no doubt owing to their greater affordability. Even in more affluent households, many people continued to send their dirty clothes to expensive commercial laundries or relied on household servants to do their laundering for them. With the economic downturn of the Great Depression in the 1930s, however, and the greater availability of **electric power**, the popularity of domestic washing machines grew rapidly, and electrically-powered machines such as the 1932 twin-tub model manufactured by Riby, became more common. Even so, washing machines in most households still had to have water heated separately for them, since the problem of hot water supply was not satisfactorily resolved until the end of the Second World War. During the post-war years, large appliance companies such as **Westinghouse**, **Hoover** and **Sanyo** produced sleek top-loading models that incorporated mangle-like wringers. Around the same time, automatic controls were improved, which meant that machines required less supervision. Although top-loading models remained and continue to remain the norm in North America, space-saving front-loading models, such as the Bendix *DRS* washing machine (c. 1961) became increasingly popular in Europe. With generally less space available than in the average North American home, it is not uncommon in Europe for the washing machine to be sited in the kitchen, where

Sanyo top-loading washing machine, 1955 – with an automatic control dial and a hinged lid that could be closed while in operation

↖Hoover top-loading washing machine, 1948 – it used piped hot water and had a mangle-like wringer that could be "folded down" when not in use

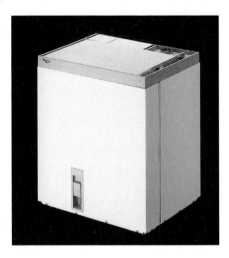

it is more likely to be seeen. For this reason, from the 1960s to present day, **styling** has played a very important role in the design of washing machines in Europe. Roberto Pezzetta's prototype *Zöe* washing machine (1992) for **Zanussi** predicts how expressive the styling of European appliances may become in the near future. Since the 1980s, there has also been a strong trend in Europe towards the design of much more energy-efficient machines, with companies like Zanussi developing appliances that adjust their power consumption according to the size of the load. New generations of machines are currently being developed in Europe that will be even more energy-efficient and will also eliminate the need for ecologically harmful detergents.

Roberto Pezzetta, *Zöe* washing machine for Zanussi, 1992 – a characterful appliance with expressive styling

↖Gastone Zanello and Zanussi Design Center, compact washing machine with 5 kg capacity for Zanussi, 1966 – received a Compasso d'Oro in 1967

←Bendix *DRS* front-loading washing machine, c. 1961

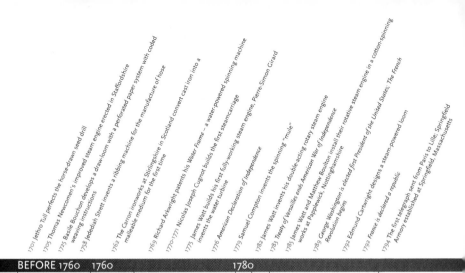

1701 Jethro Tull perfects the horse-drawn seed drill

1705 Thomas Newcomen's improved steam engine erected in Staffordshire

1725 Basile Bouchon develops a draw-loom with a perforated paper system with coded weaving instructions

1758 Jedediah Strett invents a ribbing machine for the manufacture of hose

1762 The Carron ironworks at Stirlingshire in Scotland convert cast iron into a malleable medium for the first time

1769 Richard Arkwright patents his Water Frame – a water-powered spinning machine

1770–1771 Nicolas Joseph Cugnot builds the first steamcarriage

1775 James Watt builds his first fully-working steam engine, Pierre-Simon Girard invents the water turbine

1776 American Declaration of Independence

1779 Samuel Compton invents the spinning "mule"

1782 James Watt invents his double-acting rotary steam engine

1783 Treaty of Versailles ends American War of Independence

1785 James Watt and Matthew Boulton install their rotative steam engine in a cotton-spinning works at Papplewick, Nottinghamshire

1789 George Washington is elected first President of the United States; The French Revolution begins

1792 Edmund Cartwright designs a steam-powered loom

1792 France is declared a republic

1794 The first telegraph sent from Paris to Lille; Springfield Armory established at Springfield, Massachusetts

BEFORE 1760	1760	1780

1759
Josiah Wedgwood
establishs his
own pottery

1779
Abraham Darby III constructs
the first-ever cast-iron bridge
in Coalbrookdale, England

1793
Eli Whitney designs the
cotton gin

1698
Thomas Savery
patents the first
steam engine

1781
James Watt designs
"Sun-and-Plant" wheel
for his steamengine

1764
James Hargreaves
develops the Spinning
Jenny

1783
Joseph-Michel and
Jacques-Étienne
Montgolfier
demonstrate their
hot-air balloon

1800 Richard Trevithick constructs the first light pressure steam-engine;
Eli Whitney develops a musket with interchangeable components
1801 Robert Fulton develops the first submarine, "Nautilus"
1803 Henry Shrapnel invents the first munitions shell in England
1810 Krupp factory established at Essen, Germany for the manufacture of arms
1811 John Blenkinsop patents a steam locomotive utilizing tooth-rack rail propulsion
1812 Philippe Girard develops a machine for spinning flax
1814 George Stephenson builds the first practical steam locomotive at near Newcastle
1815 Humphrey Davy designs the first safety lamp for miners
1822 First iron railway bridge constructed by George Stephenson for the Stockton-to-Darlington line
1829 William Burt granted the first US patent for a typewriter
1830 Edwin Budding designs first lawn-mower
1832 Charles Babbage builds his Difference Engine (an early precursor of the computer);
Isambard Kingdom Brunel appointed chief engineer of the Great Western Railway
1834 Cyrus Hall McCormick patents his revolutionary mechanical reaper
1837 Launching of Isambard Kingdom Brunel's ship, the *Great Western*; John Deere designs
the first commercially successful self-scouring steel plough
1839 Charles Goodyear develops a rubber "vulcanization" process

1800 **1820** **1840**

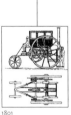

1801
Richard Trevithick
constructs a steam-
powered carriage

1819
Michael Thonet
establishes a furniture
workshop in Boppard
am Rhein, Germany

1832
Samuel Morse starts
to develop an electric
telegraph

1844
Morse's telegraph is
transmitted for the first time
between Baltimore and
Washington

1821
Michael Faraday builds
his first dynamo
(electric generator)

1835
Samuel Colt designs a revolver with
interchangable components

1805
Joseph-Marie Jacquard
invents a mechanised
loom

1829
George Stephenson's landmark engine,
The Rocket is awarded a prize of £500 in
the Rainhill Trials

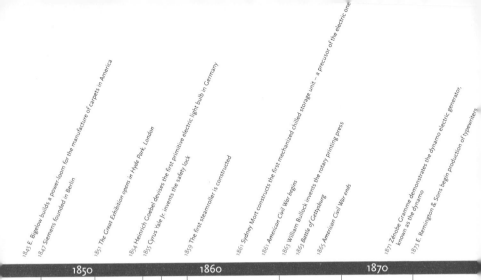

1845 E. Bigelow builds a power-loom for the manufacture of carpets in America

1847 Siemens founded in Berlin

1851 The Great Exhibition opens in Hyde Park, London

1854 Heinrich Goebel devises the first primitive electric light bulb in Germany

1855 Cyrus Yale Jr. invents the safety lock

1859 The first steamroller is constructed

1861 Sydney Mort constructs the first mechanized chilled storage unit – a precusor of the electric one

1861 American Civil War begins

1865 William Bullock invents the rotary printing press

1863 Battle of Gettysburg

1865 American Civil War ends

1871 Zénobe Gramme demonstrates the dynamo electric generator, known as the dynamo

1873 E. Remington & Sons begin production of typewriters

1850   1860   1870

1851
Isaac Singer patents his
improved sewing machine

1858
Launching of Isambard Kingdom
Brunel's ship, the *Great Eastern*

1859
Michael Thonet
designs the
Model No. 14 chair

1862
Richard Gatling
patented multi-barrel
machine gun

1873
C. L. Shoales and C. Glidden
develop their improved
typewriter

1855
General locomotive built by Rogers for
Western & Atlantic Railroad

1846
Elias Howe patents the first
practical sewing machine

1867
Nikolaus Otto and Eugen
Langen design their first
internal combustion engine

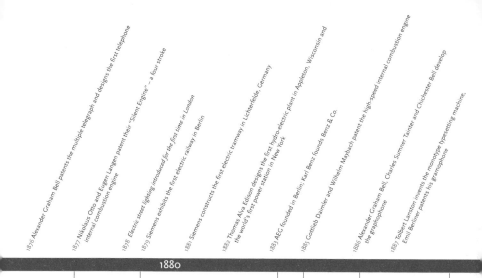

1876 Alexander Graham Bell patents the multiple telegraph and designs the first telephone

1877 Nikolaus Otto and Eugen Langen patent their "Silent Engine" – a four stroke internal combustion engine

1878 Electric street lighting introduced for the first time in London

1879 Siemens exhibits the first electric railway in Berlin

1881 Siemens constructs the first electric tramway in Lichterfelde, Germany

1882 Thomas Alva Edison designs the first hydro-electric plant in Appleton, Wisconsin and the world's first power station in New York

1883 AEG founded in Berlin; Karl Benz founds Benz & Co.

1885 Gottlieb Daimler and Wilhelm Maybach patent the high-speed internal combustion engine

1886 Alexander Graham Bell, Charles Sumner Tainter and Chichester Bell develop the graphophone

1887 Tolbert Lanston invents the monotype typesetting machine; Emil Berliner patents his gramophone

1879
Thomas Alva Edison
designs the incandescent
electric light

1884
Lewis Edison Waterman
patents his improved
fountain pen

1886
Karl Benz launches the
three-wheeled *Motorwagen* –
the world's first practical car
powered by an internal
combustion engine

1888
John Kemp Starley
designs his Rover
"Safety" bicycle

1877
Thomas Alva Edison designs his
phonograph

1885
George Eastman starts
manufacturing coated
photographic paper

1888
John Boyd Dunlop
develops the first
pneumatic tyre

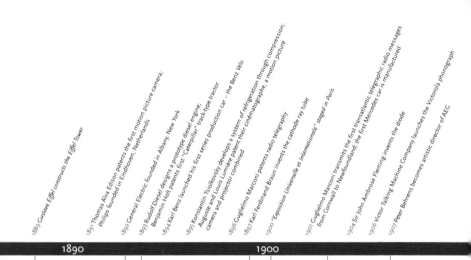

1890   1900

1889 Gustave Eiffel constructs the Eiffel Tower

1891 Thomas Alva Edison patents the first motion picture camera; Philips founded in Eindhoven, Netherlands

1892 General Electric founded in Albany, New York

1893 Rudolf Diesel designs a prototype diesel engine; Benjamin Holt patents first "Caterpillar" track-type tractor

1894 Karl Benz launched his first series production car – the Benz Velo

1895 Konstantin Tsiolkovsky develops a system of refrigeration through compression; Auguste and Louis Lumière patent their cinématographe, a motion picture camera and projector combined

1896 Guglielmo Marconi patents radio telegraphy

1897 Karl Ferdinand Braun invents the cathode ray tube

1900 "Exposition Universelle et Internationale" staged in Paris

1901 Guglielmo Marconi transmits the first transatlantic telegraphic radio messages from Cornwall to Newfoundland; the first Mercedes car is manufactured

1904 Sir John Ambrose Fleming invents the diode

1906 Victor Talking Machine Company launches the Victorola phonograph

1907 Peter Behrens becomes artistic director of AEG

1889
Linotype machine introduced for composing type

1892
Lingner Werke designs the Odol bottle

1892
Sir James Dewar invents the first vacuum flask

1893
Sears, Roebuck & Co. is founded in Chicago and publishes its first mail-order catalogue

1898
Count Ferdinand von Zeppelin constructs his first airship

1900
George Eastman introduces the Kodak Brownie camera

1903
The Wright Brothers succeed in flying the first powered aircraft, the Flyer No. 3

1907
Deutscher Werkbund founded in Munich

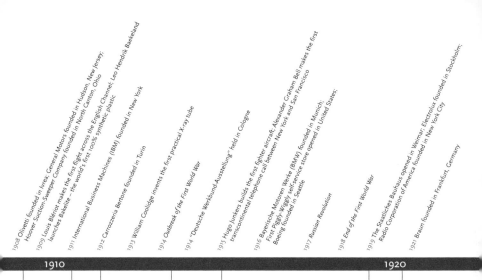

1908 Olivetti founded in Ivrea; General Motors founded in Hudson, New Jersey; Hoover Suction-Sweeper Company founded in North Canton, Ohio

1909 Louis Blériot makes the first flight across the English Channel; Leo Hendrik Baekeland launches Bakelite - the world's first 100% synthetic plastic

1911 International Business Machines (IBM) founded in New York

1912 Carrozzeria Bertone founded in Turin

1913 William Coolidge invents the first practical X-ray tube

1914 Outbreak of the First World War

1914 "Deutsche Werkbund-Ausstellung" held in Cologne

1915 Hugo Junkers builds the first fighter aircraft; Alexander Graham Bell makes the first transcontinental telephone call between New York and San Francisco

1916 Bayerische Motoren Werke (BMW) founded in Munich; First Piggly-Wiggly self-service store opened in United States; Boeing founded in Seattle

1917 Russian Revolution

1918 End of the First World War

1919 The Staatliches Bauhaus opened in Weimar; Electrolux founded in Stockholm; Radio Corporation of America founded in New York City

1921 Braun founded in Frankfurt, Germany

1910　　　　　　　　　　　　　　　　　　　**1920**

1912
Pyrex developed by
Corning Glass Works

1915
Major Wilson and Sir William
Tritton design the *Big Willie* tank

1920
Snap-on Wrench Company
founded in Milwaukee

1911
International Business
Machine (IBM) founded
in New York

1914
Bell Telephone
Company introduces
the candlestick
telephone

1919
Citroën launches its *Type A*
automobile

1908
The Ford Motor Company
produces the first *Model T* car

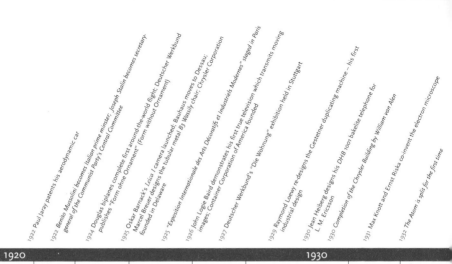

1922 Paul Jaray patents his aerodynamic car

1922 Benito Mussolini becomes Italian prime minister; Joseph Stalin becomes secretary-general of the Communist Party's Central Committee

1924 Douglas biplanes complete first around-the-world flight; Deutscher Werkbund publishes "Form ohne Ornament" (Form without Ornament)

1925 Oskar Barnack's *Leica I* camera launched; Bauhaus moves to Dessau; Marcel Breuer designs the tubular metal *B3 Wassily chair*; Chrysler Corporation founded in Delaware

1925 "Exposition Internationale des Arts Décoratifs et Industriels Modernes" staged in Paris

1926 John Logie Baird demonstrates his first true television which transmits moving images; Container Corporation of America founded

1927 Deutscher Werkbund's "Die Wohnung" exhibition held in Stuttgart

1929 Raymond Loewy re-designs the Gestetner duplicating machine – his first industrial design

1930 Jean Heiberg designs his DHB 1001 bakelite telephone for L. M. Ericsson

1930 Completion of the Chrysler Building by William von Alen

1931 Max Knott and Ernst Riska co-invent the electron microscope

1932 The Atom is split for the first time

1920

1930

1922
Gustav Dalén
designs the *Aga* cooker

1922
Austin *Seven* launched

1925
Bang & Olufsen
founded in Quistrup,
Denmark

1926
Mart Stam constructs the
first cantilevered metal chair

1928
Walter Maria Kersting
develops the first version
of his *Volksempfänger* radio

1931
Hermann Gretsch designs
his *1382* service for Arzberg

1932
Aino Marsio-Aalto designs
pitcher and glasses for Karhula

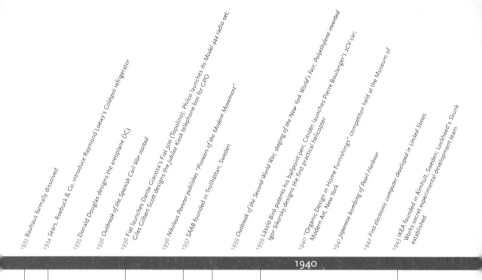

1933 Bauhaus formally dissolved

1934 Sears, Roebuck & Co. introduce Raymond Loewy's Coldspot refrigerator

1935 Donald Douglas designs the aeroplane DC3

1936 Outbreak of the Spanish Civil War started

1936 Fiat launches Dante Giacosa's Fiat 500 (Topolino); Philco launches its Model 444 radio set;
Giles Gilbert Scott designs the Jubilee Kiosk telephone box for GPO

1936 Nikolaus Pevsner publishes "Pioneers of the Modern Movement"

1937 SAAB founded in Trollhättan, Sweden

1939 Outbreak of the Second World War; staging of the New York World's Fair; Polyethylene invented

1939 László Bíró patents his ballpoint pen; Citroën launches Pierre Boulanger's 2CV car;
Igor Sikorsky designs the first practical helicopter

1940 "Organic Design in Home Furnishings" competition held at the Museum of
Modern Art, New York

1941 Japanese bombing of Pearl Harbour

1942 First electronic computer developed in United States

1943 IKEA founded in Älmhult, Sweden; Lockheed's Skunk
Works secret experimental development team
established

1940

1934
Carl Breer's *Airflow* car for
Chrysler is launched

1939
Chester Carlson produces first
xerographic image with his
prototype photocopier

1943
Sixten Sason designs his bullet-
shaped *Model 248* vacuum cleaner
for Electrolux

1936
Reginald Mitchell designs
a prototype of the *Spitfire*
monoplane

1938
Volkswagen *Beetle*
launched

1933
Wilhelm Kage designs
his *Praktika* tableware
for Gustavsberg

1937
Raymond Loewy designed
his S-1 locomotive for
Pennsylvania Railroad

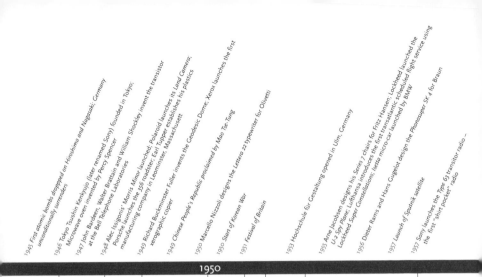

1945 First atomic bombs dropped on Hiroshima and Nagasaki; Germany unconditionally surrenders

1946 Tokyo Tsushin Kenkyujo (later renamed Sony) founded in Tokyo; Microwave oven invented by Percy Spencer

1947 John Bardeen, Walter Brattain and William Shockley invent the transistor at the Bell Telephone Laboratories

1948 Alec Issigonis' Morris Minor launched; Polaroid launches its Land Camera; Porsche launches the 356 roadster; Earl Tupper establishes his plastics manufacturing company in Leominster, Massachusett

1949 Richard Buckminster Fuller invents the Geodesic Dome; Xerox launches the first xerographic copier

1950 Chinese People's Republic proclaimed by Mao Tse-Tung

1950 Marcello Nizzoli designs the Lettera 22 typewriter for Olivetti

1950 Start of Korean War

1951 Festival of Britain

1953 Hochschule für Gestaltung opened in Ulm, Germany

1955 Arne Jacobsen designs his Series 7 chairs for Fritz Hansen; Lockheed launched the U-2 Spy Plane; Lufthansa introduces the first transatlantic scheduled flight service using Lockheed Super Constellations; Isetta micro-car launched by BMW

1956 Dieter Rams and Hans Gugelot design the Phonosuper SK 4 for Braun

1957 Launch of Sputnik satellite

1957 Sony launches the Type 63 transistor radio – the first "shirt pocket" radio

1950

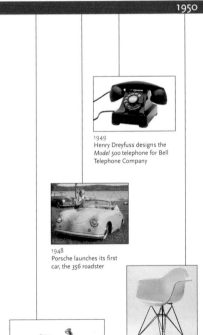

1949
Henry Dreyfuss designs the *Model 500* telephone for Bell Telephone Company

1954
Douglas Scott designs his *Routemaster* bus for London Transport

1957
Max Bill designs a wall clock for Junghans

1948
Porsche launches its first car, the 356 roadster

1952
Boeing begins production of the *B52 Stratofortress*

1951
Tetra Pak founded in Lund, Sweden and first Tetra Pak packaging launched

1946
Piaggio launches the *Vespa* scooter

1950
Charles & Ray Eames' fibreglass shell chair put into production

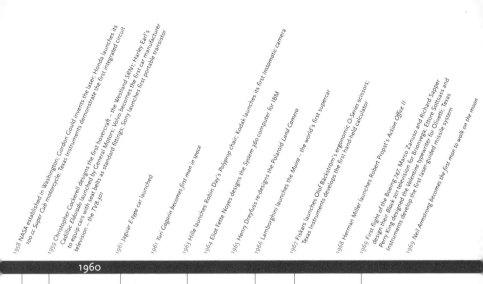

1958 NASA established in Washington; Gordon Gould invents the laser; Honda launches its 100 cc Super Cub motorcycle; Texas Instruments demonstrate the first integrated circuit

1959 Christopher Cockerell designs the first hovercraft – the Westland SRN1; Harley Earl's Cadillac Eldorado launched by General Motors; Volvo becomes the first car manufacturer to equip cars with seat belts as standard fittings; Sony launches first portable transistor television – the TV8-301

1961 Jaguar E-type car launched

1961 Yuri Gagarin becomes first man in space

1963 Hille launches Robin Day's Polyprop chair; Kodak launches its first Instamatic camera

1964 Eliot Fette Noyes designs the System 360 computer for IBM

1965 Henry Dreyfuss re-designs the Polaroid Land Camera

1966 Lamborghini launches the Miura – the world's first supercar

1967 Fiskars launches Olof Bäckström's ergonomic O-Series scissors; Texas Instruments develops the first hand-held calculator

1968 Herman Miller launches Robert Propst's Action Office II

1969 First flight of the Boeing 747; Marco Zanuso and Richard Sapper design their Black 201 television for Brionvega; Ettore Sottsass and Perry King designed the Valentine typewriter for Olivetti; Texas Instruments develop the first laser-guided missile system

1969 Neil Armstrong becomes the first man to walk on the moon

1960

1969
First flight of Concorde –
the world's first supersonic
commercial airliner

1961
Eliot Fette Noyes designs
Selectric "golfball"
typewriter for IBM

1963
Porsche launches the 911 car

1959
Alec Issigonis' Morris Mini
launched

1967
Stelton introduces Arne Jacobsen's Cylinda-Line
stainless-steel hollow-ware range

1966
Marco Zanuso and Richard Sapper
design their Grillo telephone for
Siemens

1958
Ettore Sottsass designs the
Elea 9003 main frame computer
for Olivetti

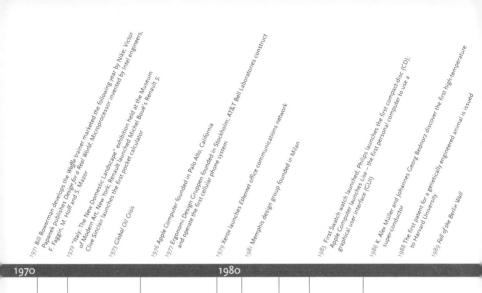

1971 Bill Bowerman develops the *Waffle* trainer marketed the following year by Nike; Victor Papanek publishes *Design for a Real World*; Microprocessor invented by Intel engineers, F. Faggin, M. Hoff and S. Mazor

1972 "Italy: The New Domestic Landscape" exhibition held at the Museum of Modern Art, New York; Renault launched Michel Boué's Renault 5; Clive Sinclair launches the first pocket calculator

1973 Global Oil Crisis

1976 Apple Computer founded in Palo Alto, California

1977 Ergonomi Design Gruppen founded in Stockholm; AT&T Bell Laboratories construct and operate the first cellular phone system

1979 Xerox launches Ethernet office communications network

1981 Memphis design group founded in Milan

1983 First Swatch watch launched; Philips launches the first compact-disc (CD); Apple Computer launches *Lisa* – the first personal computer to use a graphical user interface (GUI)

1986 K. Alex Müller and Johannes Georg Bednorz discover the first high-temperature super-conductor

1988 The first patent for a genetically engineered animal is issued to Harvard University

1989 *Fall of the Berlin Wall*

1970 | 1980

1974
Volkswagen launches Giorgetto Giugiaro's *Golf Mk I*

1982
Sony launches the *Watchman* micro-television

1972
Jakob Jensen designs the *Beogram 4000* for Bang & Olufsen

1979
Sony launches the first *Walkman* personal stereo

1981
NASA launches the first Space Shuttle

1984
Apple Computer launches the *Macintosh I* personal computer

1971
Lamborghini *Countach* launched

1980
Nikon launches Giorgetto Giugiaro's *F3* camera

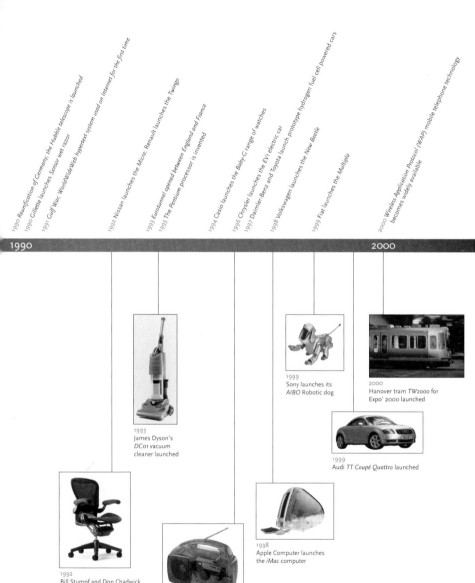

1990 *Reunification of Germany; the Hubble telescope is launched*

1990 *Gillette launches Sensor wet razor*

1991 *Gulf War; WorldWideWeb hypertext system used on Internet for the first time*

1992 *Nissan launches the Micra; Renault launches the Twingo*

1993 *Eurotunnel opened between England and France*

1993 *The Pentium processor is invented*

1994 *Casio launches the Baby-G range of watches*

1996 *Chrysler launches the EV1 electric car*

1997 *Daimler-Benz and Toyota launch prototype hydrogen fuel cell powered cars*

1998 *Volkswagen launches the New Beetle*

1999 *Fiat launches the Multipla*

2000 *Wireless Application Protocol (WAP) mobile telephone technology becomes widely available*

1990

2000

1999
Sony launches its
AIBO Robotic dog

2000
Hanover tram *TW2000* for
Expo' 2000 launched

1993
James Dyson's
DC01 vacuum
cleaner launched

1999
Audi *TT Coupé Quattro* launched

1998
Apple Computer launches
the *iMac* computer

1992
Bill Stumpf and Don Chadwick
design their *Aeron* office chair
for Herman Miller

1996
Trevor's Baylis' *Freeplay*
clockwork radio wins the Product
Category of BBC Design Awards

APPENDIX

SELECTED BIBLIOGRAPHY

ACKNOWLEDGEMENTS

PHOTOGRAPHIC CREDITS

AUTHORS' BIOGRAPHIES

Selected Bibliography

Authors' note: The majority of the research material used for this project was culled directly from primary sources – corporate archives and corporate websites in particular.

Abercrombie, S., *George Nelson: The Design of Modern Design*, MIT Press, Cambridge, Massachusetts 1995
Adams, R., *King C. Gillette, The Man & His Wonderful Shaving Device*, Little, Brown & Co., Boston 1978
Alexander, W. & Street, A., *Metals in the Service of Man*, Penguin Books, London 1985 (re-edition)
Banham, R., *Theory and Design in the First Machine Age*, Architectural Press, London/New York 1960
Baren, M., *How It All Began*, Smith Settle, Otley 1992
Bayley, S. (ed.), *The Conran Directory of Design*, Conran Octopus, London 1985
Bröhan, T. & Berg, T., *Avantgarde Design 1880–1930*, Benedikt Taschen Verlag, Cologne 1994
Buchanan, R. & Margolin, V. (eds.), *Discovering Design: Explorations in Design Studies*, The University of Chicago Press, Chicago 1995
Byars, M. (ed.), *The Design Encyclopedia*, Laurence King, London 1994
Cowley, D. (ed.), *Understanding Brands by 10 People Who Do*, Kogan Page, London 1991
Dale, H. & Dale, R., *The Industrial Revolution*, The British Library, London 1992
Dale, R., *Early Cars*, The British Library, London 1994
Dale, R., *Early Railways*, The British Library, London 1994
Dale, R. & Weaver, R., *Home Entertainment*, The British Library, London 1993
Dale, R. & Weaver, R., *Machines in the Home*, The British Library, London 1992
de Noblet, J. (ed.), *Industrial Design, Reflection of a Century*, Flammarion/APCI, Paris 1993
Droste, M., Ludewig, M. & Bauhaus Archiv, *Marcel Breuer*, Benedikt Taschen Verlag, Cologne 1990
Droste, M. & Bauhaus Archiv, *Bauhaus 1919–1933*, Benedikt Taschen Verlag, Cologne 1992
Fiell, C. & P., *Design of the 20th Century*, Benedikt Taschen Verlag, Cologne 1999
Fiell, C. & P., *Modern Furniture Classics since 1945*, Thames & Hudson, London 1991
Fiell, C. & P., *1000 Chairs*, Benedikt Taschen Verlag, Cologne 1997
Forty, A., *Objects of Desire, Design & Society 1750–1980*, Thames & Hudson, London 1986
Geddes, N. B., *Horizons*, Little Brown, Boston 1932
Giedion, S., *Mechanization Takes Command: A Contribution to Anonymous History*, Oxford University Press, New York 1948
Gloag, J., *Plastics and Industrial Design*, Scientific Book Club, London 1945
Heskett, J., *Industrial Design*, Thames & Hudson, London 1980
Hiesinger, K. & Marcus, G., *Design Since 1945*, Thames & Hudson, London 1983
Hiesinger, K. & Marcus, G., *Landmarks of Twentieth-Century Design*, Abbeville Press, New York 1993
Jervis, S., *The Penguin Dictionary of Design and Designers*, Penguin Books, London 1984
Kaplan, W. (ed.), *Designing Modernity, The Arts of Reform & Persuasion, 1885–1945*, Thames & Hudson/Wolfsonian, London/Miami 1995

Katz, S., *Plastics, Designs and Materials*, Studio Vista, London 1978

Loewy R., *Industrial Design*, Fourth Estate, London 1980

Lucie-Smith, E., *A History of Industrial Design*, Phaidon, Oxford 1983

McDermott, C., *Design Museum Book of 20th Century Design*, Carlton, London 1997

McDermott, C. (ed.), *The Product Book*, British Design & Art Direction, London + Rotovision SA, Crans-Près-Céligny 1999

McKibben, G., *Cutting Edge, Gillette's Journey to Global Leadership*, Harvard Business School Press, Boston 1998

Mende, H. U. von & Dietz, M., *Kleinwagen–Small Cars–Petites Voitures*, Benedikt Taschen Verlag, Cologne 1994

Morgan, A. L. (ed.), *Contemporary Designers*, St. James Press, London 1985

O'Brien, R., *Machines*, Time Inc., New York 1964

Ostergard, D., *Bent Wood and Metal Furniture: 1850–1946*, The American Federation of Arts, New York 1987

Perris, G., *The Industrial History of Modern England*, Kegan Paul, London 1914

Rams, D., *Less but Better*, Jo Klatt + Design Verlag, Hamburg 1995

Read, H., *Art and Industry; The Principles of Industrial Design*, Faber & Faber, London 1956 (4th edition)

Schaefer, H., *The Roots of Modern Design*, Studio Vista, London 1970

Schwartz, F., *The Werkbund: Design Theory and Mass Culture before the First World War*, Yale University Press, Yale 1996

Sharp, D., *The Penguin Dictionary of Chemistry*, Penguin Books, Middlesex 1983

Vegesack, A. von, *Deutsche Stahrohr-Möbel: 650 Modelle aus Katalogen*, Bangert Verlag, Munich 1986

Whiteley, N., *Design for Society*, Reaktion Books, London 1993

Wichmann, H., *Die Neue Sammlung, Ein neuer Museumstyp des 20. Jahrhunderts*, Prestel, Munich 1985

Woodham, J., *Twentieth Century Design*, Oxford University Press, Oxford 1997

Exhibition Catalogues & Year Books

APCI/Centre Georges Pompidou, *Design Français 1960–1990 Trois Décennies*, Éditions du Centre Pompidou, Paris 1988

The Art Journal, *The Great Exhibition, London 1851*, George Virtue, London 1851

The Design Council, *Svensk Form, A Conference about Swedish Design*, Design Council, London 1981

Galleria del Design e dell'Arredamento Cantù, *Design Italiano – Compasso d'Oro ADI*, Milan 1998

Society of Industrial Designers, New York, *US Industrial Design*, Studio Publications & Thomas Crowell Co, USA 1951

We would like to take this opportunity to thank all at Taschen – especially our editors Susanne Husemann, Uta Hoffmann and Karen Williams – for the successful realization of yet another *Klotz* project. We also wish to thank Nick Bell, Sacha Davison and Christopher Brawn of UNA for their wonderful graphic design work and for their steady nerve while under fire. Lastly – and as ever – we thank our daughters Emelia and Clementine for their good humour and tolerance of publishing deadlines.

We are also enormously grateful to the many individuals, manufacturers, distributors, design offices, auction houses, public institutions and picture libraries who have lent their assistance and provided images. Thanks must also go to Paul Chave for the new photography generated specially for this project.

Acknowledgements

Special thanks to:

A&E Design, Stockholm
Advanced Vehicle Design, Altrincham
AEG Aktiengesellschaft, Frankfurt
Airstream, Jackson Center
Alessi, Crusinallo
Alex Moulton Ltd., Bradford-on-Avon
Alfa Romeo Museo, Arese
Emilio Ambasz, New York
Animal, Wareham
Apple Computer, Cupertino
Aprilia SpA, Novale
Archivo Storico Olivetti, Ivrea
Artemide, Milan
Arzberg (Winterling AG), Kirchenlamitz
Atomic, Altenmarkt
Authentics, Holzgerlingen
Automobili Lamborghini, S. Agata Bolognese
Aviation Picture Library, London
Barry Friedman Ltd., New York
Bang & Olufsen, Struer
Bauhaus-Archiv, Berlin
Stephen Bayley, London
Caroline & Denis Bellessort, London
Bertone, Turin
Biro Bic Ltd., London
Black & Decker, Slough
B-Line Srl, Grisignano di Zocco
BMW Museum, Munich
Boeing Company, Seattle
Bombardier Inc., Montreal
Braun GmbH, Kronberg im Taunus
BT Archives, London

Burton Snowboards, Burlington
Canon Europa NV, Amstelveen
Callaway Golf Europe, Chessington
Case Corp., Racine
Caterpillar Inc., Peoria
Christies Images, London
Citroën AG, Cologne
Cooper-Hewitt Museum, New York
Corning Inc., Corning
Daimler-Chrysler, Detroit
Danese, Grumello del Monte
JC Decaux, Neuilly
Deere & Co., Moline
Design Council, London
Design Council Archives, Brighton
Design Research Unit, London
Di Palma, Milan
Ducati, Bologna
Dyson Appliances Ltd., Malmesbury
Ecco Design, New York
Electrolux, Stockholm
Erco, London
Ergonomi Design Gruppen, Bromma
L. M. Ericsson, Stockholm
Euclid-Hitachi Inc., Cleveland, Ohio
Fiat UK, Slough
Fiskars, Billnäs
Flos UK Ltd., London
Ford Motor Company, Brentwood
Franz Schneider Brakel GmbH, Brakel
Frazer Designers, London
Freeplay Energy, Cirencester
Frogdesign, Altensteig
General Electric Co., Fairfield

General Motors Corp., Detroit
Ghia, Turin
Gillette Company, Boston
Giugiaro Design, Moncalieri
GK Design Group Inc., Toyko
Gustavsberg Fabrike, Gustavsberg
Hansgrohe, Schiltach
Harley-Davidson Motor Company,
Milwaukee
Haslam & Whiteway, London
Keith Helfet, Coventry
Henry Dreyfuss Associates,
Woodridge
Herman Miller Inc., Zeeland
Hollington Associates, London
Knud Holscher, Copenhagen
Honda UK Ltd., Reading
Hoover Company, North Canton
IBM Corp., Armonk
IDEO, Palo Alto
IDEO, Tokyo
Iittala Glass Museum, Iittala
Ikea, Älmhult
Imperial War Museum, London
InterDesign, Stockholm
Jaguar Ltd., Coventry
JCB Ltd., Rochester
Junghans Uhren GmbH, Schramberg
Kartell, Milan
Kawasaki Motors UK, Bourne End
Kompan A/S, Ringe
Leica Camera AG, Solms
Lewis Moberley, London
London Transport Museum, London
Louis Poulsen, Copenhagen
Ross & Miska Lovegrove, London

MAG Instruments Inc., Ontario Cal.
Master Cutlery Corp., Chiba
Alberto Meda, Milan
Minale Tattersfield, London
Jasper Morrison, London
Motivation, Bristol
National Motor Museum, Beaulieu
Nautilus, Independence
Necchi, Pavia
Die Neue Sammlung, Munich
Nike UK Ltd., London
Nikon UK Ltd., Kingston-upon-
Thames
Nissan Motor Co., Tokyo
NPK Industrial Design, Leiden
Oakley Inc., Foothill Ranch
Opel AG, Rüsselsheim
Parker Pen Company, Newhaven
Pentagram Design, London
Pethick & Money Ltd., London
Philips NV, Eindhoven
Pi Design, Lucerne
Pininfarina SpA, Turin
Polaroid Corporation, Cambridge,
Mass.
Porsche Cars (GB) Ltd., Reading
Porsche Design, Zell am See
Queensberry Hunt Levien, London
Radio Corporation of America, New
York
Raymond Loewy International, Lon-
don
Remarkable Pencils Ltd., London
Renault Presse, Billancourt
Robert Bosch GmbH, Stuttgart
Robert Krups GmbH, Solingen

Rosenthal AG, Selb
Rubbermaid Inc., Wooster
Saab UK Ltd., Marlow
Sadler Design, Milan
Sanyo UK Ltd., Watford
Science & Society Picture Library, London
Seymour Powell, London
Shimano Europe, Nunspeet
Siemens AG, Munich
Silhouette International, Linz
Smart Cars UK, London
Snap-on Inc., Kenosha
Solari, Udine
Sony Corporation, Tokyo
Springfield Armory Museum, Springfield
Stanley Tools, New Britain
Steelcase Inc., Grand Rapids
Stelton, Hellerup
Sunseeker International Ltd., Poole
Swatch International, Zurich
TAG Heuer, Marin
Tangerine, London
Team Design, Esslingen
Tetra Pak Ltd., Uxbridge
Texas Instruments, Dallas
Torsten Bröhan GmbH, Düsseldorf
Toshiba Corp., Tokyo
Toyota Motor Corp., Tokyo
Tupperware Europe, Aalst
Vent Design, Campbell
Victorinox, Ibach
Vignelli Associates, London
Vignelli Associates, New York
Vin-Mag, London

Vitrac (Pool) Design, Paris
Volkswagen AG/Audi, Wolfsburg
Volvo Corp., Gotenburg
The Wedgwood Museum, Barlaston
Robert Welch, Chipping Campden
Werkbund-Archiv, Berlin
Weston Medical Ltd., Eye
The Wolfsonian (FIU), Miami
Xerox Corp., Stamford
Zanussi Design Center, Pordenone

The Authors

Charlotte J. Fiell (b. 1965) studied at the British Institute, Florence and at Camberwell School of Arts & Crafts, London, where she received a BA (Hons.) in the History of Drawing and Printmaking with Material Science. She later trained with Sotheby's Educational Studies, also in London.

Peter M. Fiell (b. 1958) trained with Sotheby's Educational Studies in London and later received an MA in Design Studies from Central St Martin's College of Art & Design, London.

Together the Fiells run a design consultancy in London specializing in the sale, acquisition, study and promotion of design artefacts. They have lectured widely, curated a number of exhibitions and written numerous articles and books on design and designers, including TASCHEN's *Charles Rennie Mackintosh*, *William Morris*, *1000 Chairs* and *Design of the 20th Century*. They have also edited the six-volume *Decorative Art* series published by TASCHEN GmbH.

The Fiells can be contacted at Fiell@btinternet.com.